BTEC National N II
Mathematics
for Technicians

Second Edition

A. Greer CEng, MRAeS
FORMERLY SENIOR LECTURER

G.W. Taylor BSc (Eng), CEng, MIMechE
FORMERLY PRINCIPAL LECTURER

Gloucestershire College
of Arts and Technology

Text © A. Greer and G. W. Taylor 1982, 1994
Illustrations © Nelson Thornes Ltd 1982, 1994

First published in 1982 by:
Stanley Thornes (Publishers) Ltd

Reprinted in 2001 by:
Nelson Thornes Ltd
Delta Place
27 Bath Road
CHELTENHAM
GL53 7TH
United Kingdom

01 02 03 04 05 / 10 9 8

A catalogue record for this book is available from the British Library

ISBN 0 7487 1701 3

Typeset by
Tech-Set Ltd, Gateshead, Tyne and Wear
Printed in Great Britain by
TJ International Ltd, Padstow, Cornwall

CONTENTS

ALGEBRA

TRIGONOMETRY

v

AUTHORS' NOTE

This Second Edition covers all the topics in the BTEC Mathematics for Engineering Module at National Level. Also included are some basic topics, such as the use of calculators, which should prove extremely useful for students with exemption from BTEC First Mathematics.

This book is for students of all disciplines using the standard Mathematics unit. Suitable examples have been included to cover applications for Engineering, Building Construction, and Science Technologies.

Readers will welcome the extensive revision of basic algebra, indices and solving equations. In addition the treatment of both trigonometry and statistics assumes no previous experience of these topics.

A Greer
G W Taylor

Gloucester

1. APPROXIMATION AND ACCURACY

Outcome:

1. Understand decimal places, significant figures and rounded numbers.
2. Understand accuracy of numbers.
3. Perform addition and subtraction — also multiplication and division — of non-discrete numbers stating the answers in the accepted form.
4. Realise the mistake of giving answers to an accuracy greater than that of the given data.
5. Understand implied accuracy, and what is meant by truncation.

DECIMAL PLACES (abbreviated to 'd.p.')

These refer to the number of figures which follow (i.e. after or to the right of) the decimal point.

Thus: 35.1 has one d.p. and 2.402 has three d.p.

SIGNIFICANT FIGURES (abbreviated to 's.f.')

These are the number of figures, counted from the left to the right, starting with the first *non-zero* figure.

Thus: 2700, 35.0, 0.89 and 0.0082 each have two s.f.
354, 7.21 and 0.000 234 each have three s.f.
5 782 100 and 0.537 91 each have five s.f.

Note that zero figures at the right hand end are not included in the count, as in 2700, 35.0 and 5 782 100 mentioned above.

ROUNDED NUMBERS

These are obtained by the process of 'rounding' or 'rounding-off' and enable a degree of accuracy to be stated.

Thus: Rounding 31.63 gives 31.6 correct to three s.f.
and   Rounding 31.68 gives 31.7 correct to three s.f.

Rule for Rounding

Working from right to left figures are discarded in turn:

If the discarded figure is less than 5 the preceding figure is *not* altered — in the first example given above the 3 is discarded and the 6 is unaltered. If the discarded figure is greater than (or equal to) 5 then the preceding figure is increased by one — in the second example give above the 8 is discarded and the preceding figure 6 is increased to 7.

Thus 0.472 becomes 0.47 correct to two s.f.
       or 0.5 correct to one s.f.

Also 24.0926 becomes 24.093 correct to three d.p.
       or 24.09 correct to two d.p.
       or 24.1   correct to one d.p.

Now 0.008 246 may be stated as
   0.008 25 correct to three s.f.; alternatively to five d.p.
or 0.0082 correct to two s.f.; alternatively to four d.p.
or 0.008   correct to one s.f.; alternatively to three d.p.

Care must be taken to ensure that rounding is carried out directly from the original number — thus in the above case 0.008 246 rounds to 0.0082 correct to two s.f.

It would be wrong to round successively to three s.f. and then to two s.f.: the second rounding of 0.008 25 would give an incorrect 0.0083!

Note: when rounding whole number amounts, 'discarded' figures preceding the decimal point must be replaced by zeros.

Thus: 1479 becomes 1480 (*not* 148) correct to three s.f.
      or 1500 (*not* 15) correct to two s.f.
      or 1000 (*not* 1) correct to one s.f.

Consider an attendance of 54 276 at a league soccer match. This is as precise as could have been obtained, no doubt from the turnstiles at entry.

Now 54 276 may be stated as 54 280 correct to four s.f.
        or 54 300 correct to three s.f.
        or 54 000 correct to two s.f.
     or even 50 000 correct to one s.f.

As far as press reports are concerned 54 300 may well be considered to be good enough. So when the papers state an attendance of 54 300 a three s.f. accuracy is implied (although it would not be mentioned!).

Exercise 1.1

Write down the following numbers correct to the number of significant figures stated:

1) 24.865 82   (a) to 6   (b) to 4   (c) to 2

2) 0.008 357 1   (a) to 4   (b) to 3   (c) to 2

3) 4.978 48   (a) to 5   (b) to 3   (c) to 1

4) 21.987 to 2

5) 35.603 to 4

6) 28 387 617   (a) to 5   (b) to 2

7) 4.149 76   (a) to 5   (b) to 4   (c) to 3

8) 9.2048 to 3

Write down the following numbers correct to the number of decimal places stated:

9) 2.138 87   (a) to 4   (b) to 3   (c) to 2

10) 25.165   (a) to 2   (b) to 1

11) 0.003 988   (a) to 5   (b) to 4   (c) to 3

12) 7.2039   (a) to 3   (b) to 2   (c) to 1

13) 0.7259   (a) to 3   (b) to 2

ACCURACY OF NUMBERS

A small firm employs 13 persons — this number is *exact*, or *discrete*, and therefore cannot be an approximation.

However in technology the vast majority of numbers are not discrete. Measurements usually fall into this category — such as a person's height of 177.8 cm (an approximation to 4 s.f., or to the nearest $\frac{1}{10}$ cm).

Now consider an electrical resistance measured as 52 ohm, to the nearest ohm, and considered correct to 2 s.f. The number 52 could have been obtained by rounding any number between the lowest value of 51.5 and up to the highest value of 52.5. These extremes are $52-0.5$ and $52+0.5$ and the value of the resistance would be stated as 52 ± 0.5 ohm. This shows, at a glance, the maximum possible error is 0.5 greater, or 0.5 smaller, than 52. This does not mean that the error is bound to be as great at 0.5 ohm — it may be considerably less, but it certainly cannot be any more.

Similarly 7.49 gram is considered accurate to 3 s.f., or to $\frac{1}{100}$ gram: this would come from rounding a number between the extremes of 7.485 and 7.495 and would be given as 7.49 ± 0.005 gram.

Also 0.1370 mV given accurate to the nearest $\frac{1}{10\,000}$ mV is also considered accurate to 4 s.f. You should note how the extra zero is included to indicate accuracy greater than would be implied by 0.137 mV which is only correct to $\frac{1}{1000}$ mV and only accurate to 3 s.f. Now 0.1370 mV would come from rounding between the extremes of 0.136 95 mV and 0.137 05 mV and would be stated as $0.1370 \pm 0.000\,05$ mV.

ACCURACY IN ADDITION AND SUBTRACTION

A contractor has five vehicles. If he sells two he is left with three. So in this case of $5-2=3$ the answer 3 is exact, and there is no error, since the numbers 5 and 2 are discrete.

However, let us consider the result of adding electrical resistances of 52 ± 0.5 ohm and 36 ± 0.5 ohm.

We may state this problem as $(52 \pm 0.5) + (36 \pm 0.5)$.

$$
\begin{aligned}
\text{Now the greatest answer} &= \text{ the greatest value of } 52 \pm 0.5 \\
&\quad + \text{ the greatest value of } 36 \pm 0.5 \\
&= (52 + 0.5) + (36 + 0.5) \\
&= (52 + 36) + (0.5 + 0.5) \\
&= (88 + 1.0) \text{ or } 89
\end{aligned}
$$

Similarly the smallest answer is $(88-1.0)$ or 87.

So the final result of adding the resistances lies between 87 and 89 ohm, and would be given as 88 ± 1.0 ohm, which has a maximum error of 1.0 ohm.

In general when adding and subtracting numbers the maximum error of the result may be found by adding the maximum errors of the original numbers.

Thus $623 + 56.3$ implies $(623 \pm 0.5) + (56.3 \pm 0.05)$
giving a result $(623 + 56.3) \pm (0.5 + 0.05)$
or 679.3 ± 0.55

Now $27.24 - 9.3$ implies $(27.24 \pm 0.005) - (9.3 \pm 0.05)$
giving a result $(27.24 - 9.3) \pm (0.005 + 0.05)$
or 17.94 ± 0.055

Note that although the original numbers are being subtracted, the maximum error of 0.055 is still obtained by *adding* the given individual maximum errors.

Exercise 1.2

What are the greatest and least values of:

1) 64 ± 0.5

2) 2469 ± 5

3) 3.07 ± 0.005

4) 0.6 ± 0.05

5) $(26 \pm 0.5) + (3.4 \pm 0.05)$

6) $(0.56 \pm 0.005) + (0.7 \pm 0.05)$

7) $(5.6 \pm 0.05) - (2.9 \pm 0.05)$

8) $(0.78 \pm 0.005) - (0.034 \pm 0.0005)$

9) For the answers to Questions 5–8, state the maximum error.

10) A measurement has been taken of 39.07 km to the nearest $\frac{1}{100}$ km. State this in conventional form.

11) A rectangle has sides of length 0.372 m and 1.238 m both measured to the nearest millimetre. State both these measurements in conventional form. Find also the maximum and minimum values of the perimeter of the rectangle and also the maximum error.

12) Find the maximum error of $(2.3 \pm 0.05) - (0.76 \pm 0.003) + (64 \pm 0.5)$.

13) A current has been measured correct to $\frac{1}{100}$ A, but has been listed incorrectly as 7 A. How should it have been listed and what is the maximum error?

14) Careless recording gave the length of the sides of a triangle as 3 mm, 4 mm and 5 mm when in actual fact they had been measured to the nearest $\frac{1}{100}$ mm. What are the greatest and least values of the perimeter of the triangle?

ACCURACY IN MULTIPLICATION AND DIVISION

Consider finding the area of a rectangle with sides measured as 67 mm and 62 mm. Each of the numbers will be considered accurate to two significant figures and so the problem may be stated as:

$$\text{Area} = (67 \pm 0.5) \times (62 \pm 0.5)$$

now the greatest area = (the greatest value of 67 ± 0.5)

$$\times \text{(the greatest value of } 62 \pm 0.5)$$

$$= 67.5 \times 62.5$$

$$= 4218.75$$

and   the smallest area = (the smallest value of 67 ± 0.5)

$$\times \text{(the smallest value of } 62 \pm 0.5)$$

$$= 66.5 \times 61.5$$

$$= 4089.75$$

If we examine these two extremes we see that only the four thousand figure is guaranteed in the value of the area.

Now it is generally accepted that:

> When multiplying and dividing numbers the answer should *not* be given to an accuracy greater than the least accurate of the given numbers.

If we simply calculate 67×62 we get 4154.

Here both the given lengths are to the same accuracy, so in this particular case either may be taken to be the least accurate, namely to two significant figures. Thus, after rounding, we have the area as $4200 \, \text{mm}^2$, which would be an acceptable result.

You will possibly grumble at this answer and point out that even this is not strictly correct — and you would be right! Anyway, you now realise how wrong it is to give results to an accuracy which cannot be justified by the given data: we are all guilty of this from time to time so beware!

Consider also $\dfrac{5.73 \times 21}{0.6243}$ the result of which is 192.743 87 from a

calculator. The least accurate of the three given numbers is 21 which has a two significant figure accuracy. Hence the answer must not be stated any more accurately than this: namely 190 obtained by rounding 192.743 87 correct to two significant figures.

IMPLIED ACCURACY

Suppose we decided to check graphically, by counting the squares, the area of the rectangle in the preceding section. Sides of 67 mm and 62 mm would be measured out, possibly with a rule, on squared paper. It is likely that we should try for, and achieve, a measuring accuracy to one tenth of a millimetre — thus the area we would be checking would be 67.0 mm by 62.0 mm. Now, since the calculated result of 67×62 is 4154, the area being checked would be 4150 mm^2 after rounding to *three* significant figures which is the accuracy of 67.0 mm and 62.0 mm.

With this in mind, unless there is a very good reason otherwise, a three significant figure accuracy is generally accepted on this type of data. We are really covering up for our inability to state the original data to its correct accuracy e.g. 67 mm which we should have given as 67.0 mm.

Consider, also, the problem of calculating the angles of a triangle with the sides given as 6, 8 and 9 m respectively. An accuracy of one significant figure (to which these dimensions are given) would only allow us to give the calculated angles such as 10°, 20°, 30°, ... etc. Again, we would assume the dimensions should have been given as 6.00, 8.00 and 9.00 m, and thus give the results to a three significant figure accuracy.

TRUNCATION OR CUTTING-OFF

Some calculators 'truncate' or 'cut-off' figures in their displays after computation. For instance, in an eight figure display, the result of 5 divided by 3 would be shown as 1.666 666 66 (most modern machines would round the result to show 1.666 666 67). If truncation does occur each time successive computations are performed then an accumulating error may be introduced.

Exercise 2.1 on p. 17 will give you practice in using accuracy.

2. THE SCIENTIFIC CALCULATOR

Outcome:

1. Understand the need for a rough check answer.
2. Obtain a rough check answer for any calculation.
3. Use the calculator to find the values of arithmetic expressions involving addition, subtraction, multiplication, division and use the memory.
4. Extend operations to include reciprocals, square roots and numbers in standard form.
5. Evaluate arithmetical expressions involving whole number, fractional and negative indices.
6. Use the calculator to evaluate formulae.
7. Evaluate polynomials using nested multiplication.
8. Use the calculator to find the values of trigonometrical functions.

ROUGH CHECKS

When using a calculator it is essential for you to do a rough check in order to obtain an approximate result. Any error, however small it may seem, in carrying out a sequence of operations will result in a wrong answer.

Suppose, for instance, that you had £1000 in the bank and then withdrew £97.82. The bank staff then used a calculator to find how much money you had left in your account—they calculated that £1000 less £978.2 left only £21.80 credited to you. You would be extremely annoyed and probably point out to them that a rough check of £1000 less £100 would leave £900, and that if this had been done much embarrassment would have been avoided.

The 'small' mistake was to get the decimal point in the wrong place when recording the money withdrawn, which is typical of errors we all make from time to time. You should get in the habit of doing a rough check on any calculation *before* using your machine. The advantage of a rough check answer before the actual calculation avoids the possibility of forgetting it in the excitement of obtaining a machine result. Also your rough check will not be influenced by the result obtained on your calculator.

8

KEYBOARD LAYOUT

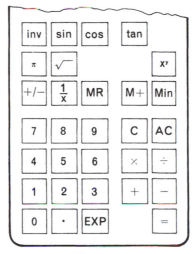

Fig. 2.1

The following are the figure keys:

| 0 | 1 | 2 | 3 | 4 | 5 | 6 | 7 | 8 | 9 |

The other keys are summarised as follows:

| . | decimal point key. |

| EXP | EXPONENTIAL KEY—allows entry of numbers in standard form (e.g. 1.46×10^3). |

| C | clear key—enables an incorrect entry to be erased (either a figure or an operation) if pressed immediately afterwards. |

| AC | all clear key—clears machine of all numbers—except the memory. |

| × ÷ + − = | arithmetical operation keys. |

| Min | memory in key—enters a number on display into the memory, erasing any number previously in the memory. |

| M+ | add to memory key—enables a number to be added to previous content of memory. |

MR	memory recall key—enables content of memory to be shown on display.
+/−	change sign key—(e.g. $+2$ to -2, or -1.5 to $+1.5$).
$\dfrac{1}{x}$	reciprocal key (e.g. $\frac{1}{2}$ to 2, or 4 to 0.25).
π	'pi' key—gives the numerical value of π.
$\sqrt{}$	square root key.
x^y	power key—gives the values of y to the index x.
sin cos tan	trigonometric function keys—gives the sine, cosine or tangent of an angle.
inv	inverse trigonometric function key—gives the angle when given the value of a trigonometrical ratio, e.g. $\operatorname{inv}\sin\theta$ (also called $\operatorname{arc}\sin\theta$ or $\sin^{-1}\theta$).

Part of a typical keyboard layout on an electronic calculator is shown in Fig. 2.1. Calculators vary in layout and operation, just as motor cars do from different manufacturers, but the methods of using each type are similar. Each calculator is supplied with an instruction booklet and you should work through this carrying out any worked examples which are given.

In this chapter we have outline procedures which are generally common to all calculators—if they are not exactly as your machine requires, you will have to make allowance according to the instruction booklet.

WORKED EXAMPLES

After first switching on the calculator, or commencing a fresh problem, you should press the AC key. This ensures that all figures entered previously have been erased and will not interfere with new data to be entered.

The memory is not cleared but this is done automatically when a new number is entered in the memory using the Min key.

EXAMPLE 2.1

Evaluate $18.24 + 4.39 - 9.72$

A rough check gives: $18 + 5 - 10 = 13$

The sequence used on the calculator is similar to the order in which the problem is given. This is shown by:

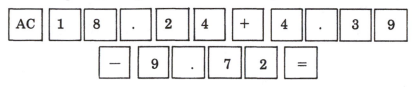

The display gives $\boxed{12.91}$ and since all the data is given to two decimal places the answer, correct to two decimal places is 12.91

It should be noted that the order of operations is not important — try for yourself the sequence $18.24 - 9.72 + 4.39$ and you will obtain the same answer.

EXAMPLE 2.2

Evaluate $\dfrac{20.3 \times 3092}{1.563}$

A rough check gives: $\dfrac{20 \times 3000}{2} = 30\,000$

The sequence of operations is:

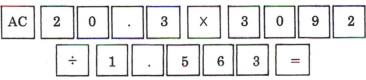

The display gives $\boxed{40\,158.413}$ which seems rather way out when compared with the approximate answer 30 000 obtained from the rough check. This is not unexpected, however, since we did allow for 2 in the denominator instead of 1.563 which would make the rough check answer smaller. It does confirm that the true answer is of the correct order.

It would not be correct to state the answer as 40 158.413 given on the display, since the least accurate of the given numbers, 20.3, has three significant figures and this is the accuracy to which we should state the answer.

Thus the answer is 40 200 correct to three significant figures.

EXAMPLE 2.3

Find the value of $6\,857\,000 \times 119\,000 \times 85.3$

For the rough check numbers which contain as many figures as these are better considered in standard form:

$$(6.857 \times 10^6) \times (1.19 \times 10^5) \times (8.53 \times 10)$$

or approximately:

$$
\begin{aligned}
(7 \times 10^6) \times (1 \times 10^5) \times (10 \times 10) &= 7 \times 1 \times 10 \times 10^{12} \\
&= 70 \times 10^{12} \\
&= 7 \times 10^{13}
\end{aligned}
$$

The sequence of operations is:

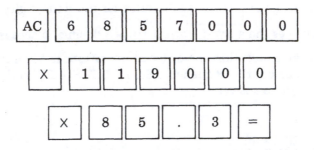

The display will show $\boxed{6.96033 \quad 13}$ which represents $6.960\,33 \times 10^{13}$. The least number of significant figures in the given numbers is three. Thus the answer is 6.96×10^{13}.

An alternative method is to enter the numbers in standard form using the $\boxed{\text{EXP}}$ (exponential) key. The sequence would then be:

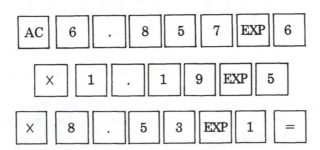

giving the same result.

The sequence used in a problem such as this would be personal choice, but if the problem includes numbers with powers of 10 the latter sequence is better.

EXAMPLE 2.4

Evaluate $\dfrac{(5.745 \times 10^3)(56.7 \times 10^{-4})}{0.0343 \times 10^6}$

The rough check gives: $\dfrac{(6 \times 10^3)(6 \times 10^{-3})}{3 \times 10^4} = \dfrac{6 \times 6}{3} \times 10^{3-3-4}$

$$= 12 \times 10^{-4}$$

The following sequence includes use of the $\boxed{+/-}$ (change sign) key to enter the negative index:

$$\boxed{AC}\ \boxed{5}\ \boxed{.}\ \boxed{7}\ \boxed{4}\ \boxed{5}\ \boxed{EXP}\ \boxed{3}$$

$$\boxed{\times}\ \boxed{5}\ \boxed{6}\ \boxed{.}\ \boxed{7}\ \boxed{EXP}\ \boxed{4}\ \boxed{+/-}$$

$$\boxed{\div}\ \boxed{0}\ \boxed{.}\ \boxed{0}\ \boxed{3}\ \boxed{4}\ \boxed{3}\ \boxed{EXP}\ \boxed{6}\ \boxed{=}$$

The display will show $\boxed{.9.496\ 84 \qquad -04}$ and since the least number of significant figures in the given numbers is three, then the answer is 9.50×10^{-4}.

Note that we give the answer stating 9.50 and not merely 9.5 which would imply only two significant figure accuracy.

EXAMPLE 2.5

Evaluate $\dfrac{0.674}{1.239} - \dfrac{0.564 \times 1.89}{0.379}$

The rough check will need a little more care when approximating numbers—it is not possible to give rules but as you gain experience you will have no difficulty.

Thus the rough check gives $\dfrac{0.5}{1} - \dfrac{0.5 \times 2}{0.5} = 0.5 - 2 = -1.5$

The sequence of operations includes use of the memory in which intermediate results may be kept for use later in the sequence. You should note also how a number may be subtracted from the contents of the memory by using the $\boxed{M+}$ (add to memory) key after changing the sign of the number.

The sequence of operations is:

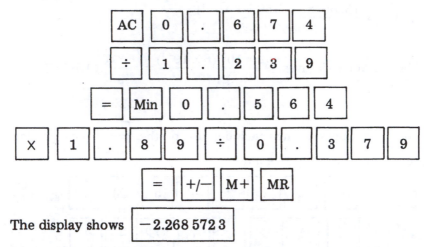

The display shows $\boxed{-2.268\,572\,3}$

This answer is considerably higher than that obtained from the rough check, but it is of the correct order, i.e. *not* 22.7 *or* 0.227

Hence the required answer is -2.27 correct to three significant figures, which is the least accuracy of the given data.

EXAMPLE 2.6

Find the value of: $9.7 + \dfrac{55.15}{29.6 - 8.64}$

The rough check gives: $10 + \dfrac{60}{30 - 9} \approx 10 + \dfrac{60}{20} = 13$

It is possible to work this problem out on the calculator by rearranging and using the memory. However no rearrangement is necessary if we make use of the $\boxed{\dfrac{1}{x}}$ (the reciprocal) key.

This key enables us to find the reciprocal of a number—for example, the reciprocal of 2 is $\frac{1}{2}$ or 0.5

Let us consider $9.7 + \dfrac{1}{\left(\dfrac{29.6 - 8.64}{55.15}\right)}$

This may be written as:

$$9.7 + 1 \div \left(\frac{29.6 - 8.64}{55.15}\right) = 9.7 + 1 \times \left(\frac{55.15}{29.6 - 8.64}\right) = \frac{55.15}{29.6 - 8.64}$$

If we make use of this knowledge then the sequence of operations is:

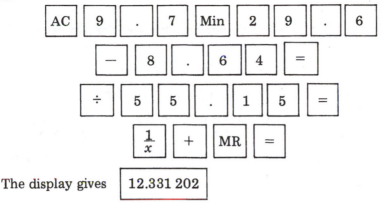

The display gives | 12.331 202

It is always difficult to assess accuracy of the answer to a calculation which involves addition and subtraction. Although the 9.7 has only two significant figures, the addition of the other portion of the calculation will increase the figures before the decimal point to two.

Thus the answer may be given as 12.3

In most engineering problems you will not often be wrong if you give answers to three significant figures—this is consistent with the accuracy of much of the data (such as ultimate tensile strengths of materials). There are exceptions, of course, such as certain machine shop problems which may require a much greater degree of accuracy.

'WHOLE NUMBER' POWERS

The obvious method, but not necessarily the quickest, to find the fifth power of 5.6 (for example) is to multiply 5.6 by itself four times. To avoid entering the number five times use may be made of the memory as the following sequence shows:

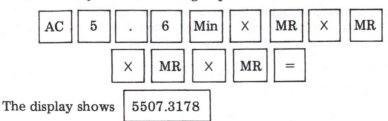

The display shows | 5507.3178

Thus $(5.6)^5 = 5500$ correct to two significant figures.

USE OF THE 'CONSTANT MULTIPLIER' FACILITY

You should check with the calculator booklet that this is possible with your machine.

Suppose that we have the following to evaluate:

$$3.1 \times 7.89, \quad 3.1 \times 6.2, \quad 3.1 \times 3.45, \quad 3.1 \times 9.8, \quad 3.1 \times 10.9$$

If each calculation is carried out individually, then the number 3.1 will have to be entered for each calculation. This may be avoided by two consecutive pressings of the ⊠ key. Thus the sequence would be:

AC	3	.	1	×	×	7	.	8	9	=

giving 24.459

6	.	2	=

giving 19.22

3	.	4	5	=

giving 10.695

9	.	8	=

giving 30.38

1	0	.	9	=

giving 33.79

If in the above sequence, entry of each of the numbers preceding the equals operation were omitted, then the calculator would automatically substitute 3.1 for each of them.

Thus $(3.1)^4 = 3.1 \times 3.1 \times 3.1 \times 3.1$ may be found by the sequence:

AC	3	.	1	×	×	=	=	=

giving 92.3521

Exercise 2.1

Evaluate the following, taking care to give the answers to an accuracy determined by the given data.

1) $45.6 + 3.5 - 21.4 - 14.6$

2) $-23.94 - 6.93 + 1.92 + 17.60$

3) $\dfrac{40.72 \times 3.86}{5.73}$

4) $\dfrac{4.86 \times 0.008\,34 \times 0.640}{0.860 \times 0.934 \times 21.7}$

5) $\dfrac{57.3 + 64.29 + 3.17}{64.2}$

6) $\dfrac{32.2}{6.45 + 7.29 - 21.3}$

7) $\dfrac{1}{\frac{1}{3} + \frac{1}{4} + \frac{1}{5}}$ to 2 d.p.

8) $\dfrac{3.76 + 42.4}{1.60 + 0.86}$

9) $\dfrac{4.82 + 7.93}{-0.730 \times 6.92}$

10) $9.38(4.86 + 7.60 \times 1.89^3)$

11) $4.93^2 - 6.86^2$

12) $(4.93 + 6.86)(4.93 - 6.86)$

13) $\dfrac{1}{6.3^2 + 9.6^2}$

14) $\dfrac{3.864^2 + 9.62}{3.74 - 8.62^2}$

15) $\dfrac{9.5}{(6.4 \times 3.2) - (6.7 \times 0.9)}$

16) $1 - \dfrac{5.0}{3.6 + 7.49}$

17) $\frac{1}{6} - \frac{1}{5}(4.6)^2$

18) $\dfrac{6.4}{20.2}\left(3.94^2 - \dfrac{5.7 + 4.9}{6.7 - 3.2}\right)$

19) $\dfrac{3.64^3 + 5.6^2 - (1/0.085)}{9.76 + 3.4 - 2.9}$

20) $\dfrac{6.54(7.69 \times 10^{-5})}{0.643^2 - 79.3(3.21 \times 10^{-4})}$

SQUARE ROOT AND 'PI' KEYS

$\boxed{\sqrt{}}$ gives the square root of any number in the display.

$\boxed{\pi}$ gives the numerical value of π to whatever accuracy the machine is designed.

EXAMPLE 2.7

The period, T seconds (the time for a complete swing), of a simple pendulum is given by the formula $T = 2\pi\sqrt{\dfrac{l}{g}}$ where l m is its length and g m/s² is the acceleration due to gravity.

Find the value of T if $l = 1.37$ m and $g = 9.81$ m/s^2.

Substituting the given values into the formula we have

$$T = 2\pi \sqrt{\frac{1.37}{9.81}}$$

The rough check gives: $T = 2 \times 3\sqrt{\frac{1}{9}} = 2 \times 3 \times \frac{1}{3} = 2$ s

The sequence of operations would be:

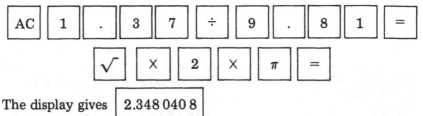

The display gives $\boxed{2.348\,040\,8}$

Thus the value of T is 2.35 seconds, correct to three significant figures.

THE POWER KEY

$\boxed{x^y}$ gives the value of x to the index y.

EXAMPLE 2.8

The relationship between the luminosity, I, of a metal filament lamp and the voltage, V, is given by the equation $I = aV^4$ where a is a constant. Find the value of I if $a = 9 \times 10^{-7}$ and $V = 60$

Substituting the given values into the equation we have

$$I = (9 \times 10^{-7})60^4$$

The rough check gives

$$I = (10 \times 10^{-7})(6 \times 10)^4 = 10^{-6} \times 6^4 \times 10^4 = 10^{-2} \times 36 \times 36$$

and if we approximate by putting 30×40 instead of 36×36

then $\qquad I = 10^{-2} \times 30 \times 40 = 10^{-2} \times 1200 = 12$

The sequence of operations would be:

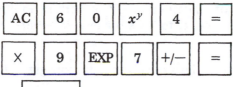

The display gives $\boxed{11.664}$

Thus the value of I is 11.7 correct to three significant figures.

EXAMPLE 2.9

The law of expansion of a gas is given by the expression $pV^{1.2} = k$ where p is the pressure, V is the volume, and k is a constant. Find the value of k if $p = 0.8 \times 10^6$ and $V = 0.2$

Substituting the given values into the formula we have

$$k = (0.8 \times 10^6) \times 0.2^{1.2}$$

The rough check gives

$$k = 1 \times 10^6 \times (\tfrac{2}{10})^{1.2} = 10^6 \times \frac{2^{1.2}}{10^{1.2}} = 10^6 \times \frac{3}{30} = 0.1 \times 10^6$$

Since it is difficult to assess the approximate value of a decimal number to an index, it becomes simpler to express the decimal number as a fraction using whole numbers. In this case it is convenient to express 0.2 as $\tfrac{2}{10}$. We can guess the rough value of $2^{1.2}$, since we know that $2^1 = 2$ and $2^2 = 4$. Similarly we judge the value of $10^{1.2}$ as being between $10^1 = 10$ and $10^2 = 100$. The more practice you have in doing calculations of this type, the more accurate your guess will be.

The sequence of operations would be:

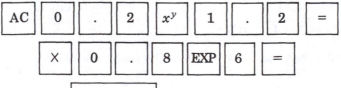

The display gives $\boxed{115\,964.74}$

Thus the required value of k is 116 000 or 0.116×10^6 correct to three significant figures.

EXAMPLE 2.10

The intensity of radiation, R, from certain radioactive materials at a particular time, t, follows the law $R = 95t^{-1.8}$. If $t = 5$ find the value of R.

Substituting $t = 5$ into the given equation gives $R = 95 \times 5^{-1.8}$

For a rough check on the value of an expression containing a negative index it helps to rearrange so that the index becomes positive.

The rough check gives: $R = 100 \times 5^{-2} = 100 \times \dfrac{1}{5^2} = 100 \times \dfrac{1}{25} = 4$

The sequence of operations would be:

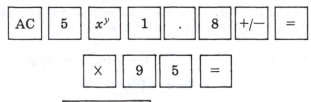

The display gives $\boxed{5.242\,972\,8}$

Hence the value of R is 5.24 correct to three significant figures.

EXAMPLE 2.11

The discharge $\dot{Q}\,\text{m}^3/\text{s}$ of water over a $90°$ vee notch is given by $\dot{Q} = 1.42\,H^{5/2}$. Find $\dot{Q}$ when $H = 0.3\,\text{m}$.

Substituting $H = 0.3$ into the given formula $\dot{Q} = 1.42(0.3)^{5/2}$.

The fractional index should be expressed as a decimal so that we may use the $\boxed{x^y}$ key. Since $\frac{5}{2} = 2.5$ then equation may be stated as $\dot{Q} = 1.42(0.3)^{2.5}$.

Remembering that $0.3 \approx \frac{1}{3}$ then the rough check gives

$$\dot{Q} \approx 1.5(\tfrac{1}{3})^{2.5} \approx \frac{1.5}{3^{2.5}} \approx \frac{1.5}{15} \approx 0.1$$

We estimated the value of $3^{2.5}$ as 15 since it lies between $3^2 = 9$ and $3^3 = 27$

The sequence of operations is:

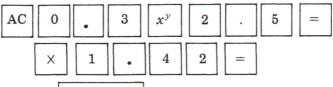

The display shows $\boxed{0.069\,998\,9}$

Hence $\dot{Q} = 0.0700\,\text{m}^3/\text{s}$ correct to three s.f.

NESTING (OR STACKING) A POLYNOMIAL

A polynomial in x is an expression containing a sum of terms, each term being a power of x.

A typical polynomial is

$$ax^4 + bx^3 + cx^2 + dx + e$$

where a, b, c, d and e are constants.

The polynomial may be factorised successively as follows:

$$ax^4 + bx^3 + cx^2 + dx + e$$
$$= (ax + b)^3 + cx^2 + dx + e$$
$$= \{(ax + b)x + c\}\,x^2 + dx + e$$
$$= [\{(ax + b)x + c\}x + d]x + e$$

The opening bracket symbols $[\{($ are usually omitted and all the remaining closure brackets are written in the same form. Thus the polynomial looks like:

$$ax + b)x + c)x + d)x + e$$

This is called the nested (or stacked) form of the polynomial. When evaluating its value we must always work from *left* to *right*.

EXAMPLE 2.12

Find the value of $y = 5x^3 - 7x^2 + 8x - 5$ when $x = 3.32$

The nested form gives $y = 5x - 7)x + 8)x - 5$

When $x = 3.32$ then $y = 5 \times 3.32 - 7)3.32 + 8)3.32 - 5$

Working from *left* to *right* the sequence of operations is:

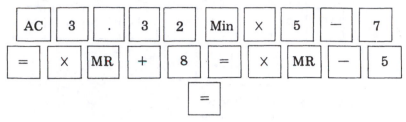

The display gives | 127.375 04 |

Thus $y = 127$ correct to three significant figures.

Note that if the polynomial is not nested it would be necessary to evaluate

$$5(3.32)^3 - 7(3.32)^2 + 8(3.32) - 5$$

Try this for yourself and you will appreciate the advantage of nesting.

TRIGONOMETRICAL FUNCTIONS

Use of | sin |, | cos |, | tan | and | inv | keys.

EXAMPLE 2.13

Find angle A if $\sin A = \dfrac{3.68 \sin 42°}{5.26}$,

Rough check: It is always difficult to find an approximate answer for a calculation involving trigonometrical functions. However, we may use a 'backwards substitution' method which is carried out after an answer has been obtained.

On most calculators there is a sliding switch which may be positioned at either RAD (radians), DEG (degrees) or GRAD (grades—a grade being one-hundredth of a right angle, used more on the continent). In this example the angles are in degrees and so we set the sliding switch to the DEG position.

A sequence of operations commencing with

| AC | 3 | . | 6 | 8 | × | 4 | 2 | sin | = |

may not function correctly as it merely gives the value of sin 42°. Try it on your own machine. We must, therefore, alter the order in which the operations are carried out. Thus the sequence used should be:

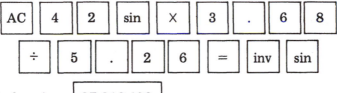

The display gives $\boxed{27.913\,433}$

Thus angle $A = 27.9°$ correct to three significant figures.

Answer check. As the result of our calculations we have

$$\sin 27.9° = \frac{3.68 \sin 42°}{5.26}$$

Now we may rearrange this expression to give

$$\sin 42° = \frac{5.26 \sin 27.9°}{3.68}$$

Thus if we find the value of the right-hand side of this new expression, it should give the value of sin 42° if the value of angle $A = 27.9°$ is correct.

The sequence of operations is similar to that given for the original calculation. Try it for yourself and thus check the result.

Exercise 2.2

Evaluate the expression in Questions 1–8 giving the answers correct to three significant figures.

1) 2.32^4 2) 1.52^6 3) 0.523^5 4) 7.9^{-2}

5) 4.59^{-3} 6) 0.321^{-4} 7) $12.1^{1.5}$ 8) $6.83^{2.32}$

Evaluate the expressions in Questions 9–16 giving the answers correct to two decimal places.

9) $0.879^{3.1}$ 10) $5.56^{0.62}$ 11) $14.7^{0.347}$ 12) $3.9^{-0.5}$

13) $6.64^{3/2}$ 14) $13.6^{2/5}$ 15) $1.23^{7/3}$ 16) $0.334^{3/5}$

In Questions 17–30 the accuracy to which you give the answer will depend on the given figures in each set of data.

17) Evaluate $4\pi r^2$ when $r = 6.1$

18) Evaluate $5\pi(R^2 - r^2)$ when $R = 1.32$ and $r = 1.24$

19) In a beam the stress, σ, due to bending is given by the expression $\sigma = \dfrac{My}{I}$. Find σ if $M = 12 \times 10^6$, $y = 60$, and $I = 11.5 \times 10^6$

20) The polar second moment of area, J, of a hollow shaft is given by the equation $J = \dfrac{\pi}{32}(D^4 - d^4)$. Find J if $D = 220$ and $d = 140$

21) The velocity, v, of a body performing simple harmonic motion is given by the expression $v = \omega\sqrt{A^2 - x^2}$. Find v if $\omega = 20.9$, $A = 0.060$, and $x = 0.020$.

22) The natural frequency of oscillation, f, of a mass, m, supported by a spring of stiffness, λ, is given by the formula

$f = \dfrac{1}{2\pi}\sqrt{\dfrac{\lambda}{m}}$. Find f if $\lambda = 5000$ and $m = 1.5$

23) The volume rate of flow, $\dot{Q}$, of water through a venturi meter is given by

$$\dot{Q} = A_2 \sqrt{\dfrac{2gH}{1 - \left(\dfrac{A_2}{A_1}\right)^2}}$$

Find $\dot{Q}$ if $A_1 = 0.0201$, $A_2 = 0.005\,03$, $g = 9.81$, and $H = 0.554$

24) A non-dimensional constant used in connection with rectangular weirs is given by $\dfrac{H^{3/2}g^{1/2}}{n}$. Find the value of this constant when $H = 4.56$, $g = 32.2$ and $n = 8.42 \times 10^{-6}$

25) The specific speed of a centrifugal pump is given by the expression $\dfrac{N\sqrt{Q}}{H^{3/4}}$. Find the specific speed when $N = 2000$, $Q = 140$ and $H = 7.0$

26) The specific speed of a hydraulic turbine is given by $\dfrac{N\sqrt{P}}{H^{5/4}}$.
Find the specific speed if $N = 150$, $P = 30 \times 10^3$ and $H = 20$

27) Find the value of a if $a = \dfrac{80.6 \sin 55°}{\sin 70°}$

28) If $\cos C = \dfrac{a^2 + b^2 - c^2}{2ab}$, find the value of the angle C when
$a = 19.37\,\text{mm}$, $b = 26.42\,\text{mm}$ and $c = 22.31\,\text{mm}$

29) The following formula is used when measuring the angle of a
vee notch: $\sin\dfrac{\theta}{2} = \dfrac{\frac{1}{2}(D-d)}{H-h-\frac{1}{2}(D-d)}$. Find θ when $D = 38.10\,\text{mm}$,
$d = 25.40\,\text{mm}$, $H = 100.92\,\text{mm}$ and $h = 78.03\,\text{mm}$

30) When checking the major diameter of a metric thread the
following formula is used: $M = D - \dfrac{5p}{\tan 30°} + d\left(1 + \dfrac{1}{\sin 30°}\right)$. Find
M when $p = 5.30\,\text{mm}$, $D = 18.34\,\text{mm}$ and $d = 15.38\,\text{mm}$

31) Use the method of nesting to evaluate the following:
(a) $5x^2 + 4x - 15$ when $x = 3.8$ correct to 1 d.p.
(b) $7x^3 + 3x^2 - 7x + 5$ when $x = 0.35$ correct to 2 d.p.
(c) $x^4 + 3x^3 - 8x^2 + x - 3$ when $x = 3.75$ correct to 3 s.f.
(d) $x^3 + 5x^2 - 6x + 9$ when $x = 1.2$ correct to 1 d.p.

3. UNITS

Outcome:

1. Recognise SI basic units.
2. Understand preferred units.
3. Recognise the Imperial system of units.
4. Convert from one system to the other.

SI UNITS

The Système International d'Unités (the international system of units) usually abbreviated to SI, is based upon six fundamental units as follows:

Length — the metre (abbreviation: m)

Mass — the kilogram (kg)

Time — the second (s)

Electric current — the ampere (A)

Luminous intensity — the candela (cd)

Temperature — the kelvin (K)

For many applications some of the above units are too small or too large and hence multiples and sub-multiples of these units are often needed. These multiples and sub-multiples are given special names as follows:

Multiplication Factor		Prefix	Symbol
1 000 000 000 000	$= 10^{12}$	tera	T
1 000 000 000	$= 10^{9}$	giga	G
1 000 000	$= 10^{6}$	mega	M
1 000	$= 10^{3}$	kilo	k
100	$= 10^{2}$	hecto	h
10	$= 10^{1}$	deca	da
0.1	$= 10^{-1}$	deci	d
0.01	$= 10^{-2}$	centi	c
0.001	$= 10^{-3}$	milli	m
0.000 001	$= 10^{-6}$	micro	μ
0.000 000 001	$= 10^{-9}$	nano	n
0.000 000 000 001	$= 10^{-12}$	pico	p
0.000 000 000 000 001	$= 10^{-15}$	femto	f
0.000 000 000 000 000 001	$= 10^{-18}$	atto	a

Preferred Units are those which have multiplication factors 10^{12}, 10^9, 10^6, 10^3, 10^{-3}, 10^{-6}, 10^{-9}, 10^{-12}, 10^{-15} and 10^{-18}. Thus 5000 metres should be written as 5 kilometres (5×10^3 metres) and *not* 50 hectometres (50×10^2 metres).

Choice of which unit to use will often be yours, but there are usually some other guidelines such as the accepted unit for a certain measurement. For example distances between towns on a road map are given in kilometres.

Length

The SI base unit of length is the metre, which is suitable for workshop or plan sites, but too large for plate thickness and too small for geographical distances, so the use of multiples and submultiples is needed. For

large distances, the kilometre	$1\,km = 10^3\,m$
small lengths, the millimetre	$1\,mm = 10^{-3}\,m$
tiny lengths, the micrometre (or micron)	$1\,\mu m = 10^{-6}\,m$

The above are a selection of preferred units — one non-preferred unit in common use for small distances is the centimetre where $1\,cm = 10\,mm = 10^{-2}\,m$.

Area

The basic unit of area is the square metre (m^2). Also for

large areas, the square kilometre	$1\,km^2 = 10^6\,m^2$
field areas, the square hectometre (or hectare)	$1\,hm^2 \text{ (or } 1\,ha) = 10^4\,m^2$
lesser areas, the square decametre (or are)	$1\,dam^2 \text{ (or } 1\,a) = 10^2\,m^2$
small areas, the square centimetre	$1\,cm^2 = 10^{-4}\,m^2$
tiny areas, the square millimetre	$1\,mm^2 = 10^{-6}\,m^2$

Although non-preferred, the units hectare, are and square centimetre are in common use.

Volume (or Capacity)

The basic unit of volume is the cubic metre (m^3) — this is suitable for most large volumes. For smaller volumes we use:

everyday measure, the cubic
decimetre (or litre*) $1 \, \mathrm{dm}^3 \; (\text{or} \; 1 \, \ell) \; = \; 10^{-3} \mathrm{m}^3$

small measures, the cubic
centimetre (or millilitre*) $1 \, \mathrm{cm}^3 \; (\text{or} \; 1 \, \mathrm{m}\ell) \; = \; 10^{-6} \mathrm{m}^3$

tiny volumes, the cubic
millimetre $1 \, \mathrm{mm}^3 \; = \; 10^{-9} \mathrm{m}^3$

*The litre (ℓ) has become the common unit for liquid measure, and to four-figure accuracy may be treated as $1 \, \mathrm{dm}^3$, but for any precise calculation the relationship between the litre and cubic decimetre is $1 \, \text{litre} = 1.000 \, 028 \, \mathrm{dm}^3$.

Mass

The SI base unit of mass is the kilogram, suitable for everyday use. For

large mass, the tonne (the metric ton) $1 \, \mathrm{t} = 10^3 \, \mathrm{kg}$

small mass, the gram $1 \, \mathrm{g} = 10^{-3} \, \mathrm{kg}$

tiny mass, the milligram $1 \, \mathrm{mg} = 10^{-3} \mathrm{g} = 10^{-6} \mathrm{kg}$

minute mass, the microgram $1 \, \mu\mathrm{g} = 10^{-6} \mathrm{g} = 10^{-9} \mathrm{kg}$

The tonne (metric ton) is very nearly equal to an Imperial ton (see p. 30). The symbol, t, is used only for the tonne; the Imperial ton should be written in full.

EXAMPLE 3.1

A measurement is 18 350 000 metres. Express this measurement in suitable preferred units.

Now

$$18 \, 350 \, 000 \, \mathrm{m} \; = \; 18 \, 350 \times 10^3 \, \mathrm{m} \; = \; 18 \, 350 \; \text{kilometres (km)}$$
$$= \; 18.35 \times 10^6 \, \mathrm{m} \; = \; 18.35 \; \text{megametres (Mm)}$$

EXAMPLE 3.2

An extremely small aperture has width 0.000 000 82 metres. Express this measurement in suitable preferred units.

Now

$$0.000 \, 000 \, 82 \, \mathrm{m} \; = \; 0.000 \, 82 \times 10^{-3} \, \mathrm{m} \; = \; 0.0082 \; \text{millimetres (mm)}$$
$$= \; 0.82 \times 10^{-6} \, \mathrm{m} \qquad = \; 0.82 \; \text{microns} \; (\mu\mathrm{m})$$
$$= \; 820 \times 10^{-9} \qquad\quad = \; 820 \; \text{nanometres (nm)}$$

EXAMPLE 3.3

How many hectares are there in a field 1 kilometre long and 400 metres wide?

Now   1 kilometre = 1000 metres

Thus   Field area = 1000×400 = 400 000 m^2

$$= 40 \times 10^4 \, m^2$$

$$= 40 \text{ hectares (ha)}$$

Exercise 3.1

Express more briefly each of the following with a suitable preferred unit.

1) 8000 m   2) 15 000 kg   3) 3800 km

4) 1 800 000 kg   5) 0.007 m   6) 0.000 001 3 m

7) 0.028 kg   8) 0.000 36 km   9) 0.000 064 kg

10) 0.0036 A

IMPERIAL SYSTEM OF UNITS

The old British units are still in use. The more common ones are listed below together with their SI conversion factors.

Length   Conversion factors

12 inches (in) = 1 foot (ft)   1 in = 25.4 mm

3 feet (ft) = 1 yard (yd)   1 ft = 0.305 m

1760 yards (yd) = 1 mile   1 mile = 1.61 km

Area

1 acre = 4840 yd^2 1 acre = 0.405 ha

Volume (Capacity)

8 pints = 1 gallon (gal)   1 gal = 4.55 ℓ

Mass

$$16 \text{ ounces (oz)} = 1 \text{ pound (lb)} \qquad 1 \text{ lb} = 0.454 \text{ kg}$$

$$112 \text{ pounds (lb)} = 1 \text{ hundredweight} \qquad 1 \text{ ton} = 1020 \text{ kg}$$
$$\text{(cwt)}$$

$$20 \text{ cwt} = 1 \text{ ton} \qquad\qquad \text{or } 1 \text{ ton} \approx 1000 \text{ kg}$$

$$\approx 1 \text{ tonne (t)}$$

EXAMPLE 3.4

Express 5.24 yards in metres.

Since there are 3 ft/yd and 0.305 m/ft

then
$$5.24 \text{ yd} = 5.24 \text{ yd} \times \left(\frac{3 \text{ ft}}{1 \text{ yd}}\right) \times \left(\frac{0.305 \text{ m}}{1 \text{ ft}}\right)$$

$$= 4.79 \text{ m correct to 3 s.f.}$$

EXAMPLE 3.5

Change 72 kilograms into hundredweight.

Since there are $\dfrac{1}{0.454}$ lb/kg and $\dfrac{1}{112}$ cwt/lb

then
$$72 \text{ kg} = 72 \text{ kg} \times \left(\frac{1 \text{ lb}}{0.454 \text{ kg}}\right) \times \left(\frac{1 \text{ cwt}}{112 \text{ lb}}\right)$$

$$= 1.42 \text{ cwt correct to 3 s.f.}$$

EXAMPLE 3.6

What is a pint of beer measured in litres? Also how many pints would I get if I ordered a litre of beer?

Since there are $\frac{1}{8}$ gal/pt and 4.55 ℓ/gal

then
$$1 \text{ pt} = 1 \text{ pt} \times \left(\frac{1 \text{ gal}}{8 \text{ pt}}\right) \times \left(\frac{4.55 \text{ ℓ}}{1 \text{ gal}}\right)$$

giving
$$1 \text{ pt} = 0.57 \text{ litres correct to 2 s.f.}$$

Now if I said that $1 ℓ = \frac{1}{0.57}$ pt from the result above, I would arrive at $1 ℓ = 1.75$ pt but I cannot say that this is correct to 3 s.f., since the 0.57 was rounded correct to only 2 s.f. Thus I must either content myself by saying that 1 litre = 1.8 pints or start afresh with conversion factors correct to 3 s.f.

Exercise 3.2

Give the answers to three significant figures unless there is a good reason for doing otherwise.

1) How many ft are there in 23 cm?

2) How many m are there in 880 yd?

3) Express 2.16 in as a number of mm.

4) How many inches are there in 20.1 mm?

5) Give 4.2 square feet as square metres.

6) How many in^2 are there in 823 cm^2?

7) A volume is 8 ft^3 — give this in m^3.

8) How many litres in 25 gallons?

9) How many cm^3 are there in 3 in^3?

10) 2 cubic metres contain how many cubic inches?

11) How many pints are there in 25 litres?

12) State a mass of 150 kg in pounds.

13) How many kg are there in 25 cwt?

14) A drawing dimension is given as 0.906 ± 0.0012 inches. Convert this to mm.

15) Three important dimensions of a 6 BA thread are: diameter 2.8 mm, pitch 0.53 mm, and depth of thread 0.0125 mm. Convert these to inches.

16) A tank is to hold 150 gallons.
(a) Find its volume in ft^3.
(b) How many litres does the tank hold?
(c) What is the volume in m^3?

17) How much must be ground off a plug gauge 1.625 inches in diameter so that it can be used for checking a hole of 41 mm diameter? Give your answer to the nearest 0.001 in.

18) Express a speed of 60 mile/hour as
(a) ft/s   (b) m/s   (c) km/h.

19) A vehicle travels 30 miles to the gallon of petrol. How many kilometres will it travel on 4 litres of petrol?

4. TABLES AND CHARTS

Outcome:

1. *Use conversion tables — interpolate.*
2. *Use parallel scale conversion charts.*
3. *Read nomographs.*
4. *Construct calibration curves.*

5. *Construct Gantt charts: use for forecasting and optimisation.*
6. *Recognise spreadsheets and appreciate their uses.*

CONVERSION TABLES — ILLUSTRATING USE OF INTERPOLATION

An example of the above is a table connecting temperature measurement in degrees Fahrenheit (°F) with degrees Celsius (°C).

EXAMPLE 4.1

Convert: **a)** 122.7°F to °C   **b)** 43.3°F to °C

c) 62.7°C to °F   **d)** −5.3°C to °F

An extract from the temperature conversion table is given opposite.

a) In this table the mean differences are listed under the heading 'interpolation' and are given at the bottom of the table.

If we consider 122.7°F we can find values in the body of the tables for the whole number values immediately above and below, i.e. 122°F and 123°F.

Thus   122°F = 50.0°C   and   123°F = 50.6°C

Now we can deduce that the required value for 122.7°F will lie between 50.0°C and 50.6°C. and being a little over half way may well be 50.4°C.

This process is called *interpolation* and is simplifed by use of the figures given at the foot of the table.

If we look along the top row of numbers, we can find 0.7°F and this is equivalent to 0.4°C which is shown immediately below in the second row.

°F	0	1	2	3	4	5	6	7	8	9
TEMPERATURE Degrees Fahrenheit to degrees Celsius (Centigrade)										
0	−17.8	−17.2	−16.7	−16.1	−15.6	−15.0	−14.4	−13.9	−13.3	−12.8
10	−12.2	−11.7	−11.1	−10.6	−10.0	−9.4	−8.9	−8.3	−7.8	−7.2
20	−6.7	−6.1	−5.6	−5.0	−4.4	−3.9	−3.3	−2.8	−2.2	−1.7
30	−1.1	−0.6	0	0.6	1.1	1.7	2.2	2.8	3.3	3.9
40	4.4	5.0	5.6	6.1	6.7	7.2	7.8	8.3	8.9	9.4
50	10.0	10.6	11.1	11.7	12.2	12.8	13.3	13.9	14.4	15.0
60	15.6	16.1	16.7	17.2	17.8	18.3	18.9	19.4	20.0	20.6
70	21.1	21.7	22.2	22.8	23.3	23.9	24.4	25.0	25.6	26.1
80	26.7	27.2	27.8	28.3	28.9	29.4	30.0	30.6	31.1	31.7
90	32.2	32.8	33.3	33.9	34.4	35.0	35.6	36.1	36.7	37.2
100	37.8	38.3	38.9	39.4	40.0	40.6	41.1	41.7	42.2	42.8
110	43.3	43.9	44.4	45.0	45.6	46.1	46.7	47.2	47.8	48.3
120	48.9	49.4	50.0	50.6	51.1	51.7	52.2	52.8	53.3	53.9
130	54.4	55.0	55.6	56.1	56.7	57.2	57.8	58.3	58.9	59.4
140	60.0	60.6	61.1	61.7	62.2	62.8	63.3	63.9	64.4	65.0
150	65.6	66.1	66.7	67.2	67.8	68.3	68.9	69.4	70.0	70.6
160	71.1	71.7	72.2	72.8	73.3	73.9	74.4	75.0	75.6	76.1
170	76.7	77.2	77.8	78.3	78.9	79.4	80.0	80.6	81.1	81.7
180	82.2	82.8	83.3	83.9	84.4	85.0	85.6	86.1	86.7	87.2
190	87.8	88.3	88.9	89.4	90.0	90.6	91.1	91.7	92.2	92.8

Interpolation: deg F:	0.1	0.2	0.3	0.4	0.5	0.6	0.7	0.8	0.9
deg C:	0.1	0.1	0.2	0.2	0.3	0.3	0.4	0.4	0.5

Thus　　　　122.7°F　is found as　　　50.0

+　0.4

50.4

Hence　　　　122.7°F = 50.4°C

b) Now for 18.2°F we can find that 18°F = −7.8 from the body of the tables. Also the 'interpolation' numbers give 0.2°F as being equivalent to 0.1°C.

Thus　　　　18.2°F　is found as　　−7.8

+ 0.1

−7.7

This result should be checked very carefully, bearing in mind the negative numbers. This check involves looking at the values in the table for the whole numbers immediately above and below 18.2°F. Now 18°F = −7.8 and 19°F = −7.2, and so the value −7.7 is reasonable.

Hence $$18.2\,°F = -7.7\,°C$$

c) Conversion from °C to °F may be carried out using the tables 'in reverse'. We must look in the body of the table in order to find numbers as close to the given 62.7 as possible. They are 62.2 corresponding to 144 °F and 62.8 corresponding to 145 °F. Now the interpolation numbers show that 0.5 °C is equivalent to 0.9 °F.

Thus

$$62.7\,°C = 62.2\,°C + 0.5\,°C = 144\,°F + 0.9\,°F = 144.9\,°F$$

A rough check shows this is reasonable since it lies between 144°F and 145 °F.

d) We again use the method in part c) using extreme caution with the negative numbers. Now $-5.3\,°C$ lies between -5.6 corresponding to 22 °F, and -5.0 corresponding to 23 °F. The interpolation numbers show that 0.3 °C is equivalent to 0.5 °F or 0.6 °F.

Thus

$$-5.3\,°C = -5.6\,°C + 0.3\,°C = 22\,°F + 0.5\,°F = 22.5\,°F$$

A rough check shows this answer is reasonable.

PARALLEL SCALE CONVERSION CHARTS

A system of parallel scales may be used when we wish to convert from one set of units to another related set.

An example of the above is a chart relating British Imperial pound (lb) mass units with SI kilogram (kg) mass units.

EXAMPLE 4.2

Convert: **a)** 3.4 lb to kg   **b)** 568 lb to kg

              c) 1.8 kg to lb   **d)** 0.23 kg to lb.

The conversion chart is shown below:

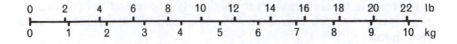

Observation of the scales shows that we cannot expect accuracy greater than 0.1 on either of them.

a) Direct reading from the adjacent scale shows that

$$3.4\,lb \;=\; 1.6\,kg$$

b) Since 568 lb is outside the range of the scales we may express it in standard form. Thus $568\,lb = 5.68 \times 10^2\,lb$.

Bearing in mind the accuracy limitations we will have to consider 5.68 as 5.7 and obtain

$$568\,lb \;=\; 5.7 \times 10^2\,lb \;=\; 2.6 \times 10^2\,kg \;=\; 260\,kg$$

correct to two significant figures only.

c) Direct reading from the adjacent scales shows that

$$1.8\,kg \;=\; 4.0\,lb$$

d) Although 0.23 kg may be found directly on the scale, accuracy will be improved if we express it as a standard number, and proceed as in part c).

Thus $0.23\,kg \;=\; 2.3 \times 10^{-1}\,kg \;=\; 5.1 \times 10^{-1}\,lb \;=\; 0.51\,lb$

NOMOGRAPHS

These are charts which represent graphically mathematical laws or relationships.

EXAMPLE 4.3

The data given on the chart shown in Fig. 4.1 refers to wires of circular cross-section supporting dead loads. The chart connects stress (MN/m^2) at failure of the wire, the load (kg) which causes the wire to break, and the diameter (mm) of the wire. Use the chart to find:

a) the stress at failure if 600 kg breaks a wire of 4 mm diameter,

b) the diameter of a wire which breaks under a load of 850 kg with a stress at failure of $1100\,MN/m^2$.

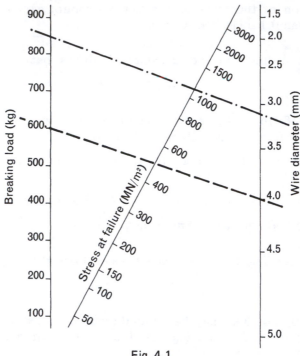

Fig. 4.1

The chart is used by placing a straight edge, such as the edge of a rule, across the three scales.

a) The dashed line shows the straight edge passing through 600 kg on the load scale and 4 mm on the wire diameter scale. The edge cuts the sloping scale at 500 MN/m², and this is the required stress at failure.

Careful judgement is needed in reading off the scales in between the marked points—especially the non-linear scales (those with unequal intervals).

b) The chain-dotted line shows the straight edge passing through 850 kg on the load scale and 1100 MN/m² on the stress at failure scale. The edge then cuts the right-hand vertical scale at 3.1 mm and this is the required diameter of the wire.

EXAMPLE 4.4

In calculations on electrical circuits we often need to find the single resistance, R_E ohms, which is equivalent to two resistances, R_A and R_B ohms connected in 'parallel' arrangement. The chart below in Fig. 4.2 relates the values of R_A, R_B, and R_E to each other. Use the chart to find:

a) R_E when $R_A = 5$ ohms and $R_B = 7$ ohms,

b) R_E when $R_A = 3.7$ ohms and $R_B = 2.4$ ohms.

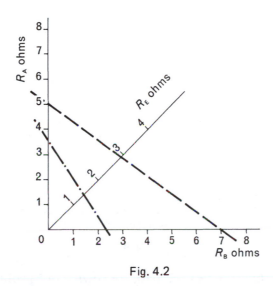

Fig. 4.2

The chart is used in a similar manner to that of the previous example by placing a straight edge across the scales.

a) The dashed line shows the straight edge passing through 5 ohms on the R_A scale and 7 ohms on the R_B scale. The intersection on the R_E scale gives 2.9 ohms and this is the required value of R_E.

b) The chain-dotted line shows the position of the straight edge for $R_A = 3.7$ ohms and $R_B = 2.4$ ohms cutting the inclined axis at 1.4 ohms, which is the required value of R_E.

EXAMPLE 4.5

The specific speed is a number used in the design of hydraulic turbines. It depends on the speed of the turbine (rev/min), the head of water available (metres), and the power output from the machine (kilowatts). The chart shown in Fig. 4.3 enables the specific speed to be found for any given speed, head, and power output. Use the chart to find the specific speed of a turbine running at 1400 rev/min, under a head of 120 metres and developing 20 megawatts of power.

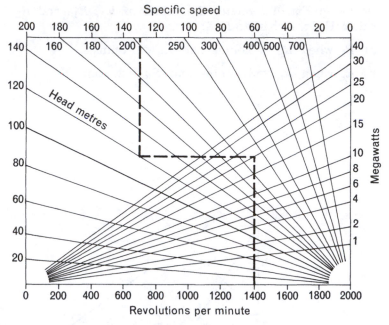

Fig. 4.3

The dashed line follows the path for solving our problem. We first locate 1400 rev/min on the bottom horizontal axis line. We then go vertically until meeting the line indicating 20 megawatts (this is one of the series of lines which radiate upwards to the right from the left-hand lower corner). We then move horizontally until we meet the line indicating 120 metres (this is one of the series of lines which start from the lower right corner and radiate upwards to the left). We then proceed vertically upwards until meeting the top horizontal line on which the required specific speeds are marked. This is 128. Accuracy is limited on this type of chart and it would be more reasonable to give an answer of 130.

CALIBRATION CURVES

These are curves—often straight lines—enabling data of one type (e.g. deflections) obtained from measuring equipment, to be converted to data of another type (e.g. loads).

A typical piece of measuring equipment is a proving ring (Fig. 4.4) which comprises a steel ring with a flat 'foot' and a flat 'hat' to support compressive loads. The dial gauge is used to measure the deflection of the ring for various loads. A typical ring is 250 mm diameter with a rectangular section 30 mm by 25 mm. Before the ring may be used to measure loads it has to be calibrated.

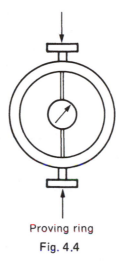

Proving ring

Fig. 4.4

EXAMPLE 4.6

A proving ring is mounted in a testing machine so that compressive loads may be applied. The load is increased steadily and at intervals dial gauge readings are taken corresponding to particular loads. Draw a suitable calibration graph for the following results:

Load (kN)	5	10	15	20	25	30	35	40	45	50
Dial gauge reading	35	66	97	130	160	190	224	254	285	317

The calibration graph is shown in Fig. 4.5.

EXAMPLE 4.7

The proving ring in the Example 4.6 was used to find the load at four places on a motor vehicle. For each measurement the vehicle was jacked up and then lowered on to the proving ring. The four dial gauge readings were 104, 110, 171, and 175. What were the corresponding loads?

The temptation here is to give the loads to an accuracy greater than that justified from the calibration graphs (Fig. 4.5). Bearing this in mind we suggest that the loads are, in order, 16, 17, 27 and 27 kN.

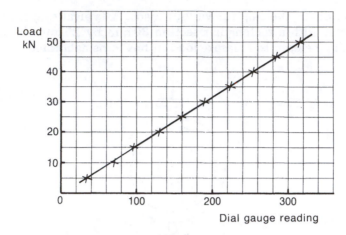

Fig. 4.5

GANTT CHARTS

There are many types of Gantt charts but all are basically horizontal bar charts on a horizontal time scale. Two examples are given below, one showing a production progress chart and the other a machine loading chart.

EXAMPLE 4.8

The planned production of an article for the first three weeks of manufacture are 150 for week 1, 200 for week 2, and 230 for week 3. The actual production of week 1 was only 130, that for week 2 was as planned, and an extra 20 were completed in week 3 making up for the shortfall in week 1. Represent this data on a production progress chart.

In the chart (Fig. 4.6) each horizontal interval represents an equal period of time, namely 1 week, but does not represent the same planned production. This is given weekly by the first figure in row two, whilst the second figure is the cumulative total. The actual production achieved per week is given by the upper bar whilst the lower bar gives the cumulative total.

	Week 1		Week 2		Week 3	
	150	150	200	350	230	580
Actual		130		200		230
					20	
Cumulative		130		330		580

Fig. 4.6

EXAMPLE 4.9

A small machine shop is manufacturing a product assembled from three components a, b and c. The available machines are a horizontal miller (HM), a vertical miller (VM), a centre lathe (CL), a universal grinder (UG) and a pedestal drill (PD). Fig. 4.7 shows the sequence of operations required to be made on each component and the number of hours (in brackets) required to complete that operation. Construct a Gantt chart showing the loading of individual machines.

Component \ Sequence	1	2	3	4	5	Total hours
a	HM(2)	UG(2)	VM(4)	UG(3)	PD(2)	13
b	CL(2)	HM(6)	CL(4)	UG(4)		16
c	CL(2)	VM(4)	HM(5)	CL(3)		14

Fig. 4.7

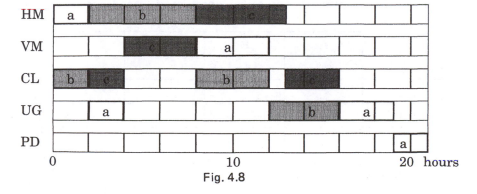

Fig. 4.8

Fig. 4.8 shows a possible loading arrangement giving a total of 21 hours to complete the operations on components for one complete product.

Fig. 4.9 shows a possible alternative which gives a reduction of 3 hours to complete one product.

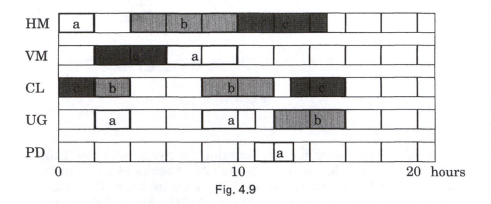

Fig. 4.9

SPREADSHEETS

A spreadsheet is a sheet of ruled paper on which information may be displayed. Each oblong is called a 'cell', and each 'cell' has an address according to its column and row, e.g. B3 in Fig. 4.10, which shows part of a spreadsheet.

A spreadsheet format may be generated on a computer screen of hardware comprising typically a keyboard processor, VDU and mouse, disk drive and printer together with a compatible software 'package'. This will enable you to enter words and numbers in particular cells, perform any arithmetic needed, and then finally obtain a printed copy of the completed spreadsheet.

	A	B	C	D
1				
2				
3		cell B3		
4				

Fig. 4.10

EXAMPLE 4.10

In June we purchased the following equipment: 30 spanners @ £3.00 each, 10 screwdrivers @ £2.30 each, 20 rules @ £1.50 each and 2 vices @ £48 each. We need to know the total cost of the spanners, etc., with and without VAT at 20%. We also need the grand total cost with and without VAT.

A suitable display would be:

	A	B	C	D	E
1	Item	Unit cost	Number bought	Total without VAT	Total with 20% VAT
2	Spanners	3.00	30	90.00	108.00
3	Screw-drivers	2.30	10	23.00	27.60
4	Rules	1.50	20	30.00	36.00
5	Vices	48.00	2	96.00	115.20
6				239.00	286.80

Fig. 4.11

An explanation of the spreadsheet shown in Fig. 4.11. The program package would enable you to enter a formula into a cell. For instance (B2)*(C2) entered into cell D2 would mean that the number in cell B2 would be multiplied by the number in cell C2 and the result stored, and displayed, in cell D2.

Other instructions would be:

(D2)*1.20 into cell E2 — total cost with 20% VAT included

Similar instructions for rows 3, 4 and 5.

Sum(D2 ... D5) in cell D6 — grand total spent without VAT

Sum(E2 ... E5) in cell E6 — grand total spent including VAT

VERSATILITY OF SPREADSHEETS

You will see from the last example how useful it is to have a tidy display of transactions which have already taken place. But fore-casting what happens in the future is often even more important

—you can enter, just as easily, the projected purchases for July, August, etc. and have the money involved displayed so that you may budget carefully.

Some software packages enable spreadsheets to display graphs, pie-charts, histograms, etc.

Other more common uses of spreadsheets are:

> Laboratory results and analysis
> Conversion tables (e.g. money exchange rates, °F to °C)
> Display of statistical information
> Mathematical calculations (e.g. iterative methods)
> Budget accounting
> Forecasting sales

In fact the use of spreadsheets is unlimited if a display is required.

Exercise 4.1

1) Use the temperature table on p. 33 to convert:

(a) 76.0°F to °C   (b) 161.5°F to °C

(c) 68.7°C to °F   (d) −11.3°C to °F

2) Use the conversion chart on p. 34 to convert:

(a) 7.7 kg to lb   (b) 332 kg to lb

(c) 18.3 lb to kg   (d) 0.42 lb to kg

3) Use Fig. 4.1 to find:

(a) the stress at failure if a 3 mm diameter wire breaks under a load of 400 kg,

(b) the diameter of a wire which breaks under a load of 728 kg if the stress at failure is 1760 MN/m²,

(c) the breaking load of a 4.43 mm diameter wire which has a stress of 273 MN/m²

4) Using Fig. 4.2 find the value of:

(a) R_E when $R_A = 6.7$ ohms and $R_B = 4$ ohms

(b) R_A when $R_B = 5.1$ ohms and $R_E = 3.1$ ohms

(c) R_B when $R_A = 2$ ohms when $R_E = 1.21$ ohms.

5) Use Fig. 4.3 to find the specific speed of a turbine having:

(a) a speed of 1200 rev/min, a head of 120 m and developing 20 megawatts of power,

(b) a head of 60 m and developing 4 megawatts of power at a speed of 600 rev/min.

6) The tension in a wire is measured by clamping a meter on to the wire. A handle is rotated and the wire is pulled sideways to a set position where the reading of the handle position corresponds to the tension in the wire. The meter is calibrated by fitting it to a wire mounted in a testing machine and then taking tensile load readings and recording them against corresponding meter readings — the results are as follows:

Load on wire (kN)	202	269	349	429	492	553	621	696	756	813
Meter reading	15	20	25	30	35	40	45	50	55	60

Load on wire (kN)	906	966	1020	1070
Meter reading	65	70	75	80

(a) Plot a suitable calibration graph.
(b) Use the graph to find the tensions in a wire if the meter readings are 481 and 630.

7) A new production line was set up and for the first four months, January to April, the estimated number of hours to be worked were 800, 1000, 1200 and 1400, whilst the actual hours worked monthly were 700, 1000, 1400 and 1000. Draw a suitable Gantt chart to show this information.

8) A household budget for January, February and March was as follows. Remaining constant for each month: Mortgage £320, Groceries £330, Car payment £220, Personal £270 and Insurance £80. The quarterly bills were Telephone £141 in January, and Electricity £86 in March.
Draw up a suitable spreadsheet format displaying the above information showing also the totals spent per month. Also list typical instructions for obtaining the total to each column.

5. LOGIC DESIGN AND NUMBER SYSTEMS

Outcome:

1. Recognise a constant, a variable and an operation.
2. Understand and use the operations NOT, AND and OR.
3. Construct and use truth tables.
4. Simplify Boolean expressions using Veitch–Karnaugh maps.
5. Understand the denary, octal, hexadecimal and binary systems of numbers.
6. Perform simple octal and binary arithmetic.
7. Appreciate a half-adder and full-adder computer circuit.

INTRODUCTION

This chapter will introduce you to the ideas behind the design of logic circuits which is the way most computer and microprocessors work. You will also become acquainted with number systems — denary, octal, hexadecimal and binary.

We make use of Boolean algebra which, together with the use of binary arithmetic, makes possible our understanding of logic circuitry. To help us handle Boolean expressions we shall use two types of tabular display — namely Truth tables and Vietch–Karnaugh maps.

BOOLEAN ALGEBRA

This topic comprises constants, variables and operations.

A CONSTANT

In normal use a constant is a fixed value or quantity, e.g. 7, 9, ...
In Boolean algebra there are only two constants, namely 0 and 1.
A typical transistor circuit may use 0 volt to represent 0, and +5 volt to represent 1.

46

A VARIABLE

In ordinary algebra a variable is a quantity which can change by taking the value of any constant. We use variables x, y ... to represent constants such as 4, 7, $6\frac{1}{2}$ ...

In Boolean algebra variables are denoted by capital letters (italic) such as A, B, X ... but since there are only two possible constants then a variable can only have only one of two values — either 0 or 1.

OPERATIONS

In ordinary algebra we add, subtract, multiply, divide and even change sign when operating on quantities.

In Boolean algebra totally different operations are used. We shall concentrate on three basic operations, namely NOT, AND and OR. There are others but these will suffice at this stage of our studies.

The operations are carried out in electric circuits by electronic units called 'gates', or alternatively by switches in specially designed circuits.

The NOT Operation

This is indicated by placing a bar ($^-$) over the constant or variable.

Alternatively due to typesetting limitations a solidus (/) may be used before the constant or variable.

Definition of NOT operation:	$\bar{1} = 0$	$\bar{0} = 1$

Gate symbol   *Truth table*   *Explanation*

A —[NOT]— $\bar{A}$

Input A	Output $\bar{A}$
0	1
1	0

When the input signal is 0 then the output is 1, and vice versa.

The AND Operation

This is shown by a dot (.) between two constants or variables.

Definition of AND operation:	$0.0 = 0$ $0.1 = 0$ $1.0 = 0$ $1.1 = 1$

Gate symbol *Truth table* *Explanation*

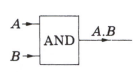

Input		Output
A	B	$A.B$
0	0	0
0	1	0
1	0	0
1	1	1

The output signal is 1 only when *all* the input signals are 1.

An alternative to the AND logic gate is a circuit containing two switches in series, a closed switch representing 1 and an open switch representing 0.

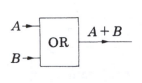

The OR Operation

This is shown by a + symbol between two constants or variables.

Definition of OR operation:	$0 + 0 = 0$ $0 + 1 = 1$ $1 + 0 = 1$ $1 + 1 = 1$

Gate symbol *Truth table* *Explanation*

Input		Output
A	B	$A + B$
0	0	0
0	1	1
1	0	1
1	1	1

The output signal is 1 when one or more of the input signals is 1.

An alternative to the OR logic gate is a circuit containing two switches in parallel, again a closed switch representing 1 and an open switch 0:

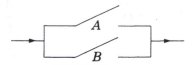

Operating on Constants

We may use the AND with more than one constant, such as
$0.1.0.\overline{1}$ and the value of this may be found by taking the terms in
pairs — brackets may be used as in ordinary algebra. Thus:

$$0.1.0.\overline{1} = 0.1.0.0 \qquad \text{since} \quad \overline{1} = 0$$

$$= (0.1).(0.0)$$

$$= 0.0 \qquad\qquad \text{since} \quad 0.1 = 0$$
$$\text{and} \quad 0.0 = 0$$

$$= 0 \qquad\qquad\quad \text{since} \quad 0.0 = 0$$

You will see that when constants are ANDed it only needs one 0
or $\overline{1}$ term to make the result zero. Remember that for a 1 result
when ANDing all the constants must be 1 or $\overline{0}$.

The diagram opposite shows
$0.1.0.\overline{1} = 0$ using two logic
gates. Note the four inputs to
the AND gate.

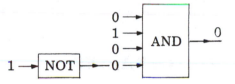

We may also use the OR with more than one constant, such as
$0 + 1 + \overline{0}$ and to find the value of this we proceed as before. Thus:

$$0 + 1 + \overline{0} = (0 + 1) + 1 \qquad \text{since} \quad \overline{0} = 1$$

$$= 1 + 1 \qquad\qquad \text{since} \quad 0 + 1 = 1$$

$$= 1 \qquad\qquad\quad \text{since} \quad 1 + 1 = 1$$

The diagram opposite shows
$0 + 1 + \overline{0} = 1$ using two logic
gates. Note the three inputs to
the OR gate.

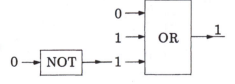

AND and OR may be used but when simplifying an expression
AND has preference over OR.

EXAMPLE 5.1

Evaluate $(\overline{1}+1).1+1.\overline{0}$

Now $(\overline{1}+1).1+1.\overline{0} = (0+1).1+1.1$ since $\overline{1} = 0$
                                                                  and $\overline{0} = 1$
                                $= 1.1+1.1$ since $0+1 = 1$
                                $= 1+1$ since $1.1 = 1$
                                $= 1$ since $1+1 = 1$

Exercise 5.1

Evaluate the following:

1) $0.1.0$ 2) $0.1.1.1$ 3) $0.\overline{1}.\overline{1}.0$

4) $1+\overline{1}$ 5) $\overline{0}+\overline{0}$ 6) $\overline{1}+0+1$

7) $1+0.\overline{1}$ 8) $\overline{0.\overline{1}+\overline{0}}$ 9) $1.0.\overline{1}+\overline{0}.1$

10) $(1+0).1$ 11) $0.(1+\overline{1})$ 12) $(1+1).(\overline{0}.1)$

13) Illustrate Questions 1–8 on diagrams with suitable logic gates. Assume that only inputs of 0 and 1 are available.

TRUTH TABLES

You have already met three truth tables which are simply a convenient way of displaying all possible values of input and corresponding output values. They may also be used to find the values of more complicated expressions.

Consider the expression $A.\overline{B}+B.\overline{A}$ whose truth table will cover all the possible combinations of constants 0 and 1 which can be given to variables A and B. Intermediate columns for $\overline{A}$, $\overline{B}$, $A.\overline{B}$ and $B.\overline{A}$ are introduced to make the working out clearer.

A	B	$\overline{A}$	$\overline{B}$	$A.\overline{B}$	$B.\overline{A}$	$A.\overline{B}+B.\overline{A}$
0	0	1	1	0	0	0
0	1	1	0	0	1	1
1	0	0	1	1	0	1
1	1	0	0	0	0	0

As you become more familiar with giving values to variables, like A and B here, you may well find it possible to do the working out mentally and dispense with a truth table — but beware mistakes are easily made!

EXAMPLE 5.2

Make a truth table for the expression $\overline{A}.B.\overline{C}$.

A	B	C	$\overline{A}$	B	$\overline{C}$	$\overline{A}.B.\overline{C}$
0	0	0	1	0	1	0
0	0	1	1	0	0	0
0	1	0	1	1	1	1
0	1	1	1	1	0	0
1	0	0	0	0	1	0
1	0	1	0	0	0	0
1	1	0	0	1	1	0
1	1	1	0	1	0	0

Note the repeat of column B for convenience between $\overline{A}$ and $\overline{C}$.

EXAMPLE 5.3

Use a truth table to verify the identity $A + \overline{A}.B = A + B$.

A	B	$\overline{A}$	$\overline{A}.B$	$A + \overline{A}.B$	$A + B$
0	0	1	0	0	0
0	1	1	1	1	1
1	0	0	0	1	1
1	1	0	0	1	1

The last two columns show that, for all possible values of variables A and B, the left hand side of the identify has the same value as the right hand side. Hence the identity $A + \overline{A}.B = A + B$ is verified.

Exercise 5.2

Verify the following identities using suitable truth tables:

1) $A + A.B = A$

2) $A.(A + B) = A$

3) $\overline{(A + B)} = \overline{A}.\overline{B}$

4) $\overline{A.B} = \overline{A} + \overline{B}$

5) $A.(B + C) = A.B + A.C$

6) $A + B.C = (A + B).(A + C)$

SIMPLIFYING BOOLEAN EXPRESSIONS

In the early 1950s Veitch and Karnaugh developed a method of tabulation in the form of a 'map'. We shall call these V–K maps.

Each V–K map contains 2^n squares for n variables.

Thus for 2 variables we have $2^2 = 4$ squares

and   for 3 variables we have $2^3 = 8$ squares

and   for 4 variables we have $2^4 = 16$ squares

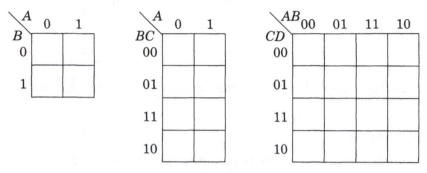

Note the change in order of values 11 and 10 in labelling the squares as compared with the value sequence on a truth table. This is because only one variable changes between adjacent squares.

The following examples will show how to plot on a V–K map.

EXAMPLE 5.4

Plot the expression $A.\overline{B} + B.C$ on a V–K map.

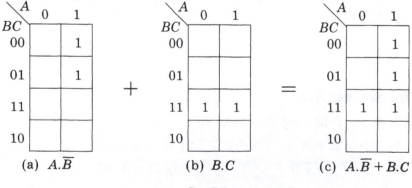

(a) $A.\overline{B}$ (b) $B.C$ (c) $A.\overline{B} + B.C$

Fig. 5.1

Since there are three variables A, B and C we require an eight square map. First consider $A.\overline{B}$. We shall plot this on its own grid (Fig. 5.1(a)) by putting a 1 in each square where the value of $A.\overline{B}$ equals 1. We can see by inspection that the required 1.1 occurs when $A = 1$ and $B = 0$. The value of the third variable C can be 0 or 1 and does not affect the value of $A.\overline{B}$ so we must plot 1 in squares for both $A = 1$, $B = 0$, $C = 0$, and $A = 1$, $B = 0$, $C = 1$. We say that C is undefined.

Fig. 5.1(b) shows the map for $B.C$ whose value is 1 when $B = 1$, $C = 1$ and either $A = 0$ or $A = 1$. Here A is undefined. Fig. 5.1(c) shows the final result of combining (a) and (b).

EXAMPLE 5.5

Plot the expression $A.\overline{B}.C + \overline{C}$ on a V-K map.

We do not need to plot $A.\overline{B}.C$ and $\overline{C}$ on individual maps — so let us look at the whole expression $A.\overline{B}.C + \overline{C}$. The OR (+) operation means that for a final 1 value then either $A.\overline{B}.C = 1$ or $\overline{C} = 1$, or both.

For $A.\overline{B}.C = 1$ then $A = 1$, $B = 0$ and $C = 1$.

For $\overline{C} = 1$ then $C = 0$ and both A and B are undefined.

The plot is shown in Fig. 5.2.

BC \ A	0	1
00	1	1
01		1
11		
10	1	1

$$A.\overline{B}.C + \overline{C}$$

Fig. 5.2

EXAMPLE 5.6

On a V-K map plot the expression $A.\overline{B}.X + B.Y + \overline{A}.B.X.Y$.

This time we have four variables so we need a grid with $2^4 = 16$ squares. Bearing in mind the OR operation, then for the whole expression to be 1 either $A.\overline{B}.X = 1$ for which $A = 1$, $B = 0$ and $X = 1$ and Y is undefined,

or $B.Y = 1$ for which $B = 1$, $Y = 1$ and both A and X are undefined,

or $\overline{A}.B.X.Y = 1$ for which $A = 0$, $B = 1$, $X = 1$ and $Y = 1$.

This last result finds a square already occupied by a 1; this does not matter as long as the square is filled already.

The final result is shown plotted in Fig. 5.3.

XY \ AB	00	01	11	10
00				
01		1	1	
11		1	1	1
10				1

$A.\overline{B}.X + B.Y + \overline{A}.B.X.Y$

Fig. 5.3

Simplified Terms for a V–K Map

These are found from a grouping of two '1 squares' if they are adjacent to each other in the same row or column, or if they are at opposite ends of the same row or column. Also each '1 square' may be used as many times as desired, but must be used at least once.

EXAMPLE 5.7

Use a V–K map to simplify the expression $A.B.C + A.\overline{B}.C$.

Fig. 5.4 shows the V-K map of $A.B.C + A.\overline{B}.C$. Now the two '1 squares' are adjacent and so may be grouped together. The grouped squares represent:

$$\left.\begin{array}{l} A = 1, \ B = 1, \ C = 1 \\ A = 1, \ B = 0, \ C = 1 \end{array}\right\} \text{ or } A.C \text{ with } B \text{ undefined.}$$

Thus $A.C$. is a simplified form and we may say that

$$A.B.C + A.\overline{B}.C = A.C$$

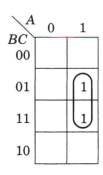

Fig. 5.4

EXAMPLE 5.8

Simplify $X.\overline{Y}.\overline{Z} + \overline{X}.\overline{Y}.Z + X.Y.\overline{Z} + \overline{X}.\overline{Y}.\overline{Z}$ using a V–K map.

Fig. 5.5 shows the V–K map of the given expression. There are two pairs of adjacent '1 squares', and one pair comprising '1 squares' at opposite ends of a column. The grouped squares represent:

$$\left. \begin{array}{l} X = 0, Y = 0, Z = 0 \\ X = 0, Y = 0, Z = 1 \end{array} \right\} \text{ or } \overline{X}.\overline{Y} \text{ with } Z \text{ undefined}$$

$$\left. \begin{array}{l} Y = 0, Z = 0, X = 0 \\ Y = 0, Z = 0, X = 1 \end{array} \right\} \text{ or } \overline{Y}.\overline{Z} \text{ with } X \text{ undefined}$$

$$\left. \begin{array}{l} X = 1, Z = 0, Y = 0 \\ X = 1, Z = 0, Y = 1 \end{array} \right\} \text{ or } X.\overline{Z} \text{ with } Y \text{ undefined}$$

Thus $\overline{X}.\overline{Y} + \overline{Y}.\overline{Z} + X.\overline{Z}$ is a simplified form and we may say that

$$X.\overline{Y}.\overline{Z} + \overline{X}.\overline{Y}.Z + X.Y.\overline{Z} + \overline{X}.\overline{Y}.\overline{Z} = \overline{X}.\overline{Y} + \overline{Y}.\overline{Z} + X.\overline{Z}$$

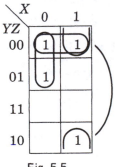

Fig. 5.5

Exercise 5.3

1) Plot the following expressions on a two variable V–K map:
(a) $A + A.B$ (b) $A + B + A.B$ (c) $X.\overline{Y} + Y.\overline{X}$

2) Plot the following expressions on a three variable V–K map:
(a) $A + \overline{B} + C$ (b) $A.\overline{B}.C + \overline{C}$ (c) $(\overline{A} + B).\overline{C}$

3) Plot the following expressions on a four variable V–K map:
(a) $W.X.Y.Z + W.Y.Z + W.Z$ (b) $A.B.\overline{C} + \overline{B}.C.D + \overline{A}$
(c) $Z + \overline{W}.\overline{Y}.\overline{X}$

4) Simplify the following expressions using suitable V–K maps:
(a) $A + \overline{A}.B$ (b) $\overline{A}.\overline{B}.C + \overline{A}.B.C$
(c) $\overline{A}.B.C + A.B.\overline{C} + A.B.C$ (d) $\overline{A}.B.\overline{C}.D + A.B.\overline{C}.D$

NUMBER SYSTEMS

The Denary System

Let us first consider the ordinary decimal (or denary) system which uses ten digits 0 to 9, and a number base of ten. For example:

$$2975 = 2000 + 900 + 70 + 5$$
$$= 2 \times 1000 + 9 \times 100 + 7 \times 10 + 5 \times 1$$
$$= 2 \times 10^3 + 9 \times 10^2 + 7 \times 10^1 + 5 \times 10^0$$

Another way of showing this is a series of divisions:

$$2975 \div 10 = 297 \text{ remainder } 5 - \text{units}$$
$$297 \div 10 = 29 \text{ remainder } 7 - \text{tens}$$
$$29 \div 10 = 2 \text{ remainder } 9 - \text{hundreds}$$
$$2 \div 10 = 0 \text{ remainder } 2 - \text{thousands}$$

$$2 \quad 9 \quad 7 \quad 5$$

Why did we choose a ten based system? Probably because we had a total of ten fingers and thumbs and these were our original calculating machine!

The Octal System

Suppose now that the inventor of a number system had lost both his thumbs in a hunting accident — so his counting would have to be done on eight fingers. The octal system has a number base of eight, and uses eight digits 0 to 7.

So let us use the series of divisions method to convert 2975 to octal:

$$2975 \div 8 = 371 \text{ remainder } 7 - \text{units}$$
$$371 \div 8 = 46 \text{ remainder } 3 - \text{eights}$$
$$46 \div 8 = 5 \text{ remainder } 6 - (8 \times 8)\text{s}$$
$$5 \div 8 = 0 \text{ remainder } 5 - (8 \times 8 \times 8)\text{s}$$

$$5 \quad 6 \quad 3 \quad 7$$

So the octal number 5637 represents $\Big\} 5 \times 8^3 + 6 \times 8^2 + 3 \times 8^1 + 7 \times 8^0$

Thus $\quad\quad\quad 2975 \text{ (denary)} = 5637 \text{ (octal)}$

Or using notation $\quad\quad 2975_{10} = 5637_8$

EXAMPLE 5.9

Convert 71 293 to octal.

$$71\,293 \div 8 = 8911 \text{ remainder } 5 \quad \text{units}$$
$$8911 \div 8 = 1113 \text{ remainder } 7 \quad \text{eights}$$
$$1113 \div 8 = 139 \text{ remainder } 1 \quad (8^2)\text{s}$$
$$139 \div 8 = 17 \text{ remainder } 3 \quad (8^3)\text{s}$$
$$17 \div 8 = 2 \text{ remainder } 1 \quad (8^4)\text{s}$$
$$2 \div 8 = 0 \text{ remainder } 2 \quad (8^5)\text{s}$$

Thus $\quad 71\,293_{10} = 2 \times 8^5 + 1 \times 8^4 + 3 \times 8^3 + 1 \times 8^2 + 7 \times 8^1 + 5$

or $\quad 71\,293_{10} = 213\,175_8$

EXAMPLE 5.10

Convert 174 023$_8$ to denary.

$$174\,023_8 = 1 \times 8^5 + 7 \times 8^4 + 4 \times 8^3 + 0 \times 8^2 + 2 \times 8^1 + 3$$

$$= 32\,768 + 28\,672 + 2048 + 0 + 16 + 3$$

$$= 63\,507$$

So $174\,023_8 = 63\,507_{10}$

Octal arithmetic

Counting in octal: 0, 1, 2, 3, 4, 5, 6, 7,   ?

There is no 8 so we continue with 10 meaning 1 eight and no units:

0, 1, 2, 3, 4, 5, 6, 7, 10, 11, 12, 13, 14, 15, 16, 17,
20, 21, 22, 23, ... 75, 76, 77,   ?

There is no 78 or 80 and we move to 100 which means one $8^2 = 64$ and no eights and no units, giving 76, 77, 100, 101, 102, ...

Adding: $1\,5\,6\,7_8$ Working from the R.H. column to
      $+\ \ \ 1\,5\,4_8$ the left, then

      $\underline{\quad 1\ 1\quad}$ ← carries $7 + 4 = 13_8$ giving 3 carry 1

      $1\,7\,4\,3_8$ $6 + 5 + 1 = 14_8$ giving 4 carry 1

                     $5 + 1 + 1 = 7_8$ giving 7 etc.

The Hexadecimal System

This is a number system with a base of sixteen. It may be used, as maybe also the octal system, in computer arithmetic in conjunction with the binary system which we shall meet next.

The hexadecimal system has a base of sixteen and thus needs sixteen digits. We have only ten digits in common use, namely 0 to 9, so we will have to invent another six — let us call these A, B, C, D, E and F. So counting in hexadecimal will be:

0, 1, 2, 3, 4, 5, 6, 7, 8, 9, A, B, C, D, E, F, 10, 11, 12, 13,
14, 15, 16, 17, 18, 19, 1A, 1B, 1C, 1D, 1E, 1F, 20, 21
23, 24, ...

We can change a hexagonal number to an ordinary denary number, and vice versa, by similar methods to those used for the octal system but using, of course, 16 instead of 8.

The Binary System

This is a number system using a base of two — this means that only two digits are needed, namely 0 and 1. You will immediately note, no doubt, why binary numbers are favoured in computer arithmetic as these are the two constants used in Boolean algebra logic.

Let us use the series of divisions method to convert 45 to binary:

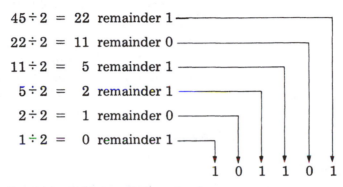

$$45 \div 2 = 22 \text{ remainder } 1$$
$$22 \div 2 = 11 \text{ remainder } 0$$
$$11 \div 2 = 5 \text{ remainder } 1$$
$$5 \div 2 = 2 \text{ remainder } 1$$
$$2 \div 2 = 1 \text{ remainder } 0$$
$$1 \div 2 = 0 \text{ remainder } 1$$

$$1 \quad 0 \quad 1 \quad 1 \quad 0 \quad 1$$

So the binary number 101101 represents:

$$1 \times 2^5 + 0 \times 2^4 + 1 \times 2^3 + 1 \times 2^2 + 0 \times 2^1 + 1 \times 2^0$$

or $\qquad 1 \times 32 + 0 \times 16 + 1 \times 8 + 1 \times 4 + 0 \times 2 + 1 \times 1$

or $\qquad 32 + 8 + 4 + 1 = 45$

Thus $\qquad\qquad 45 \text{ denary} = 101101 \text{ binary}$

or $\qquad\qquad\qquad 45_{10} = 101101_2$

Bicimals

In the denary system figures to the right of the decimal point are called decimals. In the binary system figures to the right of the bicimal point are called bicimals.

So the binary number 101.01101 represents:

$$1 \times 2^2 + 0 \times 2^1 + 1 \times 2^0 + 0 \times 2^{-1} + 1 \times 2^{-2} + 1 \times 2^{-3} + 0 \times 2^{-4}$$
$$+ 1 \times 2^{-5}$$

or $1 \times 4 + 0 \times 2 + 1 \times 1 + 0 \times \frac{1}{2} + 1 \times \frac{1}{4} + 1 \times \frac{1}{8} + 0 \times \frac{1}{16} + 1 \times \frac{1}{32}$

or $\qquad 4 + 1 + \frac{1}{4} + \frac{1}{8} + \frac{1}{32}$

or $\qquad 4 + 1 + 0.25 + 0.125 + 0.03125$

or $\qquad 5.40625$

Thus 101.01101 binary $= 5.40625$ denary

or $101.01101_2 = 5.40625_{10}$

To convert decimal 0.40625 to bicimal we use a series of multiplications:

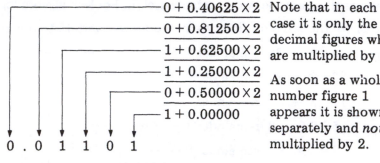

$0 + 0.40625 \times 2$	Note that in each
$0 + 0.81250 \times 2$	case it is only the
	decimal figures which
$1 + 0.62500 \times 2$	are multiplied by 2.
$1 + 0.25000 \times 2$	
	As soon as a whole
$0 + 0.50000 \times 2$	number figure 1
$1 + 0.00000$	appears it is shown
	separately and *not*
$0 . 0 \ 1 \ 1 \ 0 \ 1$	multiplied by 2.

Thus $0.40625_{10} = 0.01101_2$.

Binary Arithmetic

An understanding of binary arithmetic is useful since most computer circuits perform calculations using binary.

We shall show how to add and multiply in binary and then finish by looking at basic arithmetic logic circuits.

EXAMPLE 5.11

Add 10101_2 and 1111_2.   Working from the RH column to the left, then:

```
  1 0 1 0 1
+   1 1 1 1
  1 1 1 1 ← carries
  1 0 0 1 0 0₂
```

$1 + 1 = 10$ giving 0 carry 1

$0 + 1 + 1 = 10$ giving 0 carry 1

$1 + 1 + 1 = 11$ giving 1 carry 1

and so on.

EXAMPLE 5.12

Multiply 10011_2 by 1101_2.

```
        1 0 0 1 1
  ×       1 1 0 1
        1 0 0 1 1      (a)
        0 0 0 0 0      (b)
      1 0 0 1 1        (c)
    1 0 0 1 1          (d)
```

Sum of lines a. b. c and d.

The computer would not add the four lines a, b, c and d in one go, but would probably proceed in stages:

$$
\begin{array}{rl}
1\,0\,0\,1\,1 & \text{(a)} \\
+\,0\,0\,0\,0\,0 & \text{(b)} \\
\hline
1\,0\,0\,1\,1 & \text{(a)}+\text{(b)} \\
+\,1\,0\,0\,1\,1 & \text{(c)} \\
\hline
1\,0\,1\,1\,1\,1 & \text{(a)}+\text{(b)}+\text{(c)} \\
+\,1\,0\,0\,1\,1 & \text{(d)} \\
\hline
1\,1 & \leftarrow \text{carries} \\
\hline
1\,1\,1\,1\,0\,1\,1\,1_2 & \text{Answer: sum of a, b, c and d.}
\end{array}
$$

BASIC ARITHMETIC COMPUTER CIRCUITS

The basic arithmetic circuit is the ADDER. This will enable addition and multiplication (which, as you have seen, is merely successive addition) to be performed. There are two variants, namely the HALF-ADDER and the FULL-ADDER.

The Half-adder (Two Input)

This circuit (Fig. 5.6) has two inputs and two outputs. The inputs are two binary digits and the outputs are the sum and carry from binary addition — the corresponding truth table is also shown.

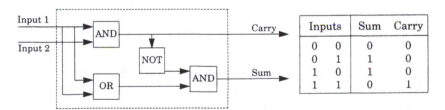

Inputs		Sum	Carry
0	0	0	0
0	1	1	0
1	0	1	0
1	1	0	1

Fig. 5.6 Half-adder circuit

The Full-adder (Three Input)

Here we have a circuit (Fig. 5.7) with three inputs — two binary and a 'carry in' from a previous stage. Again the outputs are the sum and 'carry out' from binary addition.

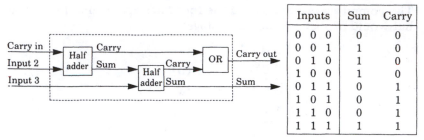

Fig. 5.7   Full-adder circuit

Exercise 5.4

1) Convert to denary
(a) 63_8 (b) 217_8 (c) 463_8

2) Convert to octal
(a) 29_{10} (b) 243_{10} (c) 1796_{10}

3) Find the sum of
(a) 637_8 and 56_8 (b) 4360_8 and 1234_8

4) Check the results of Question 3 by converting each number to denary, adding, and then converting back to octal.

5) Convert to binary
(a) 23_{10} (b) 125_{10} (c) 97_{10}

6) Convert to denary
(a) 10110   (b) 111001   (c) 1011010

7) Convert to denary
(a) 0.1101   (b) 0.0111   (c) 0.0011

8) Convert to binary
(a) $\frac{3}{8}$ (b) $\frac{5}{16}$ (c) $\frac{7}{8}$

9) Convert to binary and correct to seven bicimal places:
(a) 0.169   (b) 18.467   (c) 108.710
Hint: In (b) and (c) convert the whole numbers parts separately from the decimal parts and then combine the results for the final answer.

10) Add the following binary numbers:
(a) 1011 + 11   (b) 11011 + 1011
(c) 10111 + 11010 + 111

11) Multiply the following binary numbers:
(a) 101 × 111   (b) 1011 × 1010   (c) 11011 × 1101

6. INDICES

Outcome:

1. *Define the terms base, index and power.*
2. *Apply the rules*

$$a^m a^n = a^{m+n}$$

$$\frac{a^m}{a^n} = a^{m-n}$$

$$(a^m)^n = a^{mn}$$

3. *Deduce that* $a^0 = 1$ *for all values of* a.
4. *Deduce that* $a^{-n} = \dfrac{1}{a^n}$.
5. *Deduce that* $a^{1/n} = \sqrt[n]{a}$.
6. *Evaluate expressions which combine positive, negative and fractional indices.*

BASE, INDEX AND POWER

The quantity $2 \times 2 \times 2 \times 2$ may be written as 2^4

Now 2^4 is called the fourth power of the base 2

The figure 4, which gives the number of 2s to be multiplied together is called the index (plural: indices).

Similarly $\qquad\qquad a \times a \times a = a^3$

Here a^3 is the third power of the base a, and the index is 3

Thus in the expression x^n

x^n is called the nth power of x,

x is called the base, and

n is called the index.

Remember that, in algebra, letters such as a in the above expression merely represent numbers. Hence the laws of arithmetic apply strictly to algebraic terms as well as numbers.

RECIPROCAL

The expression $\dfrac{1}{2}$ is called the reciprocal of 2

Similarly the expression $\dfrac{1}{p}$ is called the reciprocal of p

Likewise the expression $\dfrac{1}{x^n}$ is called the reciprocal of x^n

LAWS OF INDICES

(i) Multiplication

Let us see what happens when we multiply powers of the same base together.

$$\text{Now} \qquad 5^2 \times 5^4 = (5 \times 5) \times (5 \times 5 \times 5 \times 5)$$
$$= 5 \times 5 \times 5 \times 5 \times 5 \times 5 = 5^6$$

$$\text{and} \qquad c^3 \times c^5 = (c \times c \times c) \times (c \times c \times c \times c \times c)$$
$$= c \times c \times c \times c \times c \times c \times c \times c = c^8$$

In both the examples above we see that we could have obtained the result by adding the indices together.

$$\text{Thus} \qquad 5^2 \times 5^4 = 5^{2+4} = 5^6$$
$$\text{and} \qquad c^3 \times c^5 = c^{3+5} = c^8$$

In general terms the law is

$$\boxed{a^m \times a^n = a^{m+n}}$$

We may apply this idea when multiplying more than two powers of the same base together.

$$\text{Thus} \quad 7^2 \times 7^5 \times 7^9 = 7^{2+5} \times 7^9 = 7^7 \times 7^9 = 7^{7+9} = 7^{16}$$

We see that the same result would have been obtained by adding the indices, hence

$$7^2 \times 7^5 \times 7^9 = 7^{2+5+9} = 7^{16}$$

The law is:

> When multiplying powers of the same base together, add the indices.

EXAMPLE 6.1

Simplify $m^5 \times m^4 \times m^6 \times m^2$

$$m^5 \times m^4 \times m^6 \times m^2 = m^{5+4+6+2} = m^{17}$$

(ii) Division of Powers

Now let us see what happens when we divide powers of the same base

$$\frac{3^5}{3^2} = \frac{3 \times 3 \times 3 \times 3 \times 3}{3 \times 3} = 3 \times 3 \times 3 = 3^3$$

We see that the same result could have been obtained by subtracting the indices.

Thus
$$\frac{3^5}{3^2} = 3^{5-2} = 3^3$$

In general terms the law is

$$\boxed{\frac{a^m}{a^n} = a^{m-n}}$$

or | When dividing powers of the same base subtract the index of the denominator from the index of the numerator.

EXAMPLE 6.2

a) $\dfrac{4^8}{4^5} = 4^{8-5} = 4^3$

b) $\dfrac{z^3 \times z^4 \times z^8}{z^5 \times z^6} = \dfrac{z^{3+4+8}}{z^{5+6}} = \dfrac{z^{15}}{z^{11}} = z^{15-11} = z^4$

(iii) Powers of Powers

How do we simplify $(5^3)^2$? One way is to proceed as follows:

$$(5^3)^2 = 5^3 \times 5^3 = 5^{3+3} = 5^6$$

We see that the same result would have been obtained if we multiplied the two indices together.

Thus
$$(5^3)^2 = 5^{3 \times 2} = 5^6$$

In general terms the law is

$$(a^m)^n = a^{mn}$$

or | When raising the power of a base to a power, multiply the indices together.

EXAMPLE 6.3

a) $(8^4)^3 = 8^{4 \times 3} = 8^{12}$

b) $(p^2 \times q^4)^3 = (p^2)^3 \times (q^4)^3 = p^{2 \times 3} \times q^{4 \times 3} = p^6 \times q^{12}$

c) $\left(\dfrac{a^7}{b^5}\right)^6 = \dfrac{(a^7)^6}{(b^5)^6} = \dfrac{a^{7 \times 6}}{b^{5 \times 6}} = \dfrac{a^{42}}{b^{30}}$

(iv) Zero Index

Now
$$\frac{2^5}{2^5} = \frac{2 \times 2 \times 2 \times 2 \times 2}{2 \times 2 \times 2 \times 2 \times 2} = 1$$

but using the laws of indices

$$\frac{2^5}{2^5} = 2^{5-5} = 2^0$$

Thus
$$2^0 = 1$$

Also
$$\frac{c^4}{c^4} = \frac{c \times c \times c \times c}{c \times c \times c \times c} = 1$$

But using the laws of indices

$$\frac{c^4}{c^4} = c^{4-4} = c^0$$

Thus
$$c^0 = 1$$

In general terms the law is

$$x^0 = 1$$

or | Any base raised to the index of zero is equal to 1

EXAMPLE 6.4

a) $25^0 = 1$ b) $(0.56)^0 = 1$ c) $(\frac{1}{4})^0 = 1$

(v) Negative Indices

Now $\dfrac{2^3}{2^7} = \dfrac{2 \times 2 \times 2}{2 \times 2 \times 2 \times 2 \times 2 \times 2 \times 2} = \dfrac{1}{2 \times 2 \times 2 \times 2} = \dfrac{1}{2^4}$

but using the laws of indices

$$\dfrac{2^3}{2^7} = 2^{3-7} = 2^{-4}$$

It follows that

$$2^{-4} = \dfrac{1}{2^4}$$

Also $\dfrac{d}{d^2} = \dfrac{d}{d \times d} = \dfrac{1}{d}$

but using the laws of indices

$$\dfrac{d}{d^2} = \dfrac{d^1}{d^2} = d^{1-2} = d^{-1}$$

It follows that

$$d^{-1} = \dfrac{1}{d}$$

In general terms the law is

$$\boxed{x^{-n} = \dfrac{1}{x^n}}$$

or

> The power of a base which has a negative index is the reciprocal of the power of the base having the same, but positive, index.

EXAMPLE 6.5

a) $3^{-1} = \dfrac{1}{3^1} = \dfrac{1}{3}$

b) $5x^{-3} = 5 \times x^{-3} = 5 \times \dfrac{1}{x^3} = \dfrac{5}{x^3}$

c) $(2a)^{-4} = \dfrac{1}{(2a)^4} = \dfrac{1}{2^4 \times a^4} = \dfrac{1}{16a^4}$

d) $\dfrac{1}{z^{-5}} = \dfrac{1}{\dfrac{1}{z^5}} = 1 \div \dfrac{1}{z^5} = 1 \times \dfrac{z^5}{1} = z^5$

Summary of the meaning of positive, zero, and negative indices.

$$x^4 = x \times x \times x \times x$$
$$x^3 = x \times x \times x$$
$$x^2 = x \times x$$
$$x^1 = x$$
$$x^0 = 1$$
$$x^{-1} = \frac{1}{x}$$
$$x^{-2} = \frac{1}{x \times x} = \frac{1}{x^2}$$
$$x^{-3} = \frac{1}{x \times x \times x} = \frac{1}{x^3}$$

Each line of the above sequence is obtained by dividing the previous line by x.

The above sequence may help you to appreciate the meaning of positive and negative indices, and especially the zero index. Remember:

$$(\text{elephants})^0 = 1$$

Exercise 6.1

Simplify the following, giving each answer as a power:

1) $2^5 \times 2^6$

2) $a \times a^2 \times a^5$

3) $n^8 \div n^5$

4) $3^4 \times 3^7$

5) $b^2 \div b^5$

6) $10^5 \times 10^3 \div 10^4$

7) $z^4 \times z^2 \times z^{-3}$

8) $3^2 \times 3^{-3} \div 3^3$

9) $\dfrac{m^5 \times m^6}{m^4 \times m^3}$

10) $\dfrac{x^2 \times x}{x^6}$

11) $(9^3)^4$

12) $(y^2)^{-3}$

13) $(t \times t^3)^2$

14) $(c^{-7})^{-2}$

15) $\left(\dfrac{a^2}{a^5}\right)^3$

16) $\left(\dfrac{1}{7^3}\right)^4$

17) $\left(\dfrac{b^2}{b^7}\right)^{-2}$

18) $\dfrac{1}{(s^3)^3}$

Without using tables or calculating machines find the values of the following:

19) $\dfrac{8^3 \times 8^2}{8^4}$ 20) $\dfrac{7^2 \times 7^5}{7^3 \times 7^4}$ 21) $\dfrac{2^2}{2^2 \times 2}$

22) $2^4 \times 2^{-1}$ 23) 2^{-2} 24) $\dfrac{1}{(10)^{-2}}$

25) $\dfrac{2^{-1}}{2}$ 26) $\dfrac{24^0}{7}$ 27) $(5^{-1})^2$

28) $3^{-3} \div 3^{-4}$ 29) $\dfrac{7}{24^0}$ 30) $\left(\dfrac{1}{5}\right)^{-2}$

31) 7×24^0 32) $\left(\dfrac{2}{3}\right)^{-3}$ 33) $\left(\dfrac{2}{2^{-3}}\right)^{-2}$

Fractional Indices

The cube root of 5 (written as $\sqrt[3]{5}$) is the number which, when multiplied by itself three times, gives 5.

Thus $\qquad\qquad \sqrt[3]{5} \times \sqrt[3]{5} \times \sqrt[3]{5} = 5$

But we also know that

$$5^{1/3} \times 5^{1/3} \times 5^{1/3} = 5^{1/3+1/3+1/3} = 5$$

Comparing these expressions

$$\sqrt[3]{5} = 5^{1/3}$$

Similarly the fourth root of base d (written as $\sqrt[4]{d}$) is the number which, when multiplied by itself four times, gives d.

Thus $\qquad\qquad \sqrt[4]{d} \times \sqrt[4]{d} \times \sqrt[4]{d} \times \sqrt[4]{d} = d$

But we also know that

$$d^{1/4} \times d^{1/4} \times d^{1/4} \times d^{1/4} = d^{1/4+1/4+1/4+1/4} = d$$

Comparing these expressions $\quad \sqrt[4]{d} = d^{1/4}$

In general terms the law is

$$\boxed{\sqrt[n]{x} = x^{1/n}}$$

Thus a fractional index represents a root — the denominator of the index denotes the root to be taken.

EXAMPLE 6.6

a) $7^{1/2} = \sqrt{7}$ (note that for square roots the figure 2 indicating
the root is usually omitted)

b) Find the value of $81^{1/4}$

$$81^{1/4} = \sqrt[4]{81} = 3$$

c) Find the value of $8^{2/3}$

$$8^{2/3} = 8^{(1/3) \times 2} = (8^{1/3})^2 = (\sqrt[3]{8})^2 = (2)^2 = 4$$

d) Find the value of $16^{-3/4}$

$$16^{-3/4} = \frac{1}{16^{3/4}} = \frac{1}{16^{(1/4) \times 3}} = \frac{1}{(16^{1/4})^3} = \frac{1}{(\sqrt[4]{16})^3} = \frac{1}{(2)^3}$$

$$= \frac{1}{8} = 0.125$$

e) Find the value of $(\frac{3}{2})^{-3}$

$$\left(\frac{3}{2}\right)^{-3} = \frac{(3)^{-3}}{(2)^{-3}} = (3)^{-3} \times \frac{1}{(2)^{-3}} = \frac{1}{(3)^3} \times (2)^3 = \frac{1}{27} \times 8$$

$$= \frac{8}{27} = 0.296$$

f) Find the value of $9^{2.5}$

$$9^{2.5} = 9^{5/2} = 9^{(1/2) \times 5} = (9^{1/2})^5 = (\sqrt{9})^5 = (3)^5 = 243$$

g) Find the value of $\dfrac{1}{(\sqrt{5})^{-2}}$

$$\frac{1}{(\sqrt{5})^{-2}} = (\sqrt{5})^2 = (5^{1/2})^2 = 5^{(1/2) \times 2} = 5^1 = 5$$

EXAMPLE 6.7

Write the following as powers:

a) $\sqrt{x^3} = (x^3)^{1/2} = x^{3 \times (1/2)} = x^{3/2} = x^{1.5}$

b) $\dfrac{1}{\sqrt[4]{a^5}} = \dfrac{1}{(a^5)^{1/4}} = \dfrac{1}{a^{5 \times (1/4)}} = \dfrac{1}{a^{5/4}} = \dfrac{1}{a^{1.25}} = a^{-1.25}$

c) $\dfrac{\sqrt{b}}{\sqrt[3]{b^{-2}}} = \dfrac{b^{1/2}}{(b^{-2})^{1/3}} = \dfrac{b^{1/2}}{b^{-2 \times (1/3)}} = \dfrac{b^{1/2}}{b^{-2/3}} = b^{1/2} \times b^{2/3} = b^{(1/2)+(2/3)}$

$$= b^{7/6} = b^{1.167}$$

EXAMPLE 6.8

Simplify the following:

a) $\dfrac{m^4}{m^{-3}} = m^{4-(-3)} = m^{4+3} = m^7$

b) $\dfrac{r^{-3}}{r^{2/3}} = r^{-3-(2/3)} = r^{-3.667}$

c) $\dfrac{t^{3/4}(\sqrt{t})}{t^{1/4}} = \dfrac{t^{3/4}(t^{1/2})}{t^{1/4}} = \dfrac{t^{3/4} \times t^{1/2}}{t^{1/4}} = \dfrac{t^{(3/4)+(1/2)}}{t^{1/4}} = \dfrac{t^{1.25}}{t^{0.25}} = t^{1.25-(0.25)}$

$= t^1 = t$

d) $\dfrac{z^{2.7} \times z^3}{z^{0.12}} = \dfrac{z^{2.7+3}}{z^{0.12}} = \dfrac{z^{5.7}}{z^{0.12}} = z^{5.7-0.12} = z^{5.58}$

Exercise 6.2

Write each of the following as a single power:

1) $\sqrt{x}$

2) $\sqrt[5]{x^4}$

3) $\dfrac{1}{\sqrt{x}}$

4) $\dfrac{1}{\sqrt[3]{x}}$

5) $\dfrac{1}{\sqrt[3]{x^4}}$

6) $\sqrt{x^{-3}}$

7) $\dfrac{1}{\sqrt[3]{x^{-2}}}$

8) $\dfrac{1}{\sqrt[4]{x^{-0.3}}}$

9) $(\sqrt[3]{-x})^2$

10) $\sqrt{x^{2/3}}$

11) $(\sqrt{x})^{2/3}$

12) $\left(\dfrac{1}{\sqrt[3]{x^4}}\right)^{-3/4}$

Without using tables or calculating machines find the values of the following:

13) $5^2 \times 5^{1/2} \times 5^{-3/2}$

14) $4 \div 4^{1/2}$

15) $8^{1/3}$

16) $64^{1/6}$

17) $8^{2/3}$

18) $25^{3/2}$

19) $(16^{1/4})^3$

20) $\dfrac{1}{9^{-3/2}}$

21) $\left(\dfrac{1}{4}\right)^{-1/2}$

22) $16^{0.5}$

23) $36^{-0.5}$

24) $(4^{-3})^{1/2}$

25) $\left(\dfrac{1}{4}\right)^{5/2}$

26) $\left(\dfrac{1}{16^{0.5}}\right)^{-3}$

27) $\dfrac{1}{(\sqrt{3})^{-2}}$

Simplify the following, expressing each answer as a power:

28) $\dfrac{\sqrt[3]{a}}{a^2 \times \sqrt{a}}$

29) $\dfrac{a^{-3}}{a^{2/3}}$

30) $\dfrac{x^3}{x^{-1.5}}$

31) $\dfrac{b^{5/2} \times b^{-3/2}}{b^{1/2}}$

32) $\dfrac{m^{-3/4}}{m^{-5/2}}$

33) $\dfrac{z^{2.3} \times z^{-1.5}}{z^{-3.5} \times z^2}$

34) $\dfrac{(x^{1/2})^3}{(x^3)^{1/2}}$

35) $\dfrac{\sqrt{u}}{u^3}$

36) $\dfrac{\sqrt[4]{y^3}}{\sqrt{y}}$

37) $\dfrac{(\sqrt[4]{n})^3}{\sqrt{n}}$

38) $\dfrac{\sqrt[4]{x^2}}{\sqrt[7]{x^{-2}}}$

39) $\dfrac{\sqrt[3]{t} \times \sqrt{t^3}}{t^{5/2}}$

7.

BASIC OPERATIONS

Outcome:

1. *Simplify expressions involving symbols and numbers using (i) the four arithmetic operations of addition, subtraction, multiplication and division (ii) the commutative, associative and distributive laws (iii) the precedence in the use of brackets and arithmetic operations.*

2. *Multiply an expression in brackets by a number, a symbol or by another expression in a bracket.*
3. *Factorise expressions by (i) extraction of a common factor, (ii) grouping.*
4. *Multiply, divide, add and subtract algebraic fractions.*

MULTIPLICATION AND DIVISION OF ALGEBRAIC QUANTITIES

The rules are exactly the same as those used with directed numbers.

$$(+x)(+y) = +(xy) = +xy = xy$$

$$5x \times 3y = 5 \times 3 \times x \times y = 15xy$$

$$(x)(-y) = -(xy) = -xy$$

$$(2x)(-3y) = -(2x)(3y) = -6xy$$

$$(-4x)(2y) = -(4x)(2y) = -8xy$$

$$(-3x)(-2y) = +(3x)(2y) = 6xy$$

$$\frac{+x}{+y} = +\frac{x}{y} = \frac{x}{y} \qquad\qquad \frac{-3x}{2y} = -\frac{3x}{2y}$$

$$\frac{-5x}{-6y} = +\frac{5x}{6y} = \frac{5x}{6y} \qquad\qquad \frac{4x}{-3y} = -\frac{4x}{3y}$$

When *multiplying* expressions containing the same symbols, indices are used:

$$m \times m = m^2$$

$$3m \times 5m = 3 \times m \times 5 \times m = 15m^2$$

$$(-m) \times m^2 = (-m) \times m \times m = -m^3$$

$$5m^2n \times 3mn^3 = 5 \times m \times m \times n \times 3 \times m \times n \times n \times n$$
$$= 15m^3n^4$$
$$3mn \times (-2n^2) = 3 \times m \times n \times (-2) \times n \times n = -6mn^3$$

When *dividing* algebraic expressions, cancellation between numerator and denominator is often possible. Cancelling is equivalent to dividing both numerator and denominator by the same quantity:

$$\frac{pq}{p} = \frac{\not{p} \times q}{\not{p}} = q$$

$$\frac{3p^2q}{6pq^2} = \frac{3 \times \not{p} \times p \times \not{q}}{6 \times \not{p} \times \not{q} \times q} = \frac{3p}{6q} = \frac{p}{2q}$$

$$\frac{18x^2y^2z}{6xyz} = \frac{18 \times \not{x} \times x \times \not{y} \times y \times \not{z}}{6 \times \not{x} \times \not{y} \times \not{z}} = 3xy$$

SEQUENCE OF MIXED OPERATIONS ON ALGEBRAIC QUANTITIES

Since algebraic quantities contain symbols (or letters) which represent numbers the sequence of operations is exactly the same as used with numbers.

Remember the word BODMAS which gives the initial letters of the correct sequence, i.e. Brackets, Of, Divide, Multiply, Add, Subtract.

Thus
$$2x^2 + (12x^4 - 3x^4) \div 3x^2 - x^2 = 2x^2 + 9x^4 \div 3x^2 - x^2$$
$$= 2x^2 + 3x^2 - x^2$$
$$= 5x^2 - x^2$$
$$= 4x^2$$

BRACKETS

Brackets are used for convenience in grouping terms together. When removing brackets each *term* within the bracket is multiplied by the quantity outside the bracket:

$$3(x + y) = 3x + 3y$$

$$5(2x + 3y) \;=\; 5 \times 2x + 5 \times 3y \;=\; 10x + 15y$$

$$4(a - 2b) \;=\; 4 \times a - 4 \times 2b \;=\; 4a - 8b$$

$$m(a + b) \;=\; ma + mb$$

$$3x(2p + 3q) \;=\; 3x \times 2p + 3x \times 3q \;=\; 6px + 9qx$$

$$4a(2a + b) \;=\; 4a \times 2a + 4a \times b \;=\; 8a^2 + 4ab$$

When a bracket has a minus sign in front of it, the signs of all the terms inside the bracket are changed when the bracket is removed. The reason for this rule may be seen from the following examples:

$$-3(2x - 5y) \;=\; (-3) \times 2x + (-3) \times (-5y) \;=\; -6x + 15y$$

$$-(m + n) \;=\; -m - n$$

$$-(p - q) \;=\; -p + q$$

$$-2(p + 3q) \;=\; -2p - 6q$$

When simplifying expressions containing brackets first remove the brackets and then add the like terms together:

$$(3x + 7y) - (4x + 3y) \;=\; 3x + 7y - 4x - 3y \;=\; -x + 4y$$

$$3(2x + 3y) - (x + 5y) \;=\; 6x + 9y - x - 5y \;=\; 5x + 4y$$

$$x(a + b) - x(a + 3b) \;=\; ax + bx - ax - 3bx \;=\; -2bx$$

$$2(5a + 3b) + 3(a - 2b) \;=\; 10a + 6b + 3a - 6b \;=\; 13a$$

Exercise 7.1

Remove the brackets in the following:

1) $3(x + 4)$ 2) $2(a + b)$

3) $3(3x + 2y)$ 4) $\frac{1}{2}(x - 1)$

5) $5(2p - 3q)$ 6) $7(a - 3m)$

7) $-(a + b)$ 8) $-(a - 2b)$

9) $-(3p - 3q)$ 10) $-(7m - 6)$

11) $-4(x + 3)$ 12) $-2(2x - 5)$

13) $-5(4 - 3x)$ 14) $2k(k - 5)$

15) $-3y(3x + 4)$ 16) $a(p - q - r)$

17) $4xy(ab - ac + d)$ 18) $3x^2(x^2 - 2xy + y^2)$

19) $-7P(2P^2 - P + 1)$ 20) $-2m(-1 + 3m - 2n)$

Remove the brackets and simplify the following:

21) $3(x + 1) + 2(x + 4)$ 22) $5(2a + 4) - 3(4a + 2)$

23) $3(x + 4) - (2x + 5)$ 24) $4(1 - 2x) - 3(3x - 4)$

25) $5(2x - y) - 3(x + 2y)$ 26) $\frac{1}{2}(y - 1) + \frac{1}{3}(2y - 3)$

27) $-(4a + 5b - 3c) - 2(2a + 3b - 4c)$

28) $2x(x - 5) - x(x - 2) - 3x(x - 5)$

29) $3(a - b) - 2(2a - 3b) + 4(a - 3b)$

30) $3x(x^2 + 7x - 1) - 2x(2x^2 + 3) - 3(x^2 + 5)$

THE PRODUCT OF TWO BINOMIAL EXPRESSIONS

A binomial expression consists of two terms. Thus $3x + 5$, $a + b$, $2x + 3y$ and $4p - q$ are all binomial expressions.

To find the product of $(a + b)(c + d)$ consider the diagram (Fig. 7.1).

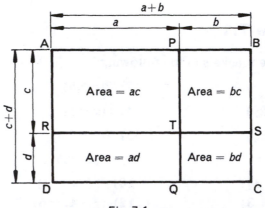

Fig. 7.1

In Fig. 7.1 the rectangular area ABCD is made up as follows:

$$ABCD = APTR + TQDR + PBST + TSCQ$$
$$\text{i.e. } (a + b)(c + d) = ac + ad + bc + bd$$

It will be noticed that the expression on the right hand side is obtained by multiplying each term in the one bracket by each term in the other bracket. The process is illustrated below:

$$(a + b)(c + d) = ac + ad + bc + bd$$

EXAMPLE 7.1

a) $(3x + 2)(4x + 5) = 3x \times 4x + 3x \times 5 + 2 \times 4x + 2 \times 5$
$$= 12x^2 + 15x + 8x + 10$$
$$= 12x^2 + 23x + 10$$

b) $(2p - 3)(4p + 7) = 2p \times 4p + 2p \times 7 - 3 \times 4p - 3 \times 7$
$$= 8p^2 + 14p - 12p - 21$$
$$= 8p^2 + 2p - 21$$

c) $(z - 5)(3z - 2) = z \times 3z + z \times (-2) - 5 \times 3z - 5 \times (-2)$
$$= 3z^2 - 2z - 15z + 10$$
$$= 3z^2 - 17z + 10$$

d) $(2x + 3y)(3x - 2y) = 2x \times 3x + 2x \times (-2y) + 3y \times 3x$
$$+ 3y \times (-2y)$$
$$= 6x^2 - 4xy + 9xy - 6y^2$$
$$= 6x^2 + 5xy - 6y^2$$

THE SQUARE OF A BINOMIAL EXPRESSION

$$(a + b)^2 = (a + b)(a + b) = a^2 + ab + ba + b^2 = a^2 + 2ab + b^2$$

The square of a binomial expression is the sum of the squares of the two terms and twice their product.

$$(a - b)^2 = (a - b)(a - b) = a^2 - ab - ba + b^2 = a^2 - 2ab + b^2$$

EXAMPLE 7.2

a) $(2x + 5)^2 = (2x)^2 + 2 \times 2x \times 5 + 5^2$
$$= 4x^2 + 20x + 25$$

b) $(3x-2)^2 = (3x)^2 + 2 \times 3x \times (-2) + (-2)^2$

$\qquad = 9x^2 - 12x + 4$

c) $(2x+3y)^2 = (2x)^2 + 2 \times 2x \times 3y + (3y)^2$

$\qquad = 4x^2 + 12xy + 9y^2$

THE PRODUCT OF THE SUM AND DIFFERENCE OF TWO TERMS

$(a+b)(a-b) = a^2 - ab + ba - b^2 = a^2 - b^2$

This result is the difference of the squares of the two terms

EXAMPLE 7.3

a) $(8x+3)(8x-3) = (8x)^2 - 3^2 = 64x^2 - 9$

b) $(2x+5y)(2x-5y) = (2x)^2 - (5y)^2 = 4x^2 - 25y^2$

Exercise 7.2

Find the products of the following:

1) $(x+1)(x+2)$

2) $(x+3)(x+1)$

3) $(x+4)(x+5)$

4) $(2x+5)(x+3)$

5) $(3x+7)(x+6)$

6) $(5x+1)(x+4)$

7) $(2x+4)(3x+2)$

8) $(5x+1)(2x+3)$

9) $(7x+2)(3x+5)$

10) $(x-1)(x-3)$

11) $(x-4)(x-2)$

12) $(x-6)(x-3)$

13) $(2x-1)(x-4)$

14) $(x-2)(3x-5)$

15) $(x-8)(4x-1)$

16) $(2x-4)(3x-2)$

17) $(3x-1)(2x-5)$

18) $(7x-5)(3x-2)$

19) $(x+3)(x-1)$

20) $(x-2)(x+7)$

21) $(x-5)(x+3)$

22) $(2x+5)(x-2)$

23) $(3x - 5)(x + 6)$ 24) $(3x + 5)(x + 6)$

25) $(3x + 5)(2x - 3)$ 26) $(6x - 7)(2x + 3)$

27) $(3x - 5)(2x + 3)$ 28) $(3x + 2y)(x + y)$

29) $(2p - q)(p - 3q)$ 30) $(3v + 2u)(2v - 3u)$

31) $(2a + b)(3a - b)$ 32) $(5a - 7)(a - 6)$

33) $(3x + 4y)(2x - 3y)$ 34) $(x + 1)^2$

35) $(2x + 3)^2$ 36) $(3x + 7)^2$

37) $(x - 1)^2$ 38) $(3x - 5)^2$

39) $(2x - 3)^2$ 40) $(2a + 3b)^2$

41) $(x + y)^2$ 42) $(P + 3Q)^2$

43) $(a - b)^2$ 44) $(3x - 4y)^2$

45) $(2x + y)(2x - y)$ 46) $(a - 3b)(a + 3b)$

47) $(2m - 3n)(2m + 3n)$ 48) $(x^2 + y)(x^2 - y)$

HIGHEST COMMON FACTOR (HCF)

The HCF of a set of algebraic expressions is the highest expression which is a factor of each of the given expressions.

The method used is similar to that for finding the HCF of a set of numbers.

EXAMPLE 7.4

Find the HCF of $ab^2c^2, a^2b^3c^3, a^2b^4c^4$.

We express each expression as the product of its factors.

Thus $ab^2c^2 = a \times b \times b \times c \times c$

and $a^2b^3c^3 = a \times a \times b \times b \times b \times c \times c \times c$

and $a^2b^4c^4 = a \times a \times b \times b \times b \times b \times c \times c \times c \times c$

We now note the factors which are common to each of the lines. Factor a is common once, factor b twice, and factor c twice. The product of these factors gives the required HCF.

Thus $\text{HCF} = a \times b \times b \times c \times c$

$$= ab^2c^2$$

EXAMPLE 7.5

Find the HCF of $\dfrac{x^3y}{m^2n^4}$, $\dfrac{x^2y^3}{m^2n^2}$, $\dfrac{x^4y^2}{mn^3}$.

Now $\qquad \dfrac{x^3y}{m^2n^4} = x \times x \times x \times y \times \dfrac{1}{m} \times \dfrac{1}{m} \times \dfrac{1}{n} \times \dfrac{1}{n} \times \dfrac{1}{n} \times \dfrac{1}{n}$

and $\qquad \dfrac{x^2y^3}{m^2n^2} = x \times x \times y \times y \times y \times \dfrac{1}{m} \times \dfrac{1}{m} \times \dfrac{1}{n} \times \dfrac{1}{n}$

and $\qquad \dfrac{x^4y^2}{mn^3} = x \times x \times x \times x \times y \times y \times \dfrac{1}{m} \times \dfrac{1}{n} \times \dfrac{1}{n} \times \dfrac{1}{n}$

Factor x is common twice, factor y once, factor $\dfrac{1}{m}$ once, and factor $\dfrac{1}{n}$ twice.

Thus $\qquad\qquad$ HCF $= x \times x \times y \times \dfrac{1}{m} \times \dfrac{1}{n} \times \dfrac{1}{n}$

$\qquad\qquad\qquad\quad = \dfrac{x^2y}{mn^2}$

An alternative method is to select the lowest power of each of the quantities which occur in **all** of the expressions, and then multiply them together.

EXAMPLE 7.6

Find the HCF of $3m^2np^3$, $6m^3n^2p^2$, $24m^3p^4$.

Dealing with the numerical coefficients 3, 6 and 24 we note that 3 is a factor of each of them. The quantities m and p occur in all three expressions, their lowest powers being m^2 and p^2. Hence,

$$\text{HCF} = 3m^2p^2$$

(Note that n does not occur in each of the three expressions and hence it does not appear in the HCF.)

FACTORISING

A factor is a common part of two or more terms which make up an algebraic expression. Thus the expression $3x + 3y$ has two terms which have the number 3 common to both of them. Thus $3x + 3y = 3(x + y)$. We say that 3 and $(x + y)$ are the factors of

$3x + 3y$. To factorise algebraic expressions of this kind, we first find the HCF of all the terms making up the expression. The HCF then appears outside the bracket. To find the terms inside the bracket divide each of the terms making up the expression by the HCF.

EXAMPLE 7.7

a) Find the factors of $ax + bx$.

The HCF of ax and bx is x.

$$\therefore \qquad ax + bx = x(a + b) \qquad \left(\text{since } \frac{ax}{x} = a \text{ and } \frac{bx}{x} = b \right)$$

b) Find the factors of $m^2 n - 2mn^2$.

The HCF of $m^2 n$ and $2mn^2$ is mn.

$$\therefore \qquad m^2 n - 2mn^2 = mn(m - 2n)$$

$$\left(\text{since } \frac{m^2 n}{mn} = m \text{ and } \frac{2mn^2}{mn} = 2n \right)$$

c) Find the factors of $3x^4 y + 9x^3 y^2 - 6x^2 y^3$.

The HCF of $3x^4 y$, $9x^3 y^2$ and $6x^2 y^3$ is $3x^2 y$.

$$\therefore \qquad 3x^4 y + 9x^3 y^2 - 6x^2 y^3 = 3x^2 y(x^2 + 3xy - 2y^2)$$

$$\left(\text{since } \frac{3x^4 y}{3x^2 y} = x^2, \; \frac{9x^3 y^2}{3x^2 y} = 3xy \text{ and } \frac{6x^2 y^3}{3x^2 y} = 2y^2 \right)$$

d) Find the factors of $\dfrac{ac}{x} + \dfrac{bc}{x^2} - \dfrac{cd}{x^3}$

The HCF of $\dfrac{ac}{x}, \dfrac{bc}{x^2}$ and $\dfrac{cd}{x^3}$ is $\dfrac{c}{x}$.

$$\therefore \qquad \frac{ac}{x} + \frac{bc}{x^2} - \frac{cd}{x^3} = \frac{c}{x}\left(a + \frac{b}{x} - \frac{d}{x^2} \right)$$

$$\left(\text{since } \frac{ac}{x} \div \frac{c}{x} = a, \; \frac{bc}{x^2} \div \frac{c}{x} = \frac{b}{x} \text{ and } \frac{cd}{x^3} \div \frac{c}{x} = \frac{d}{x^2} \right)$$

Exercise 7.3

Find the HCF of the following:

1) p^3q^2, p^2q^3, p^2q

2) $a^2b^3c^3, a^3b^3, ab^2c^2$

3) $3mn^2, 6mnp, 12m^2np^2$

4) $2ab, 5b, 7ab^2$

5) $3x^2yz, 12x^2yz, 6xy^2z^3, 3xyz^2$

Factorise the following:

6) $2x + 6$

7) $4x - 4y$

8) $5x - 5$

9) $4x - 8xy$

10) $mx - my$

11) $ax + bx + cx$

12) $\dfrac{x}{2} - \dfrac{y}{8}$

13) $5a - 10b + 15c$

14) $ax^2 + ax$

15) $2\pi r^2 + \pi rh$

16) $3y - 9y^2$

17) $ab^3 - a^2b$

18) $x^2y^2 - axy + bxy^2$

19) $5x^3 - 10x^2y + 15xy^2$

20) $9x^3y - 6x^2y^2 + 3xy^5$

21) $I_0 + I_0\alpha t$

22) $\dfrac{x}{3} - \dfrac{y}{6} + \dfrac{z}{9}$

23) $2a^2 - 3ab + b^2$

24) $x^3 - x^2 + 7x$

25) $\dfrac{m^2}{pn} - \dfrac{m^3}{pn^2} + \dfrac{m^4}{p^2n^2}$

FACTORISING BY GROUPING

To factorise the expression $ax + ay + bx + by$ first group the terms in pairs so that each pair of terms has a common factor. Thus,

$$ax + ay + bx + by = (ax + ay) + (bx + by) = a(x + y) + b(x + y)$$

Now notice that in the two terms $a(x + y)$ and $b(x + y)$, $(x + y)$ is a common factor. Hence,

$$a(x + y) + b(x + y) = (x + y)(a + b)$$

$\therefore \qquad\quad ax + ay + bx + by = (x + y)(a + b)$

Similarly,

$$np + mp - qn - qm = (np + mp) - (qn + qm)$$
$$= p(n + m) - q(n + m)$$
$$= (n + m)(p - q)$$

Exercise 7.4

Factorise the following:

1) $ax + by + bx + ay$

2) $mp + np - mq - nq$

3) $a^2c^2 + acd + acd + d^2$

4) $2pr - 4ps + qr - 2qs$

5) $4ax + 6ay - 4bx - 6by$

6) $ab(x^2 + y^2) - cd(x^2 + y^2)$

7) $mn(3x - 1) - pq(3x - 1)$

8) $k^2l^2 - mnl - k^2l + mn$

LOWEST COMMON MULTIPLE (LCM)

The LCM of a set of algebraic terms is the simplest expression of which each of the given terms is a factor.

The method used is similar to that for finding the LCM of a set of numbers.

EXAMPLE 7.8

Find the LCM of $2a$, $3ab$, and a^2b.

We express each term as a product of its factors.

Thus $\qquad\qquad\qquad 2a = 2 \times a$

and $\qquad\qquad\qquad 3ab = 3 \times a \times b$

and $\qquad\qquad\qquad a^2b = a \times a \times b$

We now note the greatest number of times each factor occurs in any one particular line.

Now factor 2 occurs once in the line for $2a$,

and factor 3 occurs once in the line for $3ab$,

and factor a occurs twice in the line for a^2b,

and factor b occurs once in either of the lines for $3ab$ or a^2b.

The product of these factors gives the required LCM.

Thus $\qquad\qquad$ LCM $= 2 \times 3 \times a \times a \times b$

$\qquad\qquad\qquad\quad = 6a^2b$

EXAMPLE 7.9

Find the LCM of $4x$, $8yz$, $2x^2y$ and yz^2.

With practice the LCM may be found by inspection, by finding the product of the highest powers of **all** factors which occur in **any** of the terms.

Thus $$\text{LCM} = 8 \times x^2 \times y \times z^2$$
$$= 8x^2yz^2$$

EXAMPLE 7.10

Find the LCM of $(a-1)$, $n(m+n)$, $(m+n)^2$.

Brackets must be treated as single factors—not the individual terms inside each bracket.

Hence $$\text{LCM} = (a-1) \times n \times (m+n)^2$$
$$= n(a-1)(m+n)^2$$

Exercise 7.5

Find the LCM for the terms in each of the following examples:

1) $2a$, $3a^2$, a, a^2 2) xy, x^2y, $2x$, $2y$

3) m^2n, mn^2, mn, m^2n^2 4) $2ab$, abc, bc^2

5) $2(x+1)$, $(x+1)$ 6) $(a+b)$, $x(a+b)^2$, x^2

7) $(a+b)$, $(a-b)$ 8) x, $(1-x)$, $(x+1)$

9) $(x-2)$, $(x+2)$, $(x-2)^2$ 10) $(x+2)$, (x^2-4)

11) $2(a-b)$, $3(a+b)$, (a^2-b^2) 12) $2(a+b)^2$, $3(a+b)$, (a^2+b^2)

HANDLING ALGEBRAIC FRACTIONS

Since algebraic expressions contain symbols (or letters) which represent numbers all the rule of operations with numbers also apply to algebraic terms, including fractions.

Thus

$$\frac{\dfrac{1}{1}}{\dfrac{1}{a}} = 1 \div \frac{1}{a} = 1 \times \frac{a}{1} = \frac{1 \times a}{1} = a$$

and

$$\frac{\dfrac{a}{b}}{\dfrac{c}{d}} = \frac{a}{b} \div \frac{c}{d} = \frac{a}{b} \times \frac{d}{c} = \frac{a \times d}{b \times c} = \frac{ad}{bc}$$

and

$$\frac{x+y}{\dfrac{1}{x-y}} = \frac{(x+y)}{\dfrac{1}{(x-y)}} = (x+y) \div \frac{1}{(x-y)} = (x+y) \times \frac{(x-y)}{1}$$

$$= (x+y)(x-y)$$

You should note in the last example how we put brackets round the $x+y$ and $x-y$ to remind us that they must be treated as single expressions—otherwise we may have been tempted to handle the terms x and y on their own.

ADDING AND SUBTRACTING ALGEBRAIC FRACTIONS

Consider the expression $\dfrac{a}{b} + \dfrac{c}{d}$ which is the addition of two fractional terms. These are called partial fractions.

If we wish to express the sum of these partial fractions as one single fraction then we proceed as follows. (The method is similar to that used when adding or subtracting number fractions.)

First find the lowest common denominator. This is the LCM of b and d which is bd. Each fraction is then expressed with bd as the denominator.

Now $\qquad \dfrac{a}{b} = \dfrac{a \times d}{b \times d} = \dfrac{ad}{bd}$ and $\dfrac{c}{d} = \dfrac{c \times b}{d \times b} = \dfrac{cb}{bd}$

and adding these new fractions we have:

$$\frac{a}{b} + \frac{c}{d} = \frac{ad}{bd} + \frac{cb}{bd} = \frac{ad+cb}{bd}$$

EXAMPLE 7.11

Express each of the following as a single fraction:

a) $\dfrac{1}{x}-\dfrac{1}{y}$ b) $a-\dfrac{1}{b}$ c) $\dfrac{1}{m}+n-\dfrac{a}{b}$

d) $\dfrac{a}{b^2}-\dfrac{1}{bc}$ e) $\dfrac{2}{x}+\dfrac{3}{x-1}$ f) $\dfrac{x}{x+1}-\dfrac{2}{x+3}$

a) The lowest common denominator is the LCM of x and y which is xy.

Therefore $\dfrac{1}{x}-\dfrac{1}{y}=\dfrac{y}{xy}-\dfrac{x}{xy}=\dfrac{y-x}{xy}$

b) $a-\dfrac{1}{b}=\dfrac{a}{1}-\dfrac{1}{b}=\dfrac{ab}{b}-\dfrac{1}{b}=\dfrac{ab-1}{b}$

c) $\dfrac{1}{m}+n-\dfrac{a}{b}=\dfrac{1}{m}+\dfrac{n}{1}-\dfrac{a}{b}=\dfrac{b}{mb}+\dfrac{nmb}{mb}-\dfrac{am}{mb}=\dfrac{b+nmb-am}{mb}$

d) $\dfrac{a}{b^2}-\dfrac{1}{bc}=\dfrac{ac}{b^2c}-\dfrac{b}{b^2c}=\dfrac{ac-b}{b^2c}$

e) $\dfrac{2}{x}+\dfrac{3}{(x-1)}=\dfrac{2(x-1)}{x(x-1)}+\dfrac{3x}{x(x-1)}=\dfrac{2(x-1)+3x}{x(x-1)}$

$$=\dfrac{2x-2+3x}{x(x-1)}=\dfrac{5x-2}{x(x-1)}$$

f) $\dfrac{x}{(x+1)}-\dfrac{2}{(x+3)}=\dfrac{x(x+3)}{(x+1)(x+3)}-\dfrac{2(x+1)}{(x+3)(x+1)}$

$$=\dfrac{x(x+3)-2(x+1)}{(x+1)(x+3)}$$

$$=\dfrac{x^2+3x-2x-2}{(x+1)(x+3)}$$

$$=\dfrac{x^2+x-2}{(x+1)(x+3)}$$

You should note that we have not attempted to multiply out the brackets in the denominator. They should be left as they are, for multiplying them out would complicate the expression rather than simplify it.

Exercise 7.6

Rearrange the following and thus express in a simplified form:

1) $\dfrac{\dfrac{1}{b}}{a}$

2) $\dfrac{\dfrac{1}{a}}{\dfrac{1}{b}}$

3) $\dfrac{\dfrac{x}{y}}{\dfrac{y}{x}}$

4) $\dfrac{\dfrac{1}{2}}{xy}$

5) $\dfrac{\dfrac{a}{b}}{a^2}$

6) $\dfrac{(a+b)}{\dfrac{1}{c}}$

7) $\dfrac{1-x}{\dfrac{1}{1+x}}$

8) $\dfrac{\dfrac{1}{a-b}}{\dfrac{a-b}{c}}$

Express with a common denominator:

9) $\dfrac{1}{x}+\dfrac{1}{y}$

10) $1+\dfrac{1}{a}$

11) $\dfrac{m}{n}-1$

12) $\dfrac{b}{c}-c$

13) $\dfrac{a}{b}-\dfrac{c}{d}$

14) $\dfrac{a}{b}-\dfrac{1}{bc}$

15) $\dfrac{1}{\dfrac{1}{a}}-\dfrac{1}{a}$

16) $\dfrac{1}{xy}+\dfrac{1}{x}+1$

17) $\dfrac{3}{x}+\dfrac{x}{4}$

18) $\dfrac{3}{c}+\dfrac{2}{d}-\dfrac{5}{e}$

19) $\dfrac{a}{b}+\dfrac{c}{d}+1$

20) $\dfrac{1}{3fg}-\dfrac{5}{6gh}-\dfrac{1}{2fh}$

21) $\dfrac{5}{y+3}+\dfrac{3}{y-5}$

22) $\dfrac{2}{x}-\dfrac{4}{x+2}$

23) $\dfrac{1}{x^2}+\dfrac{1}{x-1}$

24) $\dfrac{x}{1-x}+\dfrac{1}{1+x}$

25) $1-\dfrac{x}{x-2}$

26) $\dfrac{x+2}{x-2}+1$

EXPRESSING A SINGLE FRACTION AS PARTIAL FRACTIONS

When considering $\dfrac{x-y}{x}$ many students are tempted to cancel the

x terms and obtain $\qquad \dfrac{\cancel{x}-y}{\cancel{x}}=\dfrac{1-y}{1}=1-y$

This is completely wrong!

The correct method is to reverse the procedure used for adding algebraic terms.

Thus
$$\frac{x-y}{x} = \frac{x}{x} - \frac{y}{x} = 1 - \frac{y}{x}$$

Alternatively we may consider the numerator as enclosed in a bracket giving:

$$\frac{(x-y)}{x} = \frac{1}{x}(x-y) = \frac{1}{x} \times x - \frac{1}{x} \times y = 1 - \frac{y}{x}$$

EXAMPLE 7.12

Express as partial fractions:

a) $\dfrac{ab + bc - 1}{abc}$ b) $\dfrac{a-b}{x+y}$ c) $\dfrac{(x-1)+y}{a(x-1)}$

a) $\dfrac{ab + bc - 1}{abc} = \dfrac{ab}{abc} + \dfrac{bc}{abc} - \dfrac{1}{abc} = \dfrac{1}{c} + \dfrac{1}{a} - \dfrac{1}{abc}$

b) $\dfrac{a-b}{x+y} = \dfrac{a}{x+y} - \dfrac{b}{x+y}$

c) $\dfrac{(x-1)+y}{a(x-1)} = \dfrac{(x-1)}{a(x-1)} + \dfrac{y}{a(x-1)} = \dfrac{1}{a} + \dfrac{y}{a(x-1)}$

MIXED OPERATIONS WITH FRACTIONS

We will now combine all the ideas already used. It helps to work methodically and avoid taking short cuts by leaving out stages of simplification.

EXAMPLE 7.13

Simplify: a) $\dfrac{\dfrac{1}{x} + x}{\dfrac{1}{x}}$ b) $\dfrac{\dfrac{a}{b}}{\dfrac{1}{2ab} - \dfrac{3}{b}}$ c) $\dfrac{x - \dfrac{1}{1-x}}{\dfrac{1-x}{1+x}}$

In all these solutions the numerators and denominators are first expressed as single fractions. It is then possible to divide the numerator by the denominator.

a) $$\dfrac{\dfrac{1}{x}+x}{\dfrac{1}{x}} = \dfrac{\dfrac{1}{x}+\dfrac{x^2}{x}}{\dfrac{1}{x}} = \dfrac{\dfrac{1+x^2}{x}}{\dfrac{1}{x}} = \dfrac{(1+x^2)}{x} \div \dfrac{1}{x}$$

$$= \dfrac{(1+x^2)}{x} \times \dfrac{x}{1} = 1+x^2$$

b) $$\dfrac{\dfrac{a}{b}}{\dfrac{1}{2ab}-\dfrac{3}{b}} = \dfrac{\dfrac{a}{b}}{\dfrac{1}{2ab}-\dfrac{6a}{2ab}} = \dfrac{\dfrac{a}{b}}{\dfrac{(1-6a)}{2ab}} = \dfrac{a}{b} \div \dfrac{(1-6a)}{2ab}$$

$$= \dfrac{a}{b} \times \dfrac{2ab}{(1-6a)}$$

$$= \dfrac{2a^2}{1-6a}$$

c) $$\dfrac{x-\dfrac{1}{(1+x)}}{\dfrac{(1-x)}{(1+x)}} = \dfrac{\dfrac{x(1+x)}{(1+x)}-\dfrac{1}{(1+x)}}{\dfrac{(1-x)}{(1+x)}} = \dfrac{\dfrac{x(1+x)-1}{(1+x)}}{\dfrac{(1-x)}{(1+x)}}$$

$$= \dfrac{x(1+x)-1}{(1+x)} \div \dfrac{(1-x)}{(1+x)}$$

$$= \dfrac{x(1+x)-1}{(1+x)} \times \dfrac{(1+x)}{(1-x)}$$

$$= \dfrac{x(1+x)-1}{(1-x)}$$

Exercise 7.7

Express as partial fractions:

1) $\dfrac{a+b}{a}$

2) $\dfrac{a-b}{ab}$

3) $\dfrac{1+c}{c}$

4) $\dfrac{x^2+y}{2x}$

5) $\dfrac{a^2-ab+ac}{abc}$

6) $\dfrac{(x-1)+(x+1)}{(x+1)}$

7) $\dfrac{xy(1-a)+y^2}{x(1-a)}$ 8) $\dfrac{x+(x-y)}{x(x-y)}$ 9) $\dfrac{(a+b)-(a-b)}{(a-b)(a+b)}$

Simplify:

10) $\dfrac{\dfrac{7}{c}-\dfrac{3}{d}}{\dfrac{4}{d}}$ 11) $\dfrac{1}{1+\dfrac{1}{x}}$ 12) $\dfrac{a}{a-\dfrac{1}{a}}$

13) $\dfrac{\dfrac{1}{2a}}{4+\dfrac{3}{a}}$ 14) $\dfrac{1+\dfrac{1}{m}}{\dfrac{1}{m}-1}$ 15) $\dfrac{3+\dfrac{5}{t}}{\dfrac{1}{3t}-2}$

16) $\dfrac{1}{\dfrac{1}{R_1}+\dfrac{1}{R_2}}$ 17) $\dfrac{a-\dfrac{b}{c}}{1+\dfrac{c}{a}}$ 18) $\dfrac{\dfrac{b}{c}+\dfrac{x}{y}}{x+\dfrac{b}{y}}$

19) $\dfrac{ab-a^2}{a^2-\dfrac{a}{b}}$ 20) $\dfrac{\dfrac{a}{b}-1}{\dfrac{a-b}{a+b}}$ 21) $\dfrac{1+\dfrac{1-x}{1+x}}{\dfrac{1}{1+x}-1}$

8. SOLVING LINEAR EQUATIONS

Outcome:
1. Maintain the equality of a given equation
    whilst applying arithmetical operations.

2. Solve linear equations in one unknown.

BALANCE

Linear equations contain only the first power of the unknown quantity.

Thus $\qquad 7t - 5 = 4t + 7 \quad$ and $\quad \dfrac{5x}{3} = \dfrac{2x + 5}{2}$

are both examples of linear equations.

Consider the pair of scales in balance. $\Big\}$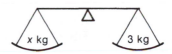

A mathematical model of this arrangement is the equation $\Big\}$ $\qquad x = 3 \qquad\qquad [1]$

Now whatever changes we make, *balance* must be maintained.

This new loading still keeps the balance. $\Big\}$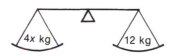

Now multiplying both sides of the equation [1] by 4, then $\Big\}$ $\qquad 4x = 4 \times 3$

giving $\qquad\qquad\qquad\qquad\qquad 4x = 12$

Remember that when 'manipulating' an equation you may 'do anything you like' to one side of an equation, providing you do the same to the other.

91

For example if we add 7
to both sides of equation [1] } $\qquad x + 7 = 3 + 7$

giving $\qquad\qquad\qquad\qquad\qquad x + 7 = 10$

Or if we divide both sides
of equation [1] by 5 then } $\qquad \dfrac{x}{5} = \dfrac{3}{5}$

We may even become ambitious
and square both sides of equation [1] } $\qquad x^2 = 3^2$

giving $\qquad\qquad\qquad\qquad\qquad x^2 = 9$

Now let us use this idea of '**doing the same thing to both sides**' to solve some equations. Before doing so we would remind you that we will use the abbreviations LHS and RHS respectively for the left and right hand sides of an equation.

EXAMPLE 8.1

Solve $\qquad\qquad\qquad 7x = 5x + 18$

We shall group all the terms containing the unknown, here x, on one side (the LHS of the given equation), and the remainder on the other.

So if we subtract $5x$
from both sides, then } $\qquad 7x - 5x = 5x + 18 - 5x$

giving $\qquad\qquad\qquad\qquad\qquad 2x = 18$

and dividing both sides
by 2 } $\qquad\qquad \dfrac{2x}{2} = \dfrac{18}{2}$

giving $\qquad\qquad\qquad\qquad\qquad x = 9$

We must now check to see if this value is correct.

Check: When $x = 9$ then LHS $= 7 \times 9 = 63$, RHS $= 5 \times 9 + 18$ $= 63$. Thus the value obtained is correct.

We say that $x = 9$ is the solution of the given equation — or we may say that $x = 9$ satisfies the given equation.

EXAMPLE 8.2

Solve $\qquad\qquad\qquad 2(4y + 3) = 3y + 8$

Removing the bracket gives $\qquad\qquad 8y + 6 = 3y + 8$

Subtracting $3y$ and 6 from both sides, then $\Big\}$ $\qquad 8y + 6 - 3y - 6 \;=\; 3y + 8 - 3y - 6$

giving $\qquad\qquad\qquad\qquad\qquad 5y \;=\; 2$

Dividing both sides by 5, then $\Big\}$ $\qquad\qquad\qquad \dfrac{5y}{5} \;=\; \dfrac{2}{5}$

from which $\qquad\qquad\qquad\qquad y \;=\; \dfrac{2}{5} \quad \text{or} \quad 0.4$

Check: When $x = 0.4$ then   LHS $= 2(4 \times 0.4 + 3) = 9.2$

$\qquad\qquad\qquad\qquad\qquad$ RHS $= 3 \times 0.4 + 8 = 9.2$

Hence the solution is correct.

EXAMPLE 8.3

Solve $\qquad\qquad\qquad \dfrac{2t}{5} + \dfrac{3}{2} \;=\; \dfrac{3t}{4} + 6$

The LCM of the denominators 5, 2 and 4 is 20.

Multiplying through by 20 $\Big\}$ $\qquad \dfrac{2t}{5} \times 20 + \dfrac{3}{2} \times 20 \;=\; \dfrac{3t}{4} \times 20 + 6 \times 20$

giving $\qquad\qquad\qquad\qquad 8t + 30 \;=\; 15t + 120$

Subtracting $15t$ and 30 from both sides, then $\Big\}$ $8t + 30 - 15t - 30 \;=\; 15t + 120 - 15t - 30$

giving $\qquad\qquad\qquad\qquad -7t \;=\; 90$

Dividing both sides by -7, then $\Big\}$ $\qquad\qquad \dfrac{-7t}{-7} \;=\; \dfrac{90}{-7}$

giving $\qquad\qquad\qquad\qquad\qquad t \;=\; -\dfrac{90}{7}$

or $\qquad\qquad\qquad\qquad\qquad t \;=\; -12.9 \text{ correct to 3 s.f.}$

Check: If $t = -12.9$ then   LHS $= \dfrac{2}{5}(-12.9) + \dfrac{3}{2} \;=\; -3.66$

$\qquad\qquad\qquad$ and   RHS $= \dfrac{3}{4}(-12.9) + 6 \;=\; -3.68$

Bearing in mind the answer was rounded to 3 s.f. we may say that the LHS = RHS, and hence that the solution is correct.

EXAMPLE 8.4

Solve $$\frac{z-4}{3} - \frac{2z-1}{2} = 4$$

The LCM of the denominators 3 and 2 is 6.

Multiplying through by 6 $\Big\}$ $\dfrac{(z-4)}{3} \times 6 - \dfrac{(2z-1)}{2} \times 6 = 4 \times 6$

giving $$2(z-4) - 3(2z-1) = 24$$

Removing the brackets $$2z - 8 - 6z + 3 = 24$$

giving $$-4z - 5 = 24$$

Adding 5 to both sides $$-4z - 5 + 5 = 24 + 5$$

giving $$-4z = 29$$

Dividing both sides by -4, then $\Big\}$ $\dfrac{-4z}{--4} = \dfrac{29}{-4}$

giving $$z = -\frac{29}{4} \quad \text{or} \quad -7.25$$

Check this value for yourself to see if it is correct.

EXAMPLE 8.5

Solve the equation $$\frac{3}{m+4} = \frac{7}{m}$$

In order to bring the unknowns m to 'the top line' we will multiply both sides of the equation by the LCM of the denominators $m(m+4)$.

Multiplying both sides by $m(m+4)$, then $\Big\}$ $\dfrac{3}{(m+4)} \times m(m+4) = \dfrac{7}{m} \times m(m+4)$

giving $$3m = 7(m+4)$$

$\therefore$ $$3m = 7m + 28$$

Subtracting $7m$ from both sides, then $\Big\}$ $3m - 7m = 7m + 28 - 7m$

giving $$-4m = 28$$

Dividing both sides by -4 $\dfrac{-4m}{-4} = \dfrac{28}{-4}$

from which $$m = -7$$

Again we will leave you to check if this answer is correct.

Exercise 8.1

Solve the equations:

1) $x + 2 = 7$

2) $t - 4 = 3$

3) $2q = 4$

4) $x - 8 = 12$

5) $q + 5 = 2$

6) $3x = 9$

7) $\dfrac{y}{2} = 3$

8) $\dfrac{m}{3} = 4$

9) $2x + 5 = 9$

10) $5x - 3 = 12$

11) $6p - 7 = 17$

12) $3x + 4 = -2$

13) $7x + 12 = 5$

14) $6x - 3x + 2x = 20$

15) $14 - 3x = 8$

16) $5x - 10 = 3x + 2$

17) $6m + 11 = 25 - m$

18) $3x - 22 = 8x + 18$

19) $0.3d = 1.8$

20) $1.2x - 0.8 = 0.8x + 1.2$

21) $2(x + 1) = 8$

22) $5(m - 2) = 15$

23) $3(x - 1) - 4(2x + 3) = 14$

24) $5(x + 2) - 3(x - 5) = 29$

25) $3x = 5(9 - x)$

26) $4(x - 5) = 7 - 5(3 - 2x)$

27) $\dfrac{x}{5} - \dfrac{x}{3} = 2$

28) $\dfrac{x}{3} + \dfrac{x}{4} + \dfrac{x}{5} = \dfrac{5}{6}$

29) $\dfrac{m}{2} + \dfrac{m}{3} + 3 = 2 + \dfrac{m}{6}$

30) $3x + \dfrac{3}{4} = 2 + \dfrac{2x}{3}$

31) $\dfrac{3}{m} = 3$

32) $\dfrac{5}{x} = 2$

33) $\dfrac{4}{t} = \dfrac{2}{3}$

34) $\dfrac{7}{x} = \dfrac{5}{3}$

35) $\dfrac{4}{7}y - \dfrac{3}{5}y = 2$

36) $\dfrac{1}{3x} + \dfrac{1}{4x} = \dfrac{7}{20}$

37) $\dfrac{x + 3}{4} - \dfrac{x - 3}{5} = 2$

38) $\dfrac{2x}{15} - \dfrac{x - 6}{12} - \dfrac{3x}{20} = \dfrac{3}{2}$

39) $\dfrac{2m - 3}{4} = \dfrac{4 - 5m}{3}$

40) $\dfrac{3 - y}{4} = \dfrac{y}{3}$

41) $x - 5 = \dfrac{3x - 5}{6}$

42) $\dfrac{x - 2}{x - 3} = 3$

43) $\dfrac{3}{x-2} = \dfrac{4}{x+4}$

44) $\dfrac{3}{x-1} = \dfrac{2}{x-5}$

45) $\dfrac{3}{2x+7} = \dfrac{5}{3(x-2)}$

46) $\dfrac{x}{3} - \dfrac{3x-7}{5} = \dfrac{x-2}{6}$

47) $\dfrac{4p-1}{3} - \dfrac{3p-1}{2} = \dfrac{5-2p}{4}$

48) $\dfrac{3m-5}{4} - \dfrac{9-2m}{3} = 0$

49) $\dfrac{x}{3} - \dfrac{2x-5}{2} = 0$

50) $\dfrac{4x-5}{2} - \dfrac{2x-1}{6} = x$

9. TRANSPOSITION OF FORMULAE

Outcome:

1. Transpose formulae in which the subject is contained in more than one term.

2. Transpose formulae which contain a root or a power.

TRANSPOSITION OF FORMULAE

In the formula $P = I^2R$, P is called the subject of the formula. It may be that we are given values of I and P and we have to find R. We can do this by substituting the given values in the formula and solving the resulting equation.

We may have several sets of corresponding values of I and P and want to find the corresponding values of R. Much time and effort will be saved if we express the formula with R as the subject because then we need only substitute the given values of I and P in the rearranged formula.

The process of rearranging a formula so that one of the other symbols becomes the subject is called *transposing the formula*. The rules used in the transposition of formulae are the same as those used in solving equations. These are given in chapter 8 and it may be a good idea to look at these to refresh your memory before starting this chapter.

EXAMPLE 9.1

Power W, current I, and voltage V, are connected by the formula $W = IV$. Transpose to make V the subject.

Divide both sides by I:
$$\frac{W}{I} = \frac{IV}{I}$$

or
$$\frac{W}{I} = V$$

∴
$$V = \frac{W}{I}$$

Check: It is possible to check whether the transposition has been made correctly by substituting numerical values.

Putting $\left.\begin{array}{l} I = 6 \\ V = 4 \end{array}\right\}$ into $W = IV$

we get $W = 6 \times 4 = 24$

Now if we use the transposed form of the given equation,

Putting $\left.\begin{array}{l} W = 24 \\ I = 6 \end{array}\right\}$ into $V = \dfrac{W}{I}$

we get $V = \dfrac{24}{6} = 4$

and this verifies the correctness of the transposition.

You may say this was unnecessary as it was obvious the transposition was correct — but this is only because the original equation was comparatively simple. Just accept that this is the procedure and use it for the more difficult formulae.

EXAMPLE 9.2

Heat energy, H, is given by $H = I^2Rt$ where I is current, R is resistance and t is time. Make R the subject of the equation.

Divide both sides by I^2t: $\dfrac{H}{I^2t} = \dfrac{I^2Rt}{I^2t}$

or $\dfrac{H}{I^2t} = R$

$\therefore$ $R = \dfrac{H}{I^2t}$

Check: Put $I = 3$, $R = 2$ and $t = 7$ and use the method given in the previous example.

EXAMPLE 9.3

The expression $R = \dfrac{V}{I}$ relates resistance R, voltage V, and current I.

Express V in terms of R and I.

Multiply both sides by I:

$$R \times I = \frac{V}{I} \times I$$

or

$$RI = V$$

$\therefore$

$$V = RI$$

Check this result for yourselves.

EXAMPLE 9.4

The tension, T, in a cord which is whirling a mass, m, round in a circular path of radius, r, and tangential velocity, v, is given by $T = \dfrac{mv^2}{r}$. Transpose for radius r.

Multiply both sides by r

$$T \times r = \frac{mv^2}{r} \cdot \times r$$

$\therefore$

$$Tr = mv^2$$

Divide both sides by T

$$\frac{Tr}{T} = \frac{mv^2}{T}$$

$\therefore$

$$r = \frac{mv^2}{T}$$

Again you may check this for yourselves.

EXAMPLE 9.5

Temperature, t, on the Celsius scale, and temperature, T, on the Absolute scale are related by $T = t + 273$. Make t the subject.

Subtract 273 from both sides $\Big\}$

$$T - 273 = t + 273 - 273$$

giving

$$T - 273 = t$$

or

$$t = T - 273$$

EXAMPLE 9.6

Pressure p, at a depth h, in a fluid of density ρ, is given by $p = p_a + \rho g h$ where p_a is atmospheric pressure and g the gravitational constant. Express h in terms of the other symbols.

Subtract p_a from both sides $\Big\}$ $p - p_a = p_a + \rho gh - p_a$

giving $p - p_a = \rho gh$

Divide both sides by ρg $\dfrac{p - p_a}{\rho g} = \dfrac{\rho gh}{\rho g}$

or $h = \dfrac{p - p_a}{\rho g}$

Check this result for yourself.

EXAMPLE 9.7

F_1 and F_2 are the tensions in the tight and slack sides of a belt drive over a pulley wheel. The velocity, v, of the belt is given by $v = \dfrac{P}{F_1 - F_2}$ where P is the power transmitted. Make F_2 the subject of this formula.

Multiply both sides by $(F_1 - F_2)$ $\Big\}$ $v(F_1 - F_2) = \dfrac{P}{(F_1 - F_2)} \times (F_1 - F_2)$

giving $v(F_1 - F_2) = P$

Divide both sides by v $\Big\}$ $\dfrac{v(F_1 - F_2)}{v} = \dfrac{P}{v}$

giving $F_1 - F_2 = \dfrac{P}{v}$

Subtract F_1 from both sides $\Big\}$ $F_1 - F_2 - F_1 = \dfrac{P}{v} - F_1$

giving $-F_2 = \dfrac{P}{v} - F_1$

Multiply both sides by (-1) $\Big\}$ $-F_2(-1) = \left(\dfrac{P}{v} - F_1\right)(-1)$

giving $F_2 = \dfrac{P}{v}(-1) - F_1(-1)$

or $F_2 = F_1 - \dfrac{P}{v}$

Check: Let us use $F_1 = 3$, $F_2 = 1$ and $P = 8$ in the given equation:

$$v = \frac{8}{3-1} = 4$$

and putting $F_1 = 3$, $P = 8$ and $v = 4$ into the transposed form, then:

$$F_2 = 3 - \frac{8}{4} = 1$$

and this verifies the transposition.

EXAMPLE 9.8

The resistance, R, of a wire after a temperature rise, t, is given by $R = R_0(1 + \alpha t)$ where R_0 is the resistance at zero temperature and α is the temperature coefficient of resistance. Transpose for t.

We may approach this problem in two ways:

a) Divide both sides by R_0 $\qquad\qquad \dfrac{R}{R_0} = \dfrac{R_0(1 + \alpha t)}{R_0}$

giving $\qquad\qquad\qquad\qquad \dfrac{R}{R_0} = 1 + \alpha t$

Subtract 1 from both sides $\Big\}$ $\qquad \dfrac{R}{R_0} - 1 = 1 + \alpha t - 1$

giving $\qquad\qquad\qquad\qquad \alpha t = \dfrac{R}{R_0} - 1$

Divide both sides by α $\qquad\qquad \dfrac{\alpha t}{\alpha} = \dfrac{(R/R_0 - 1)}{\alpha}$

So the first solution $\qquad\qquad t = \dfrac{1}{\alpha}\left(\dfrac{R}{R_0} - 1\right)$

b) Remove the bracket $\qquad\qquad R = R_0 + R_0 \alpha t$

Subtract R_0 from both sides $\Big\}$ $\qquad R - R_0 = R_0 \alpha t$

Divide both sides by $R_0 \alpha$ $\Big\}$ $\qquad \dfrac{R - R_0}{R_0 \alpha} = \dfrac{R_0 \alpha t}{R_0 \alpha}$

Thus the second solution $\qquad\qquad t = \dfrac{R - R_0}{R_0 \alpha}$

We will leave you to verify that the two solutions are equal.

EXAMPLE 9.9

The percentage profit, P, made in a transaction where the selling price is s, and the buying price was b, is given by

$$P = \frac{s-b}{b} \times 100$$

Make b the subject of this equation.

Multiply both sides by b $P \times b = \frac{(s-b)}{b} \times 100b$

and removing the bracket $Pb = 100s - 100b$

Add $100b$ to both sides $Pb + 100b = 100s - 100b + 100b$

giving $b(P + 100) = 100s$

Divide both sides $\left.\begin{array}{c}\\ \\\end{array}\right\}$
by $(P + 100)$ $\dfrac{b(P+100)}{(P+100)} = \dfrac{100s}{(P+100)}$

giving $b = \dfrac{100s}{(P+100)}$

Check this transposition for yourself.

EXAMPLE 9.10

A formula for resistances connected in parallel in an electrical circuit is $\dfrac{1}{R} = \dfrac{1}{R_1} + \dfrac{1}{R_2}$. Transpose for R.

The LCM of the RHS is $R_1 R_2$.

$\therefore$ $\dfrac{1}{R} = \dfrac{R_2 + R_1}{R_1 R_2}$

Invert both sides $R = \dfrac{R_1 R_2}{R_2 + R_1}$

Exercise 9.1

1) In thermodynamics the characteristic gas equation is $pV = nRT$ which connects pressure, p, volume, V, and temperature, T. Make T the subject of this formula.

2) A large cone has radius, R, and height, H. A small cone has radius, r, and height, h. These dimensions are related by

$$\frac{R}{r} = \frac{H}{h}$$

Rearrange this equation for h.

3) The equation of motion $v = u + at$ is for movement with constant acceleration, a. The initial velocity is u, and the final velocity is v, arrived at after time, t. Make u the subject of the equation.

4) Using $v = u + at$ again, transpose for t.

5) Temperature in degrees Fahrenheit F and in degrees Celsius C are related by $F = \dfrac{9}{5}C + 32$. Rearrange this formula for C.

6) The standard equation of a straight line is $y = mx + c$ in co-ordinate geometry. Rearrange to make x the subject of the equation.

7) A carpet of area $15\,m^2$ has a rectangle $1.5\,m$ wide and $L\,m$ long cut away. The remaining area, A, is given by $A = 15 - 1.5L$. Find L in terms of A.

8) The sum to infinity, S, of a geometric progression is given by the expression $S = \dfrac{a}{1-r}$ where a is the first term and r is the common ratio. Transpose to make r the subject.

9) The current, I, and the voltage, V, in a circuit containing two resistances R and r in series are connected by the formula $I = \dfrac{V}{R+r}$. Transpose for R.

10) The surface area, S, of a cone having height h and base radius r is given by $S = \pi r(r + h)$. Find h in terms of π, r and S.

11) The expression $H = ws(T - t)$ is used in finding total heat, H. Rearrange to make T the subject of the formula.

12) The expression $\dfrac{v^2 - u^2}{2a} = s$ is a law of motion for constant acceleration, a, the distance travelled, s, and the initial and final velocities, u and v, respectively. Transpose the equation for u^2.

13) The common difference, d, of an arithmetic progression whose sum S to n terms is given by $d = \dfrac{2(S-an)}{n(n-1)}$ where the first term is a.

Transpose this formula   (a) for S (b) for a

14) The expression $R = \dfrac{1}{1/R_1 + 1/R_2}$ is for resistances in parallel in an electrical circuit. Transpose for R_1.

15) A sphere of radius R has a cap height h cut off its top. The volume, V, of the cap is given by $V = \dfrac{\pi h^2}{3}(3R - h)$. Rearrange to make R the subject.

16) Another form of the expression for parallel resistances in an electrical circuit is $R = \dfrac{R_1 R_2}{R_1 + R_2}$. Make R_2 the subject of this equation.

17) For a statistical frequency distribution the mode, M, is given by the formula $M = L + i\left(\dfrac{d_1}{d_1 + d_2}\right)$. Transpose for d_1.

18) In co-ordinate geometry, if two straight lines have gradients m_1 and m_2 then the angle θ between the lines is given by $\tan \theta = \dfrac{m_1 - m_2}{1 + m_1 m_2}$. Find an expression for m_2 in terms of m_1 and $\tan \theta$.

19) The eccentricity, e, on an ellipse having semi-major and semi-minors axes a and b respectively is given by $e^2 = \dfrac{a^2 - b^2}{a^2}$. Make a^2 the subject.

20) The area, A, of a quadrilateral may be found by dividing the figure into two triangles giving rise to the formula $A = \dfrac{a(H+h) + bh + cH}{2}$. Transpose for H.

Formulae Containing Roots and Powers

EXAMPLE 9.11

An equation of motion for constant acceleration, a, is $v^2 = u^2 + 2as$ where the initial velocity, u, is changed to the final velocity, v, after time t. Make u the subject of the equation.

Subtract $2as$ from both sides $\qquad v^2 - 2as = u^2 + 2as - 2as$

giving $\qquad\qquad\qquad\qquad\qquad\qquad u^2 = v^2 - 2as$

and taking the square root
of both sides $\qquad\qquad\qquad\qquad \sqrt{u^2} = \sqrt{v^2 - 2as}$

giving $\qquad\qquad\qquad\qquad\qquad\qquad u = \sqrt{v^2 - 2as}$

Check this result for yourself.

EXAMPLE 9.12

The period, t, of a simple pendulum having length l is given by $t = 2\pi\sqrt{\dfrac{l}{g}}$ where g is the gravitational constant. Transpose this equation for l.

Divide both sides by 2π $\qquad\qquad\qquad \dfrac{t}{2\pi} = \sqrt{\dfrac{l}{g}}$

Square both sides $\qquad\qquad\qquad \left(\dfrac{t}{2\pi}\right)^2 = \left(\sqrt{\dfrac{l}{g}}\right)^2$

or $\qquad\qquad\qquad\qquad\qquad\qquad \dfrac{t^2}{4\pi^2} = \dfrac{l}{g}$

Multiply both sides by g $\qquad\qquad l = \dfrac{t^2 g}{4\pi^2}$

Again we leave the checking to you.

Exercise 9.2

1) The velocity, v, of a jet of water is given by the expression $v = \sqrt{2gh}$. Transpose this for head of water, h.

2) Formula $A = \pi r^2$ gives the area, A, of a circle in terms of its radius, r. Find r in terms of π and A.

3) The kinetic energy, E, of a mass m travelling at velocity v is given by $E = \frac{1}{2}mv^2$. Make v the subject of this expression.

4) The diameter, d, of a circle of area A is given by $d = 2\sqrt{\dfrac{A}{\pi}}$. Find A in terms of π and d.

5) The strain energy, U, of a material under stress is given by $U = \dfrac{f^2 V}{2E}$. Rearrange this for f.

6) In a right-angled triangle, $b = \sqrt{a^2 - c^2}$. Transpose for c.

7) The frequency, f, of a simple pendulum is given by $f = \dfrac{1}{2\pi}\sqrt{\dfrac{g}{l}}$. Make l the subject of this expression.

8) A shaft is acted on by a bending moment, M, and a torque, T. The equivalent torque, T_e, is given by $T_e^2 = M^2 + T^2$. Transpose this for M.

9) The crippling load on a strut, according to the Euler theory, is given by $P = \dfrac{\pi^2 EI}{(CL)^2}$. Make C the subject of this equation.

10) The total energy, E_t, of a mass m at a height h above a given datum and travelling with a velocity v is given by the equation $E_t = mgh + \frac{1}{2}mv^2$. Find an expression for velocity v.

11) A spherical cap having base radius r and height h has a volume V given by $V = \dfrac{h}{6}(3r^2 + h^2)$. Transpose for r.

12) A property of a solid rectangular block, called the radius of gyration, k, is given by $k = \sqrt{\dfrac{a^2 + b^2}{12}}$. Rearrange this equation for b.

13) A property of a solid cylinder, called the radius of gyration, k, is given by $k = \sqrt{\dfrac{L^2}{12} + \dfrac{R^2}{4}}$. Make L the subject of this equation.

14) A formula connected with stress in cylinders is $\dfrac{D}{d} = \sqrt{\dfrac{f+p}{f-p}}$. Transpose this for f.

15) A formula for the equivalent shear load in the design of a bolt is $Q_e = \frac{1}{2}\sqrt{P^2 + Q^2}$. Rearrange this to make an equation for P.

16) A formula for the equivalent tensile load, P_e, in the design of of a bolt, is $P_e = \frac{1}{2}(P + \sqrt{P^2 + 4Q^2})$. Transpose for Q.

10.

QUADRATIC EQUATIONS

THE PRODUCT OF TWO BINOMIAL EXPRESSIONS

A binomial expression consists of two terms. Thus $3x+5$, $a+b$, $2x+3y$ and $4p-q$ are all binomial expressions.

The product $(a+b)(c+d) = a(c+d) + b(c+d) = ac+ad+bc+bd$

It will be noticed that the expression on the right-hand side is obtained by multiplying each term in the one bracket by each term in the other bracket. The process is illustrated below:

$$(a+b)(c+d) = ac+ad+bc+bd$$

EXAMPLE 10.1

a) $(3x+2)(4x+5) = 3x \times 4x + 3x \times 5 + 2 \times 4x + 2 \times 5$

$$= 12x^2 + 15x + 8x + 10 = 12x^2 + 23x + 10$$

b) $(2p-3)(4p+7) = 2p \times 4p + 2p \times 7 - 3 \times 4p - 3 \times 7$

$$= 8p^2 + 14p - 12p - 21 = 8p^2 + 2p - 21$$

c) $(z-5)(3z-2) = z \times 3z + z \times (-2) - 5 \times 3z - 5 \times (-2)$

$$= 3z^2 - 2z - 15z + 10 = 3z^2 - 17z + 10$$

THE SQUARE OF A BINOMIAL EXPRESSION

$(a + b)^2 = (a + b)(a + b) = a^2 + ab + ba + b^2 = a^2 + 2ab + b^2$

The square of a binomial expression is the sum of the squares of the two terms and twice their product.

$(a - b)^2 = (a - b)(a - b) = a^2 - ab - ba + b^2 = a^2 - 2ab + b^2$

EXAMPLE 10.2

a) $(2x + 5)^2 = (2x)^2 + 2 \times 2x \times 5 + 5^2 = 4x^2 + 20x + 25$

b) $(3x - 2)^2 = (3x)^2 + 2 \times 3x \times (-2) + (-2)^2 = 9x^2 - 12x + 4$

THE PRODUCT OF THE SUM AND DIFFERENCE OF TWO TERMS

$$(a + b)(a - b) \ = \ a^2 - ab + ba - b^2 \ = \ a^2 - b^2$$

This result is the difference of the squares of the two terms.

EXAMPLE 10.3

a) $(8x \div 3)(8x - 3) = (8x)^2 - 3^2 = 64x^2 - 9$

b) $(2x + 5y)(2x - 5y) = (2x)^2 - (5y)^2 = 4x^2 - 25y^2$

Exercise 10.1

Find the products of the following:

1) $(x + 1)(x + 2)$

2) $(2x + 5)(x + 3)$

3) $(2x + 4)(3x + 2)$

4) $(x - 4)(x - 2)$

5) $(x - 2)(3x - 5)$

6) $(3x - 1)(2x - 5)$

7) $(x + 3)(x - 1)$

8) $(x - 5)(x + 3)$

9) $(3x - 5)(x + 6)$

10) $(6x - 7)(2x + 3)$

11) $(2p - q)(p - 3q)$

12) $(3v + 2u)(2v - 3u)$

13) $(x + 3)(x - 3)$

14) $(2x + 3)(2x - 3)$

15) $(x + 1)^2$

16) $(2x + 3)^2$

17) $(x - 1)^2$

18) $(2x - 3)^2$

19) $(2a + 3b)^2$

20) $(x + y)^2$

21) $(a - b)^2$

22) $(3x - 4y)^2$

QUADRATIC EXPRESSIONS

A quadratic expression is one in which the highest power of the symbol used is the square. Typical examples are $x^2 - 5x + 3$ or $3x^2 - 9$ in which there is no power of x greater than x^2.

You will see, from the work in the previous section, that when two binomial expressions are multiplied together the result is always a quadratic expression.

It is often necessary to try and reverse this procedure. This means that we start with a quadratic expression and wish to express this as the product of two binomial expressions—this is not always possible. For example the expressions $x^2 + 1$ or $a^2 + b^2$ cannot be factorised. You may check this for yourself after following the next section of this chapter.

FACTORISING QUADRATIC EXPRESSIONS

Consider $(7x + 4)(2x + 3)$

Now
$$(7x + 4)(2x + 3) = 14x^2 + 21x + 8x + 12$$
$$= 14x^2 + 29x + 12$$

The following points should be noted:

(1) The first terms in each bracket when multiplied together give the first term of the quadratic expression.

(2) The middle term of the quadratic expression is formed by multiplying together the terms connected by a line (see above equation) and then adding them together.

(3) The last terms in each bracket when multiplied together give the last term of the quadratic expression.

In most cases, when factorising a quadratic expression, we find all the possible factors of the first and last terms. Then, by trying various combinations, the combination which gives the correct middle term may be found.

EXAMPLE 10.4

Factorise $2x^2 + 5x - 3$

Factors of $2x^2$	Factors of -3
$2x$ $\quad$ x	-3 $\quad$ $+1$
	$+3$ $\quad$ -1

Combinations of these factors are:

$$(2x - 3)(x + 1) = 2x^2 - x - 3 \quad \text{which is incorrect,}$$
$$(2x + 1)(x - 3) = 2x^2 - 5x - 3 \quad \text{which is incorrect,}$$
$$(2x + 3)(x - 1) = 2x^2 + x - 3 \quad \text{which is incorrect,}$$
$$(2x - 1)(x + 3) = 2x^2 + 5x - 3 \quad \text{which is correct.}$$

Hence $\quad\quad\quad 2x^2 + 5x - 3 = (2x - 1)(x + 3)$

EXAMPLE 10.5

Factorise $12x^2 - 35x + 8$

Factors of $12x^2$		Factors of 8	
$12x$	x	1	8
$6x$	$2x$	8	1
$3x$	$4x$	-1	-8
		-8	-1
		2	4
		4	2
		-2	-4
		-4	-2

By trying each combination in turn the only one which will produce the correct middle term of $-35x$ is found to be $(3x - 8)(4x - 1)$.

$$\therefore \quad\quad\quad 12x^2 - 35x + 8 = (3x - 8)(4x - 1)$$

Where the Factors form a Perfect Square

A quadratic expression, which factorises into the product of two identical brackets resulting in a perfect square, may be factorised by the method used previously. However, if you can recognise that the result will be a perfect square then the problem becomes easier.

It has been shown that

$$(a + b)^2 = a^2 + 2ab + b^2 \quad \text{and} \quad (a - b)^2 = a^2 - 2ab + b^2$$

The square of a binomial expression therefore consists of:

(Square of 1st term) + (Twice product of terms)

+ (Square of 2nd term)

EXAMPLE 10.6

Factorise $9a^2 + 12ab + 4b^2$

Now $9a^2 = (3a)^2$, $4b^2 = (2b)^2$ and $12ab = 2 \times 3a \times 2b$

$\therefore \qquad 9a^2 + 12ab + 4b^2 = (3a + 2b)^2$

EXAMPLE 10.7

Factorise $16m^2 - 40m + 25$

Now $16m^2 = (4m)^2$, $25 = (-5)^2$ and $-40m = 2 \times 4m \times (-5)$

$\therefore \qquad 16m^2 - 40m + 25 = (4m - 5)^2$

The Factors of the Difference of Two Squares

It has previously been shown that

$$(a + b)(a - b) = a^2 - b^2$$

The factors of the difference of two squares are therefore the sum and the difference of the square roots of each of the given terms.

EXAMPLE 10.8

Factorise $9m^2 - 4n^2$

Now $\qquad 9m^2 = (3m)^2$, and $4n^2 = (2n)^2$

$\therefore \qquad 9m^2 - 4n^2 = (3m + 2n)(3m - 2n)$

EXAMPLE 10.9

Factorise $4x^2 - 9$

Now $\qquad 4x^2 = (2x)^2$ and $9 = (3)^2$

$\therefore \qquad 4x^2 - 9 = (2x + 3)(2x - 3)$

Exercise 10.2

Factorise:

1) $x^2 + 4x + 3$

2) $x^2 + 6x + 8$

3) $x^2 - 3x + 2$

4) $x^2 + 2x - 15$

5) $x^2 + 6x - 7$

6) $x^2 - 5x - 14$

7) $x^2 - 2xy - 3y^2$

8) $2x^2 + 13x + 15$

9) $3p^2 + p - 2$

10) $4x^2 - 10x - 6$

11) $3m^2 - 8m - 28$ 12) $21x^2 + 37x + 10$

13) $10a^2 + 19a - 15$ 14) $6x^2 + x - 35$

15) $6p^2 + 7pq - 3q^2$ 16) $12x^2 - 5xy - 2y^2$

17) $x^2 + 2xy + y^2$ 18) $4x^2 + 12x + 9$

19) $p^2 + 4pq + 4q^2$ 20) $9x^2 + 6x + 1$

21) $m^2 - 2mn + n^2$ 22) $25x^2 - 20x + 4$

23) $x^2 - 4x + 4$ 24) $m^2 - n^2$

25) $4x^2 - y^2$ 26) $9p^2 - 4q^2$

27) $x^2 - 1/9$ 28) $1 - b^2$

29) $1/x^2 - 1/y^2$ 30) $121p^2 - 64q^2$

ROOTS OF AN EQUATION

If either of two factors has zero value, then their product is zero. Thus if either $M = 0$ or $N = 0$ then $M \times N = 0$

Now suppose that either $x = 1$ or $x = 2$

$\therefore$ rearranging gives either $x - 1 = 0$ or $x - 2 = 0$

Hence $(x - 1)(x - 2) = 0$

since either of the factors has zero value.

If we now multiply out the brackets of this equation we have

$$x^2 - 3x + 2 = 0$$

and we know that $x = 1$ and $x = 2$ are values of x which satisfy this equation. The values 1 and 2 are called the solutions or *roots* of the equation $x^2 - 3x + 2 = 0$

EXAMPLE 10.10

Find the equation whose roots are -2 and 4

From the values given either $x = -2$ or $x = 4$

$\therefore$ either $x + 2 = 0$ or $x - 4 = 0$

Hence $(x + 2)(x - 4) = 0$

since either of the factors has zero value.

$\therefore$ Multiplying out gives $x^2 - 2x - 8 = 0$

EXAMPLE 10.11

Find the equation whose roots are 3 and -3

From the values given either $x = 3$ or $x = -3$

$\therefore$ either $x - 3 = 0$ or $x + 3 = 0$

Hence $(x - 3)(x + 3) = 0$

since either of the factors has zero value.

Multiplying out we have $x^2 - 9 = 0$

EXAMPLE 10.12

Find the equation whose roots are 5 and 0.

From the given values given either $x = 5$ or $x = 0$

$\therefore$ either $x - 5 = 0$ or $x = 0$

Hence $x(x - 5) = 0$

since either of the factors has zero value.

And multiplying out we have $x^2 - 5x = 0$

Exercise 10.3

Find the equations whose roots are:

1) 3, 1   2) 2, -4

3) -1, -2 4) 1.6, 0.7

5) 2.73, -1.66 6) -4.76, -2.56

7) 0, 1.4   8) -4.36, 0

9) -3.5, $+3.5$ 10) repeated, each $= 4$

QUADRATIC EQUATIONS

An equation of the type $ax^2 + bx + c = 0$, involving x in the second degree and containing no higher power of x, is called a *quadratic equation*. The constants a, b and c have any numerical values. Thus.

$$x^2 - 9 = 0 \quad \text{where } a = 1, \, b = 0 \text{ and } c = -9$$
$$x^2 - 2x - 8 = 0 \quad \text{where } a = 1, \, b = -2 \text{ and } c = -8$$
$$2.5x^2 - 3.1x - 2 = 0 \quad \text{where } a = 2.5, \, b = -3.1 \text{ and } c = -2$$

are all examples of quadratic equations. A quadratic equation may contain only the square of the unknown quantity, as in the first of the above equations, or it may contain both the square and the first power as in the other two.

1. Solution by Factors

This method is the reverse of the procedure used to find an equation when given the two roots. We shall now start with the equation and proceed to solve the equation and find the roots.

We shall again use the fact that if the product of two factors is zero then one factor or the other must be zero. Thus if $M \times N = 0$ then either $M = 0$ or $N = 0$

When the factors are easy to find the factor method is very quick and simple. However do not spend too long trying to find factors: if they are not easily found use the formula given in the next method (p. 116) to solve the equation.

EXAMPLE 10.13

Solve the equation $(2x + 3)(x - 5) = 0$

Since the product of the two factors $2x + 3$ and $x - 5$ is zero then either

$$2x + 3 = 0 \quad \text{or} \quad x - 5 = 0$$

Hence $\qquad x = -\dfrac{3}{2} \quad \text{or} \quad x = 5$

EXAMPLE 10.14

Solve the equation $\quad 6x^2 + x - 15 = 0$

Factorising gives $(2x - 3)(3x + 5) = 0$

$\therefore \qquad$ either $\quad 2x - 3 = 0 \quad \text{or} \quad 3x + 5 = 0$

Hence $\qquad\qquad\qquad x = \dfrac{3}{2} \quad \text{or} \quad x = -\dfrac{5}{3}$

EXAMPLE 10.15

Solve the equation $14x^2 = 29x - 12$

Bring all the terms to the left-hand side:

$$14x^2 - 29x + 12 = 0$$

$\therefore$
$$(7x - 4)(2x - 3) = 0$$

$\therefore$ either $7x - 4 = 0$ or $2x - 3 = 0$

Hence $x = \dfrac{4}{7}$ or $x = \dfrac{3}{2}$

EXAMPLE 10.16

Find the roots of the equation $x^2 - 16 = 0$

Factorising gives $(x - 4)(x + 4) = 0$

$\therefore$ either $x - 4 = 0$ or $x + 4 = 0$

Hence $x = 4$ or $x = -4$

In this case an alternative method may be used:

Rearranging the given equation gives $x^2 = 16$

and taking the square root of both sides gives $x = \sqrt{16} = \pm 4$

Remember that when we take a square root we must insert the $\pm$ sign, because $(+4)^2 = 16$ and $(-4)^2 = 16$

EXAMPLE 10.17

Solve the equation $x^2 - 2x = 0$

Factorising gives $x(x - 2) = 0$

$\therefore$ either $x = 0$ or $x - 2 = 0$

Hence $x = 0$ or $x = 2$

Note: The solution $x = 0$ must not be omitted as it is a solution in the same way as $x = 2$ is a solution. Equations should not be divided through by variables, such as x, since this removes a root of the equation.

EXAMPLE 10.18

Solve the equation $x^2 - 6x + 9 = 0$

Factorising gives $(x - 3)(x - 3) = 0$

$\therefore$ either $x-3 = 0$ or $x-3 = 0$

Hence $x = 3$ or $x = 3$

In this case there is only one arithmetical value for the solution. Technically, however, there are two roots and when they have the same numerical value they are said to be repeated roots.

2. Solution by Formula

In general, quadratic expressions do not factorise and therefore some other method of solving quadratic equations must be used.

Consider the expression $ax^2 + bx = a\left(x^2 + \dfrac{b}{a}x\right)$

If we add (half the coefficient of x)^2 to the terms inside the bracket we get

$$ax^2 + bx = a\left[x^2 + \frac{b}{a}x + \left(\frac{b}{2a}\right)^2\right] - a\left(\frac{b}{2a}\right)^2$$

$$= a\left(x + \frac{b}{2a}\right)^2 - \frac{b^2}{4a}$$

We are said to have completed the square of $ax^2 + bx$.

We shall now establish a formula which may be used to solve any quadratic equation.

If $ax^2 + bx + c = 0$

then $ax^2 - bx = -c$

Completing the square of the LHS gives

$$a\left(x + \frac{b}{2a}\right)^2 - \frac{b^2}{4a} = -c$$

$\therefore$ $$4a^2\left(x + \frac{b}{2a}\right)^2 - b^2 = -4ac$$

or $$4a^2\left(x + \frac{b}{2a}\right)^2 = b^2 - 4ac$$

Taking the square root of both sides

$$2a\left(x + \frac{b}{2a}\right) = \pm\sqrt{b^2 - 4ac}$$

from which $$x = \frac{-b \pm \sqrt{b^2 - 4ac}}{2a}$$

The *standard form* of the *quadratic equation* is:

$$ax^2 + bx + c = 0$$

As shown on the previous page the *solution* of this equation is:

$$x = \frac{-b \pm \sqrt{b^2 - 4ac}}{2a}$$

EXAMPLE 10.19

Solve the equation $3x^2 - 8x + 2 = 0$

Comparing with $ax^2 + bx + c = 0$, we have $a = 3$, $b = -8$ and $c = 2$

Substituting these values in the formula, we have

$$x = \frac{-(-8) \pm \sqrt{(-8)^2 - 4 \times 3 \times 2}}{2 \times 3}$$

$$= \frac{8 \pm \sqrt{64 - 24}}{6} = \frac{8 \pm \sqrt{40}}{6} = \frac{8 \pm 6.325}{6}$$

$\therefore$ either $\quad x = \dfrac{8 + 6.325}{6} \quad$ or $\quad x = \dfrac{8 - 6.325}{6}$

Hence $\quad x = 2.39 \qquad$ or $\quad x = 0.28$

It is important that we check the solutions in case we have made an error. We may do this by substituting the values obtained in the left-hand side of the given equation and checking that the solution is zero, or approximately zero.

Thus when $x = 2.39$ we have LHS $= 3(2.39)^2 - 8(2.39) + 2 \approx 0$ and when $x = 0.28$ we have LHS $= 3(0.28)^2 - 8(0.28) + 2 \approx 0$

EXAMPLE 10.20

Solve the equation $2.13x^2 + 0.75 - 6.89 = 0$

Here $a = 2.13$, $b = 0.75$, $c = -6.89$

$$x = \frac{-0.75 \pm \sqrt{(0.75)^2 - 4(2.13)(-6.89)}}{2 \times 2.13}$$

$$= \frac{-0.75 \pm \sqrt{0.5625 + 58.70}}{4.26} = \frac{-0.75 \pm \sqrt{59.26}}{4.26}$$

$$= \frac{-0.75 \pm 7.698}{4.26}$$

$\therefore$ either $x = \dfrac{-0.75 + 7.698}{4.26}$ or $x = \dfrac{-0.75 - 7.698}{4.26}$

Hence $x = 1.631$ or $x = -1.983$

Solution check

When $x = 1.631$

we have   LHS $= 2.13(1.631)^2 + 0.75(1.631) - 6.89 \approx 0$

When $x = -1.983$

we have   LHS $= 2.13(-1.983)^2 + 0.75(-1.983) - 6.89 \approx 0$

EXAMPLE 10.21

Solve the equation $x^2 + 4x + 5 = 0$

Here $a = 1$, $b = 4$ and $c = 5$

$$\therefore \quad x = \frac{-4 \pm \sqrt{4^2 - 4(1)(5)}}{2(1)} = \frac{-4 \pm \sqrt{16 - 20}}{2} = \frac{-4 \pm \sqrt{-4}}{2}$$

Now when a number is squared the answer must be a positive quantity because two quantities having the same sign are being multiplied together. Therefore the square root of a negative quantity, as $\sqrt{-4}$ in the above equation, has no arithmetical meaning and is called an imaginary quantity. The equation $x^2 + 4x + 5 = 0$ is said to have imaginary or complex roots. Equations which have complex roots are beyond the scope of this book and are dealt with in more advanced mathematics.

Exercise 10.4

Solve the following equations by the factor method:

1) $x^2 - 36 = 0$ 2) $4x^2 - 6.25 = 0$

3) $9x^2 - 16 = 0$ 4) $x^2 + 9x + 20 = 0$

5) $x^2 + x - 72 = 0$ 6) $3x^2 - 7x + 2 = 0$

7) $m^2 = 6m - 9$ 8) $m^2 + 4m + 4 = 36$

9) $14q^2 = 29q - 12$ 10) $9x + 28 = 9x^2$

Solve the following equations by using the quadratic formula:

11) $4x^2 - 3x - 2 = 0$ ✓

12) $x^2 - x + \frac{1}{4} = \frac{1}{9}$

13) $3x^2 + 7x - 5 = 0$ ✓

14) $7x^2 + 8x - 2 = 0$ ✓

15) $5x^2 - 4x - 1 = 0$ ✓

16) $2x^2 - 7x = 3$

17) $x^2 + 0.3x - 1.2 = 0$

18) $2x^2 - 5.3x + 1.25 = 0$

Solve the following equations:

19) $x(x + 4) + 2x(x + 3) = 5$

20) $x^2 - 2x(x - 3) = -20$

21) $\dfrac{2}{x + 2} + \dfrac{3}{x + 1} = 5$

22) $\dfrac{x + 2}{3} - \dfrac{5}{x + 2} = 4$

23) $\dfrac{6}{x} - 2x = 2$

24) $40 = \dfrac{x^2}{80} + 4$

25) $\dfrac{x + 2}{x - 2} = x - 3$

26) $\dfrac{1}{x + 1} - \dfrac{1}{x + 3} = 15$

PROBLEMS INVOLVING QUADRATIC EQUATIONS

EXAMPLE 10.22

The distance, s m, moved by a vehicle in time, t s, with an initial velocity, v_1 m/s, and a constant acceleration, a m/s^2, is given by $s = v_1 t + \frac{1}{2} a t^2$. Find the time taken to cover 84 m with a constant acceleration 2 m/s^2 if the initial velocity is 5 m/s.

Using $s = 84$ m, $v_1 = 5$ m/s and $a = 2$ m/s^2 we have

$$84 = 5t + \tfrac{1}{2} 2t^2$$

from which $\quad\quad\quad t^2 + 5t - 84 = 0$

Factorising gives $\quad (t + 12)(t - 7) = 0$

∴ $\quad\quad\quad$ either $\quad t + 12 = 0 \quad\quad$ or $\quad t - 7 = 0$

∴ $\quad\quad\quad$ either $\quad\quad t = -12 \quad$ or $\quad\quad t = 7$

Now the solution $t = -12$ is not acceptable since negative time has no meaning in this question. Thus the required time is 7 seconds.

Solution check

When $t = 7$ we have LHS $= 7^2 + 5 \times 7 - 84 = 0$

EXAMPLE 10.23

The diagonal of a rectangle is 15 m long and one side is 2 m longer than the other. Find the dimensions of the rectangle.

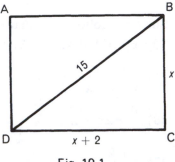

In Fig. 10.1, let the length of BC be x cm. The length of CD is then $(x + 2)$ m. $\triangle$BCD is right-angled and so by Pythagoras,

Fig. 10.1

$$x^2 + (x + 2)^2 = 15^2$$

$\therefore \quad x^2 + x^2 + 4x + 4 = 225$

$\therefore \quad 2x^2 + 4x - 221 = 0$

Here $a = 2$, $b = 4$ and $c = -221$

$$\therefore \qquad x = \frac{-4 \pm \sqrt{4^2 - 4 \times 2 \times (-221)}}{2 \times 2}$$

$\therefore \qquad x = 9.56 \quad \text{or} \quad -11.56$

Since the answer cannot be negative, then $x = 9.56$ m.

Now $\qquad\qquad x + 2 = 11.56$ m

$\therefore$ the rectangle has adjacent sides equal to 9.56 m and 11.56 m.

Solution check

When $x = 9.56$ we have LHS $= 2(9.56)^2 + 4(9.56) - 221 \approx 0$

EXAMPLE 10.24

A section of an air duct is shown by the full lines in Fig. 10.2.

a) Show that: $w^2 - 2Rw + \dfrac{R^2}{4} = 0$

b) Find the value of w when $R = 2$ m.

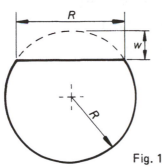
Fig. 10.2

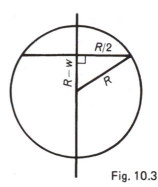
Fig. 10.3

a) Using the construction shown in Fig. 10.3 we have, by Pythagoras,

then
$$(R-w)^2+\left(\frac{R}{2}\right)^2 = R^2$$

$$\therefore \qquad R^2-2Rw+w^2+\frac{R^2}{4} = R^2$$

$$\therefore \qquad w^2-2Rw+\frac{R^2}{4} = 0$$

b) When $R = 2$, $\qquad w^2-4w+1 = 0$

Here $a = 1$, $b = -4$ and $c = 1$

$$\therefore \qquad w = \frac{-(-4)\pm\sqrt{(-4)^2-4\times1\times1}}{2\times1}$$

Hence $\qquad w = 3.732 \quad$ or $\quad 0.268\,\text{m}$

Now w must be less than $2\,\text{m}$, thus $w = 0.268\,\text{m}$

Solution check

When $w = 0.268$ we have LHS $= (0.268)^2-4(0.268)+1 \approx 0$

EXAMPLE 10.25

Specified quantities of hydrogen and iodine are mixed together in a vessel of given volume at a certain temperature. If the equilibrium constant, K_c, is given by $K_c = \dfrac{(2-x)(1-x)}{(2x)^2}$ find the quantity, $(2-x)$, of hydrogen at equilibrium, given that $K_c = 0.018$

If we substitute $K_c = 0.018$ in the given expression for K_c

then
$$0.018 = \frac{(2-x)(1-x)}{(2x)^2}$$

from which $\qquad 0.018(2x)^2 = (2-x)(1-x)$

$\therefore \qquad\qquad\qquad 0.072x^2 = 2-3x+x^2$

$\therefore \qquad\qquad 0.928x^2-3x+2 = 0$

Now the standard form of a quadratic equation is

$$ax^2+bx+c = 0$$

from which $\qquad\qquad\qquad x = \dfrac{-b\pm\sqrt{b^2-4ac}}{2a}$

Comparing the quadratic equations we have $a = 0.928$, $b = -3$ and $c = 2$

Thus $\qquad\qquad x = \dfrac{-(-3)\pm\sqrt{(-3)^2-4(0.928)2}}{2(0.928)}$

from which $\qquad x = 2.293 \quad$ or $\quad 0.940$

Now if $x = 2.293$ then the quantity of hydrogen is $(2-2.293)$ $= -0.293$ which is unacceptable since there cannot be a negative quantity. We therefore reject this answer and use the other.

Hence the required quantity of hydrogen is $(2-0.940) = 1.06$

Solution check

In this case we would check the solution by substituting $x = 0.940$ into the right-hand side of the given expression and hope that we obtain 0.018

Thus $\qquad$ RHS $= \dfrac{(2-0.940)(1-0.940)}{(2\times0.940)^2} = 0.018$

Exercise 10.5

1) The length L of a wire stretched tightly between two supports in the same horizontal line is given by

$$L = S+\dfrac{8D^2}{3S}$$

where S is the span and D is the (small) sag. If $L = 150$ and $D = 5$, find the value of S.

2) In a right-angled triangle the hypotenuse is twice as long as one of the sides forming the right angle. The remaining side is 80 mm long. Find the length of the hypotenuse.

3) The area of a rectangle is $61.75\,m^2$. If the length is 3 m more than the width find the dimensions of the rectangle.

4) The total surface area of a cylinder whose height is 75 mm is 29 000 mm^2. Find the radius of the cylinder.

5) If a segment of a circle has a radius R, a height H and a length of chord W show that

$$R = \frac{W^2}{8H} + \frac{H}{2}$$

Rearrange this equation to give a quadratic equation for H and hence find H when $R = 12$ m and $W = 8$ m.

6) Fig. 10.4 shows a template whose area is 9690 mm^2. Find the value of r.

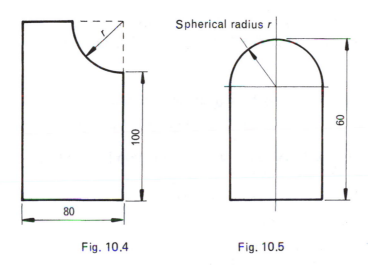

Fig. 10.4   Fig. 10.5

7) A pressure vessel is of the shape shown in Fig. 10.5, the radius of the vessel being r mm. If the surface area is 30000 mm^3 find r.

8) The total iron loss in a transformer is given by the equation $P = 0.1f + 0.006f^2$. If $P = 20$ watts, find the value of the frequency f.

9) The volume of a frustum of a cone is given by the formula $V = \frac{1}{3}\pi h(R^2 + rR + r^2)$ where h is the height of the frustum and R and r are the radii at the large and small ends respectively. If $h = 9$ m, $R = 4$ m and the volume is 337.2 m^3, what is the value of r?

10) A square steel plate is pierced by a square tool leaving a margin of 20 mm all round. The area of the hole is one third that of the original plate. What are the dimensions of the original plate?

11) The velocity, v, of a body in terms of time, t, is given by the expression $v = 3t^2 - 6t - 3$. Find the times at which the velocity is zero.

12) The value of the equilibrium constant, K_c, when phosphorus pentachloride vapour is heated is given by the expression $K_c = \dfrac{x^2}{V(1-x)}$ where, for an equilibrium mixture, V is the total volume and x is the quantity of chlorine. Find the value of x if $K_c = 0.012$ and $V = 24$

SIMULTANEOUS SOLUTION OF A QUADRATIC AND LINEAR EQUATION

EXAMPLE 10.26

Solve simultaneously the equations: $\quad y = x^2 + 3x - 4 \qquad$ [1]

$$\text{and} \quad y = 2x + 4 \qquad [2]$$

Substituting the value of y given by equation [2], that is $y = 2x + 4$ into equation [1] we have

$$2x + 4 = x^2 + 3x - 4$$

from which $\qquad x^2 + x - 8 = 0$

This equation does not factorise so we will use the formula. Here $a = 1$, $b = 1$ and $c = -8$

$$\therefore \qquad x = \frac{-1 \pm \sqrt{1^2 - 4(1)(-8)}}{2(1)} = \frac{-1 \pm \sqrt{33}}{2}$$

$$\therefore \quad \text{either} \quad x = \frac{-1 + 5.745}{2} \quad \text{or} \quad x = \frac{-1 - 5.745}{2}$$

$$\therefore \quad \text{either} \quad x = 2.372 \qquad \text{or} \quad x = -3.372$$

Now for each of these values of x there will be a corresponding value of y. We may find these values of y by substituting the values $x = 2.372$ and $x = -3.372$ into either of the given equations.

Therefore from equation [2]

$$\text{when} \quad x = 2.372, \quad y = 2(2.372) + 4 = 8.744$$

and

$$\text{when} \quad x = -3.372, \quad y = 2(-3.372) + 4 = -2.744$$

Hence the required solutions are

x	2.372	-3.372
y	8.744	-2.744

These solutions may be checked by substituting the values into equation [1]:

RHS $= (2.372)^2 + 3(2.372) - 4 = 8.742 \approx$ LHS

and

RHS $= (-3.372)^2 + 3(-3.372) - 4 = -2.745 \approx$ LHS

The inaccuracies occur because the values have been rounded off correct to the third place of decimals.

EXAMPLE 10.27

Solve simultaneously the equations: $\quad y^2 - 2y - 4 = x \qquad$ [1]

$$\text{and} \quad x + 3y - 2 = 0 \qquad [2]$$

Now equation [2] may be rearranged to give $x = -3y + 2$ and if we substitute this value of x into equation [1] we have

$$y^2 - 2y - 4 = -3y + 2$$

from which $\qquad y^2 + y - 6 = 0$

and factorising $\qquad (y + 3)(y - 2) = 0 \quad$ gives

Thus $\qquad$ either $\quad y + 3 = 0 \quad$ or $\quad y - 2 = 0$

$\therefore \qquad$ either $\qquad y = -3 \quad$ or $\qquad y = 2$

Now for each of these values of y there will be a corresponding value of x. We may find these values of x by substituting the values of y into either of the given equations. Therefore from equation [2]

when $y = -3$, $x + 3(-3) - 2 = 0 \;\; \therefore x = 11$

and when $y = 2$, $x + 3(2) - 2 = 0 \;\; \therefore x = -4$

Hence the required solutions are

x	11	-4
y	-3	2

These solutions may be checked by substituting the values into equation [1]:

LHS $= (-3)^2 - 2(-3) - 4 = 9 + 6 - 4 = 11 =$ RHS

and LHS $= (2)^2 - 2(2) - 4 = 4 - 4 - 4 = -4 =$ RHS

Exercise 10.6

Solve simultaneously, by substitution:

1) $y = 3x^2 - 3x - 1$
   $y = 4x - 3$

2) $y = 8x^2 - 2$
   $y = 1 - 5x$

3) $y = 4x^2 - 2x - 1$
   $x - y + 1 = 0$

4) $x = y^2 - 0.4y - 1.5$
   $x + 0.7y + 0.3 = 0$

5) $y = 7x^2 + 6x - 3$
   $2x + y = -1$

6) $y = x^2 + 1.213x + 0.574$
   $y = 2.213x + 0.435$

7) A cutting is in the shape of an isosceles trapezium which is $2x$ m wide at the top, $2y$ m wide at the bottom and has a vertical height of 8 m. The cross-sectional area of the cutting is 144 m^2 and its perimeter (two sloping sides and the base) is 32 m. It can be shown that

$$x + y = 18 \qquad [1]$$

and $\qquad x^2 - 2xy + 4y = -60 \qquad [2]$

Solve these equations and hence find the bottom width of the cross-section of the cutting.

8) Fig. 10.6 shows a hot water cylinder with a surface area of 138 m^2. Show that

$$3\pi r^2 + 2\pi rh = 138 \qquad [1]$$

and $\qquad r + h = 10 \qquad [2]$

By solving this pair of simultaneous equations find the values of r and h.

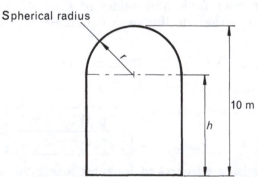

Fig. 10.6

11. LOGARITHMS AND EXPONENTIALS

Outcome:

1. *Define a logarithm to any base.*
2. *Convert from logarithmic to index form and vice versa.*
3. *Deduce the laws of logarithms.*
4. *Show that* $\log_b 1 = 0$, $\log_b b = 1$ *and* $\log_b 0 = -\infty$.
5. *Apply the laws to simplify expressions and solve equations.*
6. *Plot the graphs of the exponential functions* e^x *and* e^{-x}.

7. *Show that at any point on the curve of the exponential function* e^x *that the gradient of the curve is equal to the value of the function.*
8. *Solve problems, from the relevant technologies, of exponential growth* $(k.e^{ax})$ *and decay* $(k.e^{-ax})$ *curves.*
9. *Recognise that at any point the gradients of the functions in 8 are proportional to the value of the function.*

LOGARITHMS

If N is a number such that

we may write this in the alternative form

$$N = b^x$$
$$\log_b N = x$$

which, in words, is   'the logarithm of N, to the base b, is x'

or   'x is the logarithm of N to the base b'

The word 'logarithm' is often abbreviated to just 'log'.

It is helpful to remember that: $\text{Number} = \text{Base}^{\text{logarithm}}$

Alternatively in words:

> The log of a number is the power to which the base must be raised to give that number

Thus:

We may write	$8 = 2^3$	We may write	$81 = 3^4$
in log form as	$\log_2 8 = 3$	in log form as	$\log_3 81 = 4$

We may write	$2 = \sqrt{4}$	We may write	$\dfrac{1}{4} = \dfrac{1}{2^2}$
or	$2 = 4^{1/2}$		
or	$2 = 4^{0.5}$	or	$0.25 = 2^{-2}$
in log form as	$\log_4 2 = 0.5$	in log form as	$\log_2 0.25 = -2$

EXAMPLE 11.1

If $\log_7 49 = x$, find the value of x.

Writing the equation in index form we have $49 = 7^x$

or $7^2 = 7^x$

Since the bases are the same on both sides of the equation the indices must be the same.   Thus $x = 2$

EXAMPLE 11.2

If $\log_x 8 = 3$, find the value of x.

Writing this equation in index form we have $8 = x^3$

or $2^3 = x^3$

Since the indices on both sides of the equation are the same the bases must be the same.   Thus $x = 2$

THE VALUE OF $\log_b 1$

Let $\log_b 1 = x$

then in index form $1 = b^x$

Now the only value of the index x which will satisfy this expression is zero.

Hence $\log_b 1 = 0$

Thus: | To any base the value of log 1 is zero |

THE VALUE OF $\log_b b$

Let $\log_b b = x$

then in index form $b = b^x$

Now the only value of the index x which will satisfy the expression is unity.

Hence $\log_b b = 1$

Thus: | The value of the log of a number to the same base is unity |

THE VALUE OF $\log_b 0$

Let $$\log_b 0 = x$$

then in index form $$0 = b^x$$

Now consider, for example, the value of 10^x:

If $x = -2$ then the value of 10^x will be $10^{-2} = \dfrac{1}{10^2}$

If $x = -20$ then the value of 10^x will be $10^{-20} = \dfrac{1}{10^{20}}$

If $x = -200$ then the value of 10^x will be $10^{-200} = \dfrac{1}{10^{200}}$

Now $\dfrac{1}{10^{200}}$ is a very small number indeed and from this pattern we may deduce that if the value of the index x is an infinitely large negative number (called 'minus infinity', written as $-\infty$) then the value of 10^x would be zero. It follows that $b^{-\infty} = 0$

Hence $$\log_b 0 = -\infty$$

Thus: | To any base the log of zero is minus infinity |

THE VALUE OF $\log_b(-N)$

Let $$\log_b(-N) = x$$

then in index form $$-N = b^x$$

If we examine this expression we can see that whatever the value of the negative number N, or whatever the value of the base, b, it is not possible to find a value for the index x which will satisfy the expression.

Hence $$\log_b(-N) \text{ has no real value}$$

Thus | Only positive numbers have real logarithms |

Exercise 11.1

Express in logarithmic form:

1) $n = a^x$　　　　2) $2^3 = 8$　　　　3) $5^{-2} = 0.04$

4) $10^{-3} = 0.001$ 5) $x^0 = 1$ 6) $10^1 = 10$

7) $a^1 = a$ 8) $e^2 = 7.39$ 9) $10^0 = 1$

Find the value of x in each of the following:

10) $\log_x 9 = 2$ 11) $\log_x 81 = 4$ 12) $\log_2 16 = x$

13) $\log_5 125 = x$ 14) $\log_3 x = 2$ 15) $\log_4 x = 3$

16) $\log_{10} x = 2$ 17) $\log_7 x = 0$ 18) $\log_x 8 = 3$

19) $\log_x 27 = 3$ 20) $\log_9 3 = x$ 21) $\log_n n = x$

LAWS OF LOGARITHMS

Let $\log_b M = x$ and $\log_b N = y$

or in index form $M = b^x$ and $N = b^y$

(1) Now $MN = b^x \times b^y$

∴ $MN = b^{x+y}$

or in log form $\log_b MN = x + y$

Hence $\boxed{\log_b MN = \log_b M + \log_b N}$

In words this relationship is:

> The logarithm of two numbers multiplied together may be found by adding their individual logarithms

(2) Now $\dfrac{M}{N} = \dfrac{b^x}{b^y}$

∴ $\dfrac{M}{N} = b^{x-y}$

or in log form $\log_b \dfrac{M}{N} = x - y$

Hence $\boxed{\log_b \dfrac{M}{N} = \log_b M - \log_b N}$

In words this relationship is:

> The logarithm of two numbers divided may be found by subtracting their individual logarithms

(3) Now
$$M^n = (b^x)^n$$

$\therefore$
$$M^n = b^{nx}$$

or in log form
$$\log_b M^n = nx$$

Hence
$$\boxed{\log_b M^n = n(\log_b M)}$$

In words this relationship is:

> The logarithm of a number raised to a power may be found by multiplying the logarithm of the number by the power

LOGARITHMS TO THE BASE 10

Logarithms to the base 10 are called common logarithms and stated as $\log_{10}$ (or lg). When logarithmic tables are used to solve numerical problems, tables to this base are preferred as they are simpler to use than tables to any other base. Common logarithms are also used for scales on logarithmic graph paper and also for calculations on the measurement of sound.

LOGARITHMS TO THE BASE 'e'

In higher mathematics all logarithms are taken to the base e, where e = 2.71828. Logarithms to this base are often called natural logarithms. They are also called Naperian or hyperbolic logarithms.

Natural logarithms are stated as $\log_e$ (or ln).

CHOICE OF BASE FOR CALCULATIONS

If an electronic calculating machine of the scientific type is used, then it is just as easy to use logarithms to the base e. Some machines have keys for both $\log_e$ and $\log_{10}$ but on the more limited models only $\log_e$ is given.

The natural logarithms is found by using the $\log_e$ (or ln) key and the natural antilogarithm is found by using the e^x key.

The common logarithm is found using the $\log_{10}$ (or lg) key and the common antilogarithm is found using the 10^x key.

In the example which follows it is appreciated that use of the power key x^y will give an immediate solution, but it is instructive to work through the alternative method of solution using logarithms.

EXAMPLE 11.3

Evaluate $3.714^{2.87}$

Let $\qquad\qquad x = 3.714^{2.87}$

and taking logarithms of both sides we have

$$\log x = \log 3.714^{2.87}$$

$\therefore\qquad\qquad \log x = 2.87 \times \log 3.714$

$\therefore\qquad\qquad x = \text{antilog}\,(2.87 \times \log 3.714)$

The base of the logarithms has not yet been chosen — the sequence given will be true for any base value.

Using a calculator with natural logarithms the sequence of operations is:

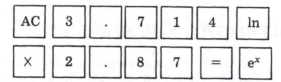

The display gives $\boxed{43.1963}$

Thus the answer is 43.2 correct to three significant figures.

INDICAL EQUATIONS

These are equations in which the number to be found is an index, or part of an index.

The method of solution is to reduce the given equation to an equation involving logarithms, as the following examples will illustrate.

EXAMPLE 11.4

If $8.79^x = 67.65$ find the value of x.

Now taking logarithms of both sides of the given equation we have

$$\log 8.79^x = \log 67.65$$

∴ $$x(\log 8.79) = \log 67.65$$

∴ $$x = \frac{\log 67.65}{\log 8.79}$$

The base of the logarithms has not yet been chosen, the above procedure being true for any base value.

The quickest way, that is without using the reciprocal $\boxed{\dfrac{1}{x}}$ key,

is to find the value of the bottom line and put this into the memory. Then find the value of the top line and divide this by the content of the memory.

The sequence, using natural logarithms, would be:

The display gives $\boxed{1.93886}$

Thus the answer is 1.94 correct to three significant figures.

EXAMPLE 11.5

Find the value of x if $1.793^{(x+3)} = 20^{0.982}$

Now taking logarithms of both sides of the given equation we have

$$\log 1.793^{x+3} = \log 20^{0.982}$$

∴ $$(x+3)(\log 1.793) = (0.982)(\log 20)$$

∴ $$x + 3 = \frac{(0.982)(\log 20)}{\log 1.793}$$

∴ $$x = \frac{(0.982)(\log 20)}{\log 1.793} - 3$$

The procedure will be similar to that used in Example 11.4. The sequence of operations, using natural logarithms, is then:

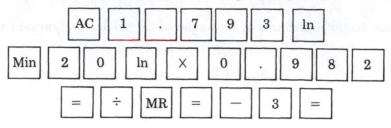

The display gives $\boxed{2.03829}$

Thus the answer is 2.04 correct to three significant figures.

Exercise 11.2

Evaluate the following:

1) $11.57^{0.3}$ 2) $15.62^{2.15}$ 3) $0.6327^{0.5}$

4) $0.06521^{3.16}$ 5) $27.15^{-0.4}$

Find the value of x in the following:

6) $3.6^x = 9.7$ 7) $0.9^x = 2.176$

8) $\left(\dfrac{1}{7.2}\right)^x = 1.89$ 9) $1.4^{(x+2)} = 9.3$

10) $21.9^{(3-x)} = 7.334$ 11) $2.79^{(x-1)} = 4.377^x$

12) $\left(\dfrac{1}{0.64}\right)^{(2+x)} = 1.543^{(x+1)}$ 13) $\dfrac{1}{0.9^{(x-2)}} = 8.45$

CALCULATIONS INVOLVING THE EXPONENTIAL FUNCTIONS, e^x and e^{-x}

EXAMPLE 11.6

Evaluate $50\,e^{2.16}$

The sequence of operations is:

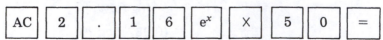

giving an answer 434 correct to three significant figures.

EXAMPLE 11.7

Evaluate $200\,e^{-1.34}$

The sequence of operations would then be:

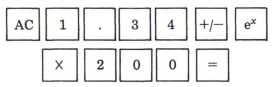

giving an answer 52.4 correct to three significant figures.

EXAMPLE 11.8

In a capacitive circuit the instantaneous voltage across the capacitor is given by $v = V(1 - e^{-t/CR})$ where V is the initial supply voltage, R ohms the resistance, C farads the capacitance, and t seconds the time from the instant of connecting the supply voltage.

If $V = 200$, $R = 10000$, and $C = 20 \times 10^{-6}$ find the time when the voltage v is 100 volts.

Substituting the given values in the equation we have

$$100 = 200(1 - e^{t/\,20 \times 10^{-6} \times 10\,000})$$

$$\therefore \quad \frac{100}{200} = 1 - e^{-t/0.2}$$

$$\therefore \quad 0.5 = 1 - e^{-5t}$$

$$\therefore \quad e^{-5t} = 1 - 0.5$$

$$\therefore \quad e^{-5t} = 0.5$$

Thus in log form

$$\log_e 0.5 = -5t$$

$$\therefore \quad t = -\frac{\log_e 0.5}{5}$$

The sequence of operation is:

$$\boxed{\text{AC}}\ \boxed{0}\ \boxed{.}\ \boxed{5}\ \boxed{\ln}\ \boxed{\div}\ \boxed{5}\ \boxed{=}\ \boxed{+/-}$$

giving an answer 0.139 seconds correct to three significant figures.

EXAMPLE 11.9

$$R = \frac{(0.42)S}{l} \times \log_e \frac{d_2}{d_1}$$

refers to the insulation resistance of a wire. Find the value of R when $S = 2000$, $l = 120$, $d_1 = 0.2$ and $d_2 = 0.3$

Substituting the given values gives

$$R = \frac{0.42 \times 2000}{120} \times \log_e \frac{0.3}{0.2}$$

$$= \frac{0.42 \times 2000}{120} \times \log_e 1.5$$

The sequence of operations would be:

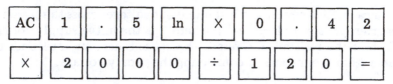

giving an answer 2.84 correct to three significant figures.

Exercise 11.3

1) Find the numbers whose natural logarithms are:

(a) 2.76   (b) 0.677   (c) 0.09

(d) −3.46   (e) −0.543   (f) −0.078

2) Find the values of:

(a) $70\,e^{2.5}$ (b) $150\,e^{-1.34}$ (c) $3.4\,e^{-0.445}$

3) The formula $L = 0.000644\left(\log_e \dfrac{d}{r} + \dfrac{1}{4}\right)$ is used for calculating the self-inductance of parallel conductors. Find L when $d = 50$ and $r = 0.25$

4) The inductance (L microhenrys) of a straight aerial is given by the formula: $L = \dfrac{1}{500}\left(\log_e \dfrac{4l}{d} - 1\right)$ where l is the length of the aerial in mm and d its diameter in mm. Calculate the inductance of an aerial 5000 mm long and 2 mm in diameter.

5) Find the value of $\log_e\left(\dfrac{c_1}{c_2}\right)^2$ when $c_1 = 4.7$ and $c_2 = 3.5$

6) If $T = R \log_e \left(\dfrac{a}{a-b} \right)$ find T when $R = 28$, $a = 5$ and $b = 3$.

7) When a chain of length $2l$ is suspended from two points $2d$ apart on the same horizontal level, $d = c \log_e \left(\dfrac{l + \sqrt{l^2 + c^2}}{c} \right)$. If $c = 80$ and $l = 200$ find d.

8) The instantaneous value of the current when an inductive circuit is discharging is given by the formula $i = Ie^{-Rt/L}$. Find the value of this current, i, when $I = 6$, $R = 30$, $L = 0.5$ and $t = 0.005$

9) In a circuit in which a resistor is connected in series with a capacitor the instantaneous voltage across the capacitor is given by the formula $v = V(1 - e^{-t/CR})$. Find this voltage, v, when $V = 200$, $C = 40 \times 10^{-6}$, $R = 100000$ and $t = 1$

10) In the formula $v = Ve^{-Rt/L}$ the values of v, V, R and L are 50, 150, 60 and 0.3 respectively. Find the corresponding value of t.

11) The instantaneous charge in a capacitive circuit is given by $q = Q(1 - e^{-t/CR})$. Find the value of t when $q = 0.01$, $Q = 0.015$, $C = 0.0001$, and $R = 7000$

EXPONENTIAL GRAPHS

Curves which have equations of the type e^x and e^{-x} are called exponential graphs.

We may plot the graphs of e^x and e^{-x} by using mathematical tables to find values of e^x and e^{-x} for chosen values of x. We should remember that any number to a zero power is unity: hence $e^0 = 1$

Drawing up a table of values we have:

x	-2	-1	0	1	2
e^x	0.14	0.37	1	2.72	7.39
$-x$	2	1	0	-1	-2
e^{-x}	7.39	2.72	1	0.37	0.14

For convenience both the curves are shown plotted on the same axes in Fig. 11.1. Although the range of values chosen for x is limited the overall shape of the curves is clearly shown.

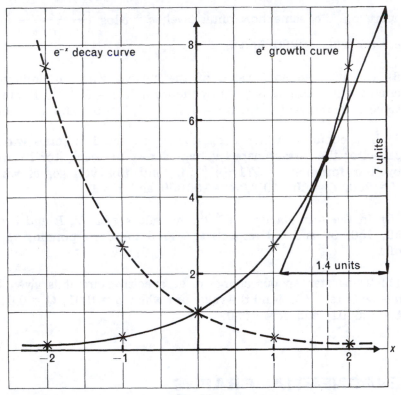

Fig. 11.1

The rate at which a curve is changing at any point is given by the gradient of the tangent at that point.

Remember the sign convention for gradients is

Positive
gradient

Negative
gradient

Now the gradient at any point on the e^x graph is positive and so the rate of change is positive. In addition the rate of change increases as the values of x increase. A graph of this type is called a *growth curve*.

The gradient at any point on the e^{-x} graph is negative and so the rate of change is negative. In addition the rate of change decreases as the values of x decrease. A graph of this type is called a *decay curve*.

AN IMPORTANT PROPERTY OF THE EXPONENTIAL FUNCTION, e^x

Suppose we choose any point, P, on the graph of the function e^x shown in Fig. 11.1 and draw a tangent to the curve at P. We may find the gradient of the curve at P by finding the gradient of the tangent using the right-angled triangle shown. The gradient at P $= \dfrac{7}{1.4} = 5$. Now this is also the value of e^x at P. The reader may like to draw the curve of e^x and check that the gradient at various points is always equal to the value of e^x at corresponding points. This illustrates the important, and unique, property of the exponential function e^x which is:

> At any point on the curve of the exponential function e^x the gradient of the curve is equal to the value of the function

EXAMPLE 11.10

The population size, N, at a certain time, t hours, after commencement of growth of a unit sized population is given by the exponential growth relationship $N = e^{0.8t}$. Show that the instantaneous rate of growth is proportional to the population size.

The values of N for corresponding values of t may be found using the scientific calculator. The table of values shows results for values of t from zero to 4 hours:

t hours	0	0.5	1	1.5	2	2.5	3	3.5	4
$N = e^{0.8t}$	1	1.49	2.23	3.32	4.95	7.39	11.0	16.4	24.5

The curve is shown plotted in Fig. 11.2.

The instantaneous rate of growth means the rate of growth at any instant. This is given by the gradient of the curve at any particular point. The gradient may be found by drawing a tangent to the curve at the point and calculating its slope by constructing a suitable right-angled triangle.

At P, using the right-angled triangle shown, the gradient $= \dfrac{5.4}{3} = 1.8$

Also the value of N is 2.23.

Hence   the ratio: $\dfrac{\text{Gradient}}{N \text{ value}} = \dfrac{1.8}{2.23} = 0.81$

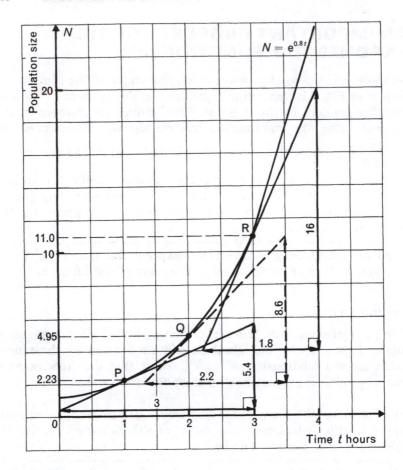

Fig. 11.2

Similarly at Q the ratio: $\dfrac{\text{Gradient}}{N \text{ value}} = \dfrac{8.6/2.2}{4.95} = 0.79$

And at R the ratio: $\dfrac{\text{Gradient}}{N \text{ value}} = \dfrac{16/1.8}{11} = 0.81$

The reader may like to plot the curve and calculate the ratio $\dfrac{\text{Gradient}}{N \text{ value}}$ for other points. We can see from these results that the value of the ratio (within the limitations of accuracy of values obtained from the graph) is constant—in this case 0.80

It is, therefore, reasonable to assume that at any point on the expontential curve the ratio:

$$\frac{\text{Gradient}}{N \text{ value}} = \text{Constant}$$

or when rearranged $\text{Gradient} = (\text{Constant})(N \text{ value})$

$\therefore$ $\text{Gradient} \propto N \text{ value}$

Thus the instantaneous rate of growth is proportional to the population size.

> From this we may conclude that a property of an exponential curve is that, at any point, the gradient is proportional to the N value (i.e. the ordinate)

EXAMPLE 11.11

The instantaneous e.m.f. in an inductive circuit is given by the expression $100\,e^{-4t}$ volts, where t is time in seconds. Plot the graph of the e.m.f. for values of t from 0 to 0.5 seconds, and use the graph to find:

a) the value of the e.m.f. when $t = 0.25$ seconds, and
b) the rate of change of the e.m.f. when $t = 0.1$ seconds.

The graph is shown plotted in Fig. 11.3, from values obtained from the sequence of operations:

| AC | | t value | | X | | 4 | | = | | +/− | | e^x | | X | | 100 | | = |

a) The point P on the curve is at 0.25 seconds shown on the t scale and the corresponding value of e.m.f. can be read directly from the vertical axis scale. The value is 37 volts.

b) The point Q on the graph is at 0.1 seconds. Now the rate of change of the curve at Q is given by the gradient of the tangent at Q. This gradient may be found by constructing a suitable right-angled triangle such as MNO in Fig. 11.3, and finding the ratio $\dfrac{\text{MO}}{\text{ON}}$. Hence the

$$\text{Gradient at } Q = \frac{\text{MO}}{\text{ON}} = \frac{94 \text{ volts}}{0.35 \text{ seconds}} = 269 \text{ volts per second}$$

According to the sign convention a line sloping downwards from left to right has a negative gradient.

Hence the gradient at Q is -269 volts per second, which means that the rate of change of the curve at Q is -269 volts per second.

This is the same as saying that the e.m.f. at $t = 0.1$ seconds is decreasing at the rate of 269 volts per second.

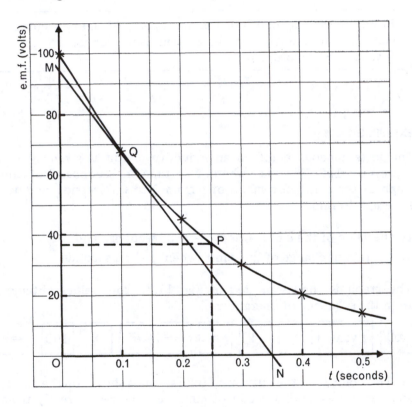

Fig. 11.3

EXAMPLE 11.12

The formula $i = 2(1-e^{-10t})$ gives the relationship between the instantaneous current i amperes and the time t seconds in an inductive circuit. Plot a graph of i against t taking values of t from 0 to 0.3 seconds at intervals of 0.05 seconds. Hence find:

a) the initial rate of growth of the current i when $t = 0$, and
b) the time taken for the current to increase from 1 to 1.6 amperes.

The curve is shown plotted in Fig. 11.4 from values obtained from the sequence of operations:

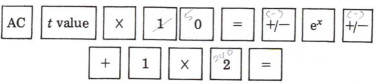

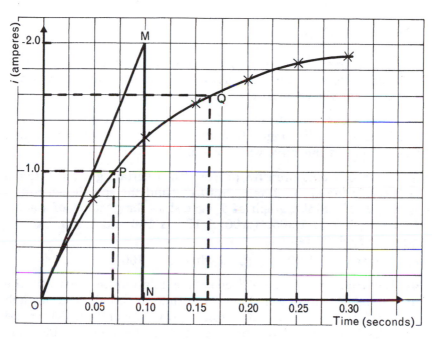

Fig. 11.4

a) When $t = 0$ the initial rate of growth will be given by the gradient of the tangent at O. The tangent at O is the line OM and its gradient may be found by using a suitable right-angled triangle at MNO and finding the ratio $\dfrac{MN}{ON}$.

Hence the initial rate of growth of $\quad i = \dfrac{MN}{ON} = \dfrac{2\ \text{amperes}}{0.1\ \text{seconds}}$

$$= 20\ \text{amperes per second}$$

b) The point P on the curve corresponds to a current of 1.0 amperes and the time at which this occurs may be read from the t scale and is 0.07 seconds.

Similarly point Q corresponds to a 1.6 ampere current and occurs at 0.16 seconds.

Hence the time between P and Q = 0.16 − 0.07 = 0.09 seconds.

This means that the time for the current to increase from 1 to 1.6 amperes is 0.09 seconds.

Exercise 11.4

1) Plot a graph of $y = e^{2x}$ for values of x from -1 to $+1$ at 0.25 unit intervals. Use the graph to find the value of y when $x = 0.3$, and the value of x when $y = 5.4$

2) Using values of x from -4 to $+4$ at one unit intervals plot a graph of $y = e^{-x/2}$. Hence find the value of x when $y = 2$, and the gradient of the curve when $x = 0$

3) For a constant pressure process on a certain gas the formula connecting the absolute temperature T and the specific entropy s is $T = 24\,e^{3s}$. Plot a graph of T against s taking values of s equal to 1.000, 1.033, 1.066, 1.100, 1.133, 1.166 and 1.200. Use the graph to find the value of:

(a) T when $s = 1.09$ (b) s when $T = 700$

4) The equation $i = 2.4e^{-6t}$ gives the relationship between the instantaneous current, i mA, and the time, t seconds. Plot a graph of i against t for values of t from 0 to 0.6 seconds at 0.1 second intervals. Use the curve obtained to find the rate at which the current is decreasing when $t = 0.2$ seconds.

5) In a capacitive circuit the voltage v and the time t seconds are connected by the relationship $v = 240(1 - e^{-5t})$. Draw the curve of v against t for values of $t = 0$ to $t = 0.7$ seconds at 0.1 second intervals. Hence find:

(a) the time when the voltage is 140 volts, and
(b) the initial rate of growth of the voltage when $t = 0$

6) The number of cells, N, in a bacterial population in time, t hours, from the commencement of growth is given by $N = 100e^{1.7t}$, find:

(a) the size of the population after 4 hours growth,
(b) the time in which the population increases tenfold from its initial value,
(c) the instantaneous rate of growth after 3 hours from the start.

7) Given that the mass, m grams, of a bacterial population after t hours from the beginning of growth is given by $m = (10^{-10})e^{1.2t}$, find:

(a) the mass of the population after 2 hours from growth commencement,
(b) the mass of the population at beginning of growth,
(c) the time when the population has doubled from its initial value.

8) The decomposition of a chemical compound, C, over a period of time, t, is given by $C = k(1 - e^{-0.2t})$. If $k = 10$ find the rate of decomposition after 15 seconds.

12.

RADIAN MEASURE

Outcome:

1. *Define the radian.*
2. *Convert measurements in degrees to radians, and vice-versa.*
3. *Express angles in multiples of π radians.*
4. *Use the relationships* $s = r\theta$ *for the length*

of an arc and $A = \dfrac{1}{2}r^2\theta$ *for the area of a sector.*

5. *Solve problems involving lengths of arcs, areas and angles in radians.*

RADIAN MEASURE

We have seen that an angle is usually measured in degrees but there is another way of measuring an angle. In this, the unit is known as the radian (abbreviation rad).

Referring to Fig. 12.1 gives

$$\text{Angle in radians} = \frac{\text{Length of arc}}{\text{Radius of circle}}$$

$$\theta \text{ radians} = \frac{l}{r}$$

$$l = r\theta$$

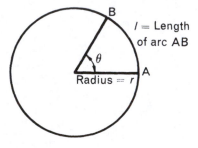

$l = $ Length of arc AB

Radius $= r$

Hence

$$\text{Length of arc} = r\theta$$

Fig. 12.1

RELATION BETWEEN RADIANS AND DEGREES

If we make the arc AB equal to a semi-circle then

$$\text{Length of arc} = \pi r$$

and

$$\text{Angle in radians} = \frac{\pi r}{r} = \pi$$

Now the angle subtended by a semi-circle $= 180°$

146

Therefore π radians $= 180°$

or 1 radian $= \dfrac{180°}{\pi} = 57.3°$

Thus to convert from degrees to radians

$$\theta° = \frac{\pi\theta}{180} \text{ radians}$$

Thus $30° = \dfrac{\pi(30)}{180} \text{ rad} = \dfrac{\pi}{6} \text{ rad}$

$90° = \dfrac{\pi}{2} \text{ rad}$ $180° = \pi \text{ rad}$

$45° = \dfrac{\pi}{4} \text{ rad}$ $270° = \dfrac{3\pi}{2} \text{ rad}$

$60° = \dfrac{\pi}{3} \text{ rad}$ $360° = 2\pi \text{ rad}$

To convert from radians to degrees

$$\theta \text{ radians} = \left(\frac{180}{\pi} \times \theta\right)°$$

DEGREES, MINUTES AND SECONDS

There are   60 seconds in 1 minute, or $60'' = 1'$

and   60 minutes in 1 degree, or $60' = 1°$

Thus 60×60 seconds in 1 degree, or $3600'' = 1°$

Modern calculating methods make the use of decimal degrees (e.g. $36.783°$) more likely than the use of minutes and seconds.

EXAMPLE 12.1

Convert $29°37'29''$ to radians stating the answer correct to 4 significant figures.

The first step is to convert the given angle into degrees and decimals of a degree.

$$29°37'29'' = 29 + \frac{37}{60} + \frac{29}{3600} = 29.625°$$

$$= \frac{\pi \times 29.625}{180} = 0.5171 \text{ radians}$$

Many scientific calculators will convert degrees, minutes and seconds into decimal degrees, and vice versa, using special keys — instructions for use of these keys will be given in the accompanying booklet.

However, the sequence given below will perform the calculation given in the solution shown above.

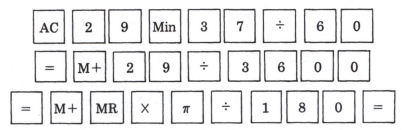

giving an answer of 0.5171 correct to 4 significant figures.

EXAMPLE 12.2

Convert 0.089 35 radians into degrees, minutes and seconds.

$$0.089\,35 \text{ radians} = \frac{0.089\,35 \times 180}{\pi} = 5.1194° = 5°7'10''$$

For calculators without decimal degree conversion facility the following sequence may be used—this is a reverse of the sequence used in the previous example.

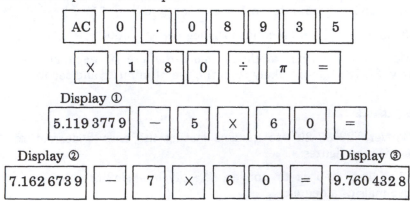

In this sequence it is necessary to record three results as they appear. The whole number, namely 5, in display ① is the number of degrees. The whole number 5 is then subtracted to leave the decimal part which is then multiplied by 60. The whole number, namely 7, in display ② is the number of minutes. The whole number, namely 7 is now subtracted to leave the decimal part which is then multiplied by 60 to give display ③—this figure is the number of seconds.

Thus the result is $5°7'10''$ to the nearest second.

COMPONENTS OF A CIRCLE

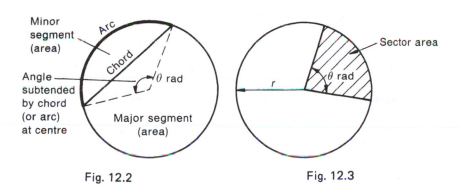

Fig. 12.2   Fig. 12.3

AREA OF A SECTOR

The area of a circle $= \pi r^2$.

So, by proportion, (Fig. 12.2),

$$\text{Area of sector} = \pi r^2 \times \frac{\theta}{2\pi} = \tfrac{1}{2}r^2\theta$$

Summary

Length of arc $= r\theta$	θ in radians	or $2\pi r\left(\dfrac{\theta°}{360}\right)$
Area of a sector $= \tfrac{1}{2}r^2\theta$	θ in radians	or $\pi r^2\left(\dfrac{\theta°}{360}\right)$

EXAMPLE 12.3

Calculate a) the length of arc of a circle whose radius is 8 m and which subtends an angle of 56° at the centre, and b) the area of the sector so formed.

a) Length of arc $= 2\pi r \times \dfrac{\theta°}{360} = 2 \times \pi \times 8 \times \dfrac{56}{360} = 7.82\,\mathrm{m}$

b) Area of sector $= \pi r^2 \times \dfrac{\theta°}{360} = \pi \times 8^2 \times \dfrac{56}{360} = 31.28\,\mathrm{m}^2$

EXAMPLE 12.4

Find the angle of a sector of radius 35 mm and area $1020\,\mathrm{mm}^2$

Now   Area of sector $= \tfrac{1}{2}r^2\theta$

and substituting the given values of

$$\text{Area} = 1020\,\mathrm{mm}^2 \quad \text{and} \quad r = 35\,\mathrm{mm}$$

we have $1020 = \tfrac{1}{2}(35)^2\theta$

from which $\theta = \dfrac{1020 \times 2}{35^2} = 1.67\,\mathrm{rad}$

$$= \dfrac{180 \times 1.67}{\pi} = 95.7°$$

EXAMPLE 12.5

Water flows in a 400 mm diameter pipe to a depth of 300 mm. Calculate the wetted perimeter of the pipe and the area of cross-section of the water.

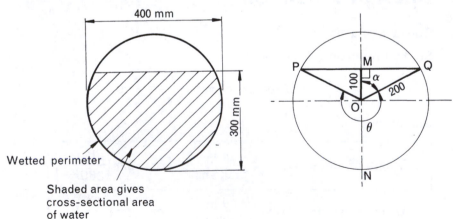

Fig. 12.4

From Fig. 12.4

the right-angled triangle MQO

$$\cos \alpha = \frac{OM}{OQ} = \frac{100}{200} = 0.5$$

$\therefore$ $\alpha = 60°$

Also $\sin \alpha = \dfrac{MQ}{OQ}$

$\therefore$ $MQ = OQ \sin \alpha = 200 \sin 60° = 173.2 \, mm$

Now $\theta + 2\alpha = 360°$

$\therefore$ $\theta = 360° - 2(60°) = 240°$

Thus

Wetted perimeter $=$ Arc PNQ

$$= 2\pi r \left(\frac{\theta}{360}\right) = 2\pi(200)\left(\frac{240}{360}\right) = 838 \, mm$$

Also

$$\begin{pmatrix} \text{Cross-sectional} \\ \text{area of water} \end{pmatrix} = \begin{pmatrix} \text{Area of} \\ \text{sector PNQ} \end{pmatrix} + \begin{pmatrix} \text{Area of} \\ \text{triangle POQ} \end{pmatrix}$$

$$= \pi r^2 \left(\frac{\theta}{360}\right) + \tfrac{1}{2}(PQ)(MO)$$

$$= \pi(200)^2\left(\frac{240}{360}\right) + \tfrac{1}{2}(2 \times 173.2)(100)$$

$$= 83\,780 + 17\,320$$

$$= 101\,000 \, mm^2$$

Exercise 12.1

1) Convert the following angles to radians stating the answers correct to 4 significant figures:

(a) 35°   (b) 83°28′   (c) 19°17′32″   (d) 43°39′49

2) Convert the following angles to degrees, minutes and seconds correct to the nearest second:

(a) 0.1732 radians   (b) 1.5632 radians   (c) 0.0783 radians

3) If r is the radius and θ is the angle subtended by an arc, find the length of arc when:

(a) $r = 2\,\text{m}, \quad \theta = 30°$ (b) $r = 34\,\text{mm}, \quad \theta = 38°40'$

4) If l is the length of an arc, r is the radius and θ the angle subtended by the arc, find θ when:

(a) $l = 9.4\,\text{m}, \quad r = 4.5\,\text{m}$ (b) $l = 14\,\text{mm}, \quad r = 79\,\text{mm}$

5) If an arc 70 mm long subtends an angle of 45° at the centre, what is the radius of the circle?

6) Find the areas of the following sectors of circles:

(a) radius 3 m, angle of sector 60°
(b) radius 27 mm, angle of sector 79°45′
(c) radius 78 mm, angle of sector 143°42′

7) Calculate the area of the part shaded in Fig. 12.5.

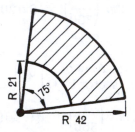

Fig. 12.5

8) A chord 26 mm is drawn in a circle of 35 mm diameter. What are the lengths of arcs into which the circumference is divided?

9) The radius of a circle is 60 mm. A chord is drawn 40 mm from the centre. Find the area of the minor segment.

10) In a circle of radius 30 mm a chord is drawn which subtends an angle of 80° at the centre. What is the area of the minor segment?

11) A flat is machined on a circular bar of 15 mm diameter, the maximum depth of cut being 2 mm. Find the area of the cross-section of the finished bar.

12) Water flows in a 300 mm diameter drain to a depth of 100 mm. Calculate the wetted perimeter of the drain and the area of cross-section of the water.

13) In marking out the plan of part of a building, a line 8 m long is pegged down at one end. Then with the line held horizontal and taut, the free end is swung through an angle of 57°. Calculate the distance moved by the free end of the line and determine the area swept out.

14) Find the area of brickwork necessary to fill in the tympanum of the segmental arch shown in Fig. 12.6.

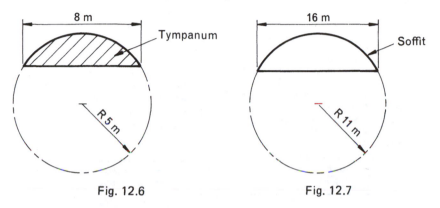

Fig. 12.6   Fig. 12.7

15) Fig. 12.7 shows a segmental arch for a bridge. Calculate the length of the soffit of the arch.

13. TRIANGLES AND WAVEFORMS

THE TRIGONOMETRICAL RATIOS OF A RIGHT-ANGLED TRIANGLE

The definitions of the three trigonometrical ratios are given below. Refer to Fig. 13.1.

$$\sin A = \frac{a}{b} = \frac{\text{Side opposite to A}}{\text{Hypotenuse}}$$

$$\cos A = \frac{c}{b} = \frac{\text{Side adjacent to A}}{\text{Hypotenuse}}$$

$$\tan A = \frac{a}{c} = \frac{\text{Side opposite to A}}{\text{Side adjacent to A}}$$

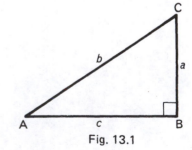

Fig. 13.1

154

EXAMPLE 13.1

Find the sides marked x in Figs. 13.2, 13.3 and 13.4.

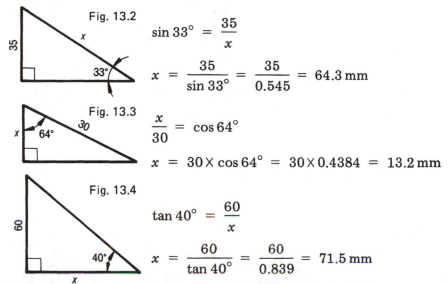

Fig. 13.2

$$\sin 33° = \frac{35}{x}$$

$$x = \frac{35}{\sin 33°} = \frac{35}{0.545} = 64.3 \text{ mm}$$

Fig. 13.3

$$\frac{x}{30} = \cos 64°$$

$$x = 30 \times \cos 64° = 30 \times 0.4384 = 13.2 \text{ mm}$$

Fig. 13.4

$$\tan 40° = \frac{60}{x}$$

$$x = \frac{60}{\tan 40°} = \frac{60}{0.839} = 71.5 \text{ mm}$$

EXAMPLE 13.2

Find the angles marked θ in Figs. 13.5 and 13.6.

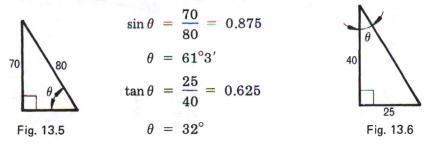

$$\sin \theta = \frac{70}{80} = 0.875$$

$$\theta = 61°3'$$

$$\tan \theta = \frac{25}{40} = 0.625$$

$$\theta = 32°$$

Fig. 13.5

Fig. 13.6

Exercise 13.1

Find the lengths of the sides marked x in Fig. 13.7.

1)   2)   3)

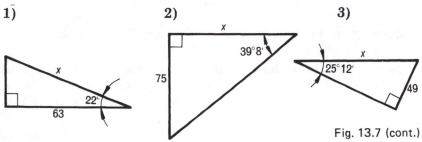

Fig. 13.7 (cont.)

4)   **5)**   **6)**

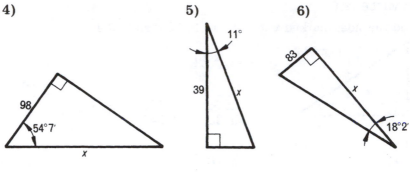

Fig. 13.7

Find the angles marked θ in Fig. 13.8.

7)   **8)**   **9)**

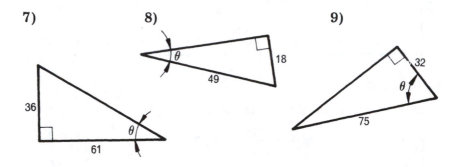

Fig. 13.8

10) The altitude of an isosceles triangle is 86 mm and each of the equal angles is 29°. Calculate the lengths of the equal sides.

TRIGONOMETRICAL IDENTITIES

A statement of the type $\operatorname{cosec} A \equiv \dfrac{1}{\sin A}$ is called an *identity*.

The sign $\equiv$ means 'is identical to'. Any statement using this sign is true for all values of the variables, i.e. the angle A in the above identity. In practice, however, the $\equiv$ sign is often replaced by the $=$ (equals) sign and the identity would be given as $\operatorname{cosec} A = \dfrac{1}{\sin A}$.

Many trigonometrical identities may be verified by the use of a right-angled triangle.

EXAMPLE 13.3

To show that $\tan A = \dfrac{\sin A}{\cos A}$

The sides and angles of the triangle may be labelled in any way providing that the 90° angle is *not* called A. In Fig. 13.9 the standard notation for a triangle has been used.

Now $\qquad\qquad \sin A = \dfrac{a}{b}$ Fig. 13.9

and $\qquad\qquad \cos A = \dfrac{c}{b}$

and $\qquad\qquad \tan A = \dfrac{a}{c}$

Hence from the given identity,

$$RHS = \frac{\sin A}{\cos A} = \frac{a/b}{c/b} = \frac{ab}{bc} = \frac{a}{c} = \tan A = LHS$$

EXAMPLE 13.4

To show that $\sin^2 A + \cos^2 A = 1$

In Fig. 7.9 $\sin A = \dfrac{a}{b}$ $\qquad \therefore \quad \sin^2 A = \left(\dfrac{a}{b}\right)^2 = \dfrac{a^2}{b^2}$

$\qquad\qquad \cos A = \dfrac{c}{b}$ $\qquad \therefore \quad \cos^2 A = \left(\dfrac{c}{b}\right)^2 = \dfrac{c^2}{b^2}$

$\therefore \qquad\qquad LHS = \sin^2 A + \cos^2 A = \dfrac{a^2}{b^2} + \dfrac{c^2}{b^2} = \dfrac{a^2 + c^2}{b^2}$

But by Pythagoras' theorem, $\quad a^2 + c^2 = b^2$

$\therefore \qquad\qquad LHS = \dfrac{b^2}{b^2} = 1 = RHS$

Thus $\qquad\qquad \sin^2 A + \cos^2 A = 1$

RECIPROCAL RATIOS

In addition to sin, cos and tan there are three other ratios that may be obtained from a right-angled triangle. These are:

cosecant   (called cosec for short)
secant   (called sec for short)
cotangent   (called cot for short)

The three ratios are defined as follows:

$$\text{cosec A} = \frac{1}{\sin A} \qquad \sec A = \frac{1}{\cos A} \qquad \cot A = \frac{1}{\tan A}$$

The reciprocal of x is $\dfrac{1}{x}$ and it may therefore be seen why the terms cosec, sec and cot are called 'reciprocal ratios', since they are equal respectively to $\dfrac{1}{\sin}$, $\dfrac{1}{\cos}$ and $\dfrac{1}{\tan}$

Formulae in technical reference books often include reciprocal ratios. It will then be necessary for you to re-write the formula before use in terms of the more familiar ratios, namely sin, cos, and tan.

For example, the formula

$$M = D - \frac{5p}{6}\cot\theta + d(\text{cosec}\,\theta + 1)$$

should be re-written as

$$M = D - \frac{5p}{6}\left(\frac{1}{\tan\theta}\right) + d\left(\frac{1}{\sin\theta} + 1\right)$$

TRIGONOMETRICAL RATIOS BETWEEN 0° AND 360°

The sine, cosine and tangent of an angle between 0° and 90° have been previously defined. We now show how to deal with angles between 0° and 360°.

From Fig. 13.10 by definition, $\sin \theta = \dfrac{PM}{OP}$

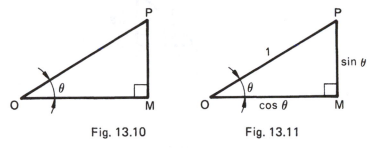

Fig. 13.10   Fig. 13.11

If we make OP = 1 unit as shown in Fig. 13.11 then $\sin \theta = PM$ and $\cos \theta = OM$.

In Fig. 7.12 the axes XOX′ and YOY′ have been drawn at right-angles to each other to form four quadrants shown in the diagram. Drawing the axes in this way allows us to use the same sign convention as when drawing a graph.

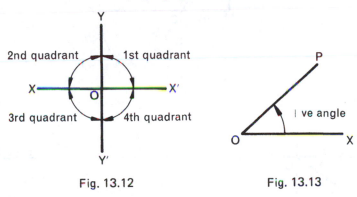

Fig. 13.12   Fig. 13.13

An angle, if positive, is measured in an anti-clockwise direction from the axis OX, which is the datum line from which all angles are measured. It is formed by rotating a line, such as OP (Fig. 7.13) in an anti-clockwise direction.

Now if we draw a circle whose radius is OP we see that OP, as it rotates, forms the angle θ. If OP is made equal to 1 unit, then by drawing the right-angled triangle OPM (Fig. 13.14) the vertical height PM gives the value of $\sin \theta$ and the horizontal distance OM gives the value of $\cos \theta$.

The idea can be extended to angles greater than 90° as shown in Fig. 13.15.

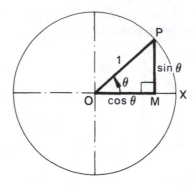

Fig. 13.14

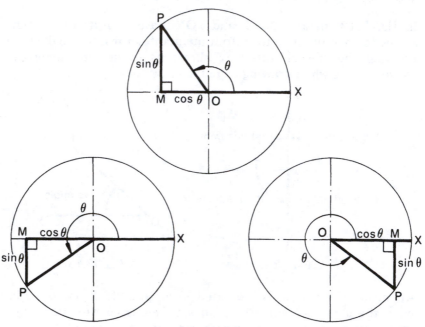

Fig. 13.15

We now make use of sign convention used when drawing a graph. This means that when the height PM lies above the horizontal axis it is a positive length and when it lies below the horizontal axis it is a negative length. Hence when PM lies above the axis XOX' $\sin \theta$ is positive and when it lies below the axis XOX' $\sin \theta$ is negative.

Similarly, when the distance OM lies to the right of the origin O, it is regarded as being a positive distance; if it lies to the left of O it is regarded as being a negative distance. Hence when OM lies to the right of O, $\cos \theta$ is positive and when it lies to the left of O, $\cos \theta$ is negative.

This gives us the signs of the ratios sine and cosine in the four quadrants, see Fig. 13.16.

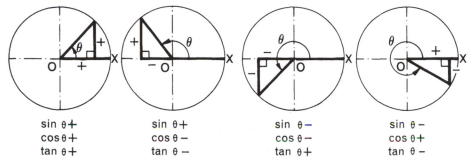

sin θ+	sin θ+	sin θ−	sin θ−
cos θ+	cos θ−	cos θ−	cos θ+
tan θ+	tan θ−	tan θ+	tan θ−

Fig. 13.16

We have shown previously that $\tan \theta \equiv \dfrac{\sin \theta}{\cos \theta}$

Remembering that like signs, when divided, give a positive result and unlike signs give a negative result, we find that $\tan \theta$ is positive in the first and third quadrants and negative in the second and fourth quadrants.

All of the above results are summarised in Fig. 13.17.

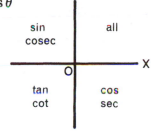

Positive trig. ratios

Fig. 13.17

EXAMPLE 13.5

Find the values of $\sin 158°$, $\cos 158°$ and $\tan 158°$

As shown in Fig. 13.18, $\sin 158°$ is given by the length PM. But from △OPM, PM gives the sine of the angle POM,

$$\therefore \quad \sin 158° = \sin P\hat{O}M$$

$$= \sin (180° - 158°)$$

$$= \sin 22°$$

$$= 0.3746$$

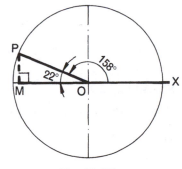

Fig. 13.18

Also from Fig. 13.18, OM gives the value of $\cos 158°$: but this is a negative length since it lies to the left of origin O,

$$\therefore \qquad \cos 158° = -\cos P\hat{O}M$$
$$= -\cos(180° - 158°)$$
$$= -\cos 22°$$
$$= -0.9272$$

and
$$\tan 158° = -\tan P\hat{O}M$$
$$= -\tan(180° - 158°)$$
$$= -\tan 22°$$
$$= -0.4040$$

EXAMPLE 13.6

Find the values of $\sin 247°$, $\cos 247°$ and $\tan 247°$.

From Fig. 13.19

$$\sin 247° = MP = -\sin P\hat{O}M$$
$$= -\sin(247° - 180°)$$
$$= -\sin 67°$$
$$= -0.9205$$

$$\cos 247° = OM = -\cos P\hat{O}M$$
$$= -\cos 67°$$
$$= -0.3907$$

$$\tan 247° = \frac{MP}{OM} = +\tan P\hat{O}M$$
$$= +\tan 67°$$
$$= 2.3559$$

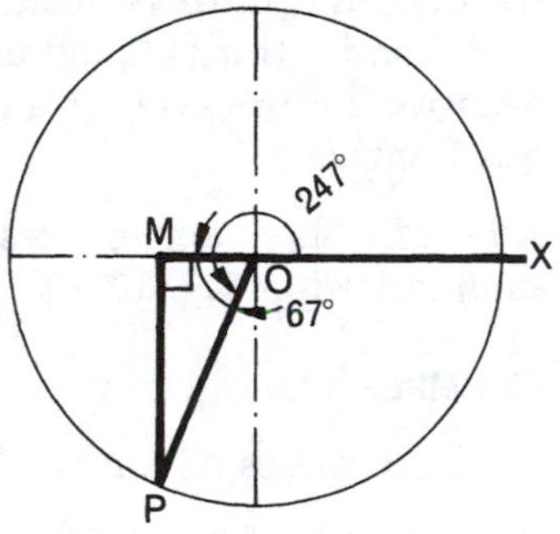

Fig. 13.19

EXAMPLE 13.7

Find all the angles between $0°$ and $360°$ whose:

a) sines are -0.4676
b) cosines are 0.3572
c) cotangents are -0.9827

a) Let $\sin \theta = -0.4676$

Since the sine is negative the angles θ lie in the 3rd and 4th quadrants. These are shown in Figs. 13.20 and 13.21.

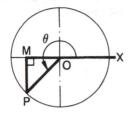

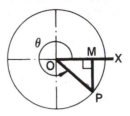

Fig. 13.20   Fig. 13.21

From right-angled triangle OPM,

$$\sin P\hat{O}M = MP = 0.4676$$

$\therefore$ $P\hat{O}M = 27°53'$ from calculator

From Fig. 7.20 $\theta = 180° + 27°53' = 207°53'$

and from Fig. 7.21 $\theta = 360° - 27°53' = 332°7'$

b) Let $\cos \theta = 0.3572$

Since the cosine is positive the angles θ lie in the 1st and 4th quadrants. These are shown in Figs. 13.22 and 13.23.

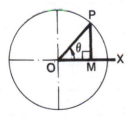

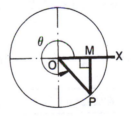

Fig. 13.22   Fig. 13.23

From the right-angled triangle OPM,

$$\cos P\hat{O}M = OM = 0.3572$$

$\therefore$ $P\hat{O}M = 69°4'$ from calculator

From Fig. 13.22 $\theta = 69°4'$

and from Fig. 13.23 $\theta = 360° - 69°4' = 290°56'$

c) Let $\cot\theta = -0.9827$

Cotangents are treated in a similar manner to tangents, i.e. since the cotangent is negative the angles θ lie in the 2nd and 4th quadrants. These are shown in Figs. 13.24 and 13.25.

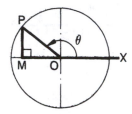   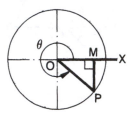

Fig. 13.24   Fig. 13.25

From the right-angled triangle OPM,

$$\cot P\hat{O}M = \frac{OM}{MP} = 0.9827$$

$\therefore$ $\qquad\qquad P\hat{O}M = 45°30'$ from calculator

From Fig. 13.24 $\theta = 180° - 45°30' = 134°30'$

and from Fig. 13.25 $\theta = 360° - 45°30' = 314°30'$

The following tables may be used for angles in any quadrant:

Quadrant	Angle	$\sin\theta =$
First	0° to 90°	$\sin\theta$
Second	90° to 180°	$\sin(180° - \theta)$
Third	180° to 270°	$-\sin(\theta - 180°)$
Fourth	270° to 360°	$-\sin(360° - \theta)$
Quadrant	$\cos\theta =$	$\tan\theta =$
First	$\cos\theta$	$\tan\theta$
Second	$-\cos(180° - \theta)$	$-\tan(180° - \theta)$
Third	$-\cos(\theta - 180°)$	$\tan(\theta - 180°)$
Fourth	$\cos(360° - \theta)$	$-\tan(360° - \theta)$

Exercise 13.2

1) Write down the values of the sine, cosine and tangent of the following angles:

(a) 121°   (b) 178°23'   (c) 102°29'   (d) 211°

(e) 239°17'   (f) 258°28'   (g) 318°27'   (h) 297°17'

2) Evaluate: $6 \sin 23° - 2 \cos 47° + 3 \tan 17°$.

3) Evaluate: $5 \sin 142° - 3 \tan 148° + 3 \cos 230°$.

4) Evaluate: $\sin A \cos B - \sin B \cos A$ given that $\sin A = \frac{3}{5}$ and $\tan B = \frac{4}{3}$. A and B are both acute angles.

5) An angle A is in the 2nd quadrant. If $\sin A = \frac{3}{5}$ find, without actually finding angle A, the value of cos A and tan A.

6) If $\sin \theta = 0.1432$ find all the values of θ from 0° to 360°.

7) If $\cos \theta = -0.8927$ find all the values of θ from 0° to 360°.

8) Find the angles in the first and second quadrants:

(a) whose sine is 0.7137   (b) whose cosine is -0.4813

(c) whose tangent is 0.9476   (d) whose tangent is -1.7642

9) Find the angles in the third or fourth quadrants:

(a) whose sine is -0.7880 (b) whose cosine is 0.5592

(c) whose tangent is -2.9042

10) If $\sin A = \dfrac{a \sin B}{b}$ find the values of A between 0° and 360° when $a = 7.26$ mm, $b = 9.15$ mm and B = 18°29'

11) If $\cos C = \dfrac{(a^2 + b^2 - c^2)}{2ab}$ find the values of C between $0°$ and $360°$ given that $a = 1.26\,\text{m}$, $b = 1.41\,\text{m}$ and $c = 2.13\,\text{m}$.

SINE, COSINE AND TANGENT CURVES

A sine curve (or sine waveform) is the result of plotting vertically the values of $\sin\theta$ against θ horizontally (often called an angle base). The values of $\sin\theta$ may be found using a scientific calculator, but an alternative method is shown in Fig. 13.26. This makes use of the ideas expressed in Figs. 13.14 and 13.15.

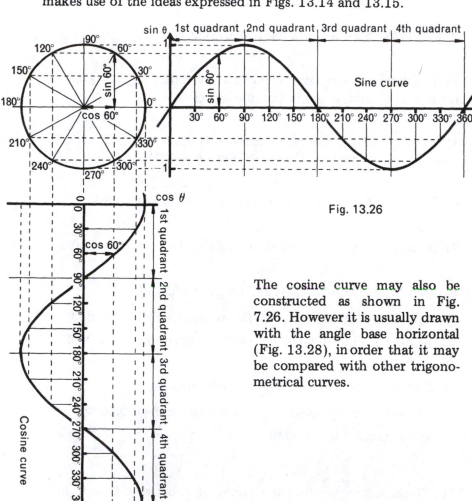

Fig. 13.26

The cosine curve may also be constructed as shown in Fig. 7.26. However it is usually drawn with the angle base horizontal (Fig. 13.28), in order that it may be compared with other trigonometrical curves.

The Sine Curve

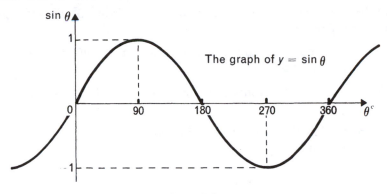

Fig. 13.27

The following features should be noted:

(1) In the first quadrant as θ increases from $0°$ to $90°$, $\sin\theta$ increases from 0 to 1

(2) In the second quadrant as θ increases from $90°$ to $180°$, $\sin\theta$ decreases from 1 to 0

(3) In the third quadrant as θ increases from $180°$ to $270°$, $\sin\theta$ decreases from 0 to -1

(4) In the fourth quadrant as θ increases from $270°$ to $360°$, $\sin\theta$ increases from -1 to 0

The Cosine Curve

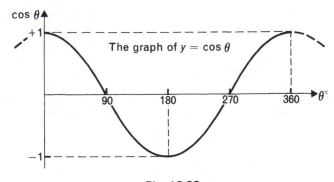

Fig. 13.28

Note that:

(1) In the first quadrant as θ increases from $0°$ to $90°$, $\cos\theta$ decreases from 1 to 0

(2) In the second quadrant as θ increases from $90°$ to $180°$, $\cos\theta$ decreases from 0 to -1

(3) In the third quadrant as θ increases from $180°$ to $270°$, $\cos\theta$ increases from -1 to 0

(4) In the fourth quadrant as θ increases from $270°$ to $360°$, $\cos\theta$ increases from 0 to 1

The Tangent Curve

The graph of $y = \tan\theta$ may be drawn using values obtained from the identity $\tan\theta = \dfrac{\sin\theta}{\cos\theta}$. A table of values may be drawn up part of which is given below.

You should remember that:

$$\frac{1}{\text{Very small number}} = \text{Very large number}$$

Thus:
$$\frac{1}{\text{Zero}} = \text{Infinity (symbol } \infty)$$

$\theta°$	0	10	20	30	40
$\sin\theta$	0	0.174	0.342	0.500	0.643
$\cos\theta$	1	0.985	0.940	0.866	0.766
$\tan\theta = \dfrac{\sin\theta}{\cos\theta}$	0	0.176	0.364	0.577	0.839

$\theta°$	50	60	70	80	90
$\sin\theta$	0.766	0.866	0.940	0.985	1
$\cos\theta$	0.643	0.500	0.342	0.174	0
$\tan\theta = \dfrac{\sin\theta}{\cos\theta}$	1.19	1.73	2.75	5.67	∞

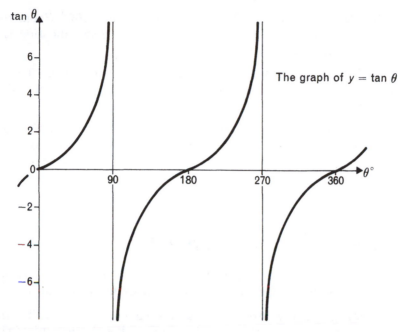

The graph of $y = \tan \theta$

Fig. 13.29

Note that:

(1) In the first quadrant as θ increases from $0°$ to $90°$, $\tan \theta$ increases from 0 to infinity

(2) In the second quadrant as θ increases from $90°$ to $180°$, $\tan \theta$ increases from minus infinity to 0

(3) In the third quadrant as θ increases from $180°$ to $270°$, $\tan \theta$ increases from 0 to infinity

(4) In the fourth quadrant as θ increases from $270°$ to $360°$, $\tan \theta$ increases from minus infinity to 0

Exercise 13.3

1) Draw the graphs of (i) $y = \sin \theta$, (ii) $y = \cos \theta$ for values of θ between $0°$ and $360°$. From the graphs find values of the sine and cosine of the angles:

(a) $38°$ (b) $72°$ (c) $142°$ (d) $108°$
(e) $200°$ (f) $250°$ (g) $305°$ (h) $328°$

2) Plot the graph of $y = 3 \sin x$ between $0°$ and $360°$. From the graph read off the values of x for which $y = 1.50$ and find the value of y when $x = 250°$.

3) Draw the graphs of $3 \cos \theta$ for values of θ from $0°$ to $360°$. Use the graph to find approximate values of the two angles for which $3 \cos \theta = 0.6$

4) By projection from the circumference of a suitably marked off circle, draw the graph of $4 \sin \theta$ for values of θ from $0°$ to $360°$. Use the graph to find approximate values of the two angles for which $4 \sin \theta = 1.6$

THE SOLUTION OF TRIANGLES

We now deal with triangles which are *not* right-angled. Every triangle consists of six elements—three angles and three sides.

If we are given any three of these six elements we can find the other three by using either the *Sine Rule* or the *Cosine Rule*. (The exception is when we are given three angles, since it is obvious that a triangle of a given shape can be of any size.)

When we have found the values of the three missing elements we are said to have 'solved the triangle'.

THE SINE RULE

The sine rule may be used when given:

(1) one side and any two angles; or

(2) two sides and an angle opposite to one of the given sides. (In this case two solutions may be found giving rise to what is called the 'ambiguous case', see Example 13.9.)

Using the notation of Fig. 13.30 the *sine rule* states:

$$\frac{a}{\sin A} = \frac{b}{\sin B} = \frac{c}{\sin C}$$

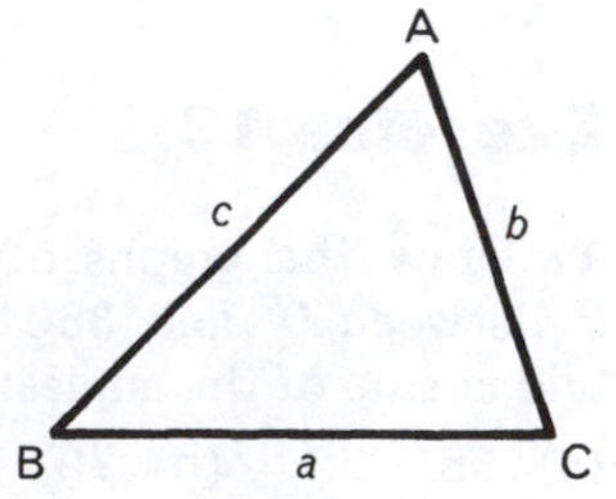

Fig. 13.30

EXAMPLE 13.8

Solve the triangle ABC given that
A = 42°, C = 72° and b = 61.8 mm.

The triangle should be drawn for
reference as shown in Fig. 13.31,
but there is no need to draw it to
scale.

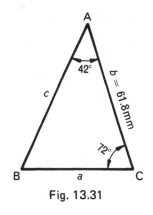

Fig. 13.31

Since $\angle A + \angle B + \angle C = 180°$

$\angle B = 180° - 42° - 72° = 66°$

The sine rules states:

$$\frac{a}{\sin A} = \frac{b}{\sin B} \qquad\qquad \frac{c}{\sin C} = \frac{b}{\sin B}$$

$$\therefore \quad a = \frac{b \sin A}{\sin B} \qquad\qquad \therefore \quad c = \frac{b \sin C}{\sin B}$$

$$= \frac{61.8 \times \sin 42°}{\sin 66°} \qquad\qquad = \frac{61.8 \times \sin 72°}{\sin 66°}$$

$$= 45.3 \text{ mm} \qquad\qquad\qquad = 64.3 \text{ mm}$$

The complete solution is:

$$\angle B = 66°, \quad a = 45.3 \text{ mm}, \quad c = 64.3 \text{ mm}$$

A rough check on sine rule calculations may be made by
remembering that in any triangle the longest side lies opposite
the largest angle and the shortest side lies opposite the smallest
angle.

Thus in the previous example:

Smallest angle = 42° = A;   Shortest side = a = 45.3 mm

Largest angle = 72° = C;   Longest side = c = 64.3 mm

The Ambiguous Case

There are two angles between 0° and 180° which have the same
sine. For instance if $\sin A = 0.5000$, then A can be either 30° or
150°. When using the sine rule to find an angle we must always
examine the problem to see if there are two possible values for the
angle.

EXAMPLE 13.9

In triangle ABC, $b = 93.23\,\text{mm}$, $c = 85.61\,\text{mm}$ and $\angle C = 37°$.

Solve the triangle.

Referring to Fig. 13.32 we have

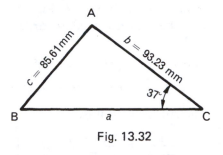

Fig. 13.32

$$\frac{b}{\sin B} = \frac{c}{\sin C}$$

$$\therefore \quad \sin B = \frac{b \sin C}{c}$$

$$= \frac{93.23 \times \sin 37°}{85.61}$$

$$= 0.6552$$

The angle B may be in either the first or second quadrants.

In the first quadrant, $\angle B = 40°56'$ (Fig. 13.33).

In the second quadrant, $\angle B = 139°4'$ (Fig. 13.34).

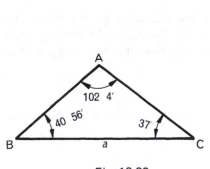

Fig. 13.33

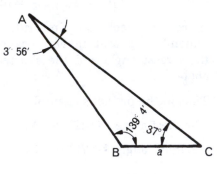

Fig. 13.34

When

$\angle B = 40°56'$

$\angle A = 180° - 40°56' - 37°$

$\quad = 102°4'$

When

$\angle B = 139°4'$

$\angle A = 180° - 139°4' - 37°$

$\quad = 3°56'$

Now

$$\frac{a}{\sin A} = \frac{c}{\sin C}$$

$$\therefore \quad a = \frac{c \sin A}{\sin C}$$

$$= \frac{85.61 \sin 102°4'}{\sin 37°}$$

$$= \frac{85.61 \sin 77°56'}{\sin 37°}$$

$$= 130 \, \text{mm}$$

Now

$$\frac{a}{\sin A} = \frac{c}{\sin C}$$

$$\therefore \quad a = \frac{c \sin A}{\sin C}$$

$$= \frac{85.61 \sin 3°56'}{\sin 37°}$$

$$= 9.77 \, \text{mm}$$

The ambiguous case may be seen clearly by constructing the given triangle geometrically as follows (Fig. 13.35).

Using a full size scale draw AC = 93.23 mm and draw CX such that ACX = 37°. Now with centre A and radius 85.61 mm describe a circular arc to cut CX at B and B'.

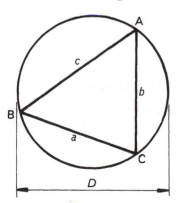

Fig. 13.35

Then ABC represents the triangle shown in Fig. 13.33 and AB'C represents the triangle shown in Fig. 13.34.

Use of the Sine Rule to Find the Diameter (*D*) of the Circumscribing Circle of a Triangle

Using the notation of Fig. 13.36.

$$\frac{a}{\sin A} = \frac{b}{\sin B} = \frac{c}{\sin C} = D$$

The rule is useful when we wish to find the pitch circle diameter of a ring of holes.

Fig. 13.36

EXAMPLE 13.10

In Fig. 13.37 three holes are
positioned by the angle and dimensions shown. Find the pitch circle
diameter.

We are given

$$\angle B = 41° \text{ and } b = 112.5\,\text{mm}$$

$$\therefore \quad D = \frac{b}{\sin B} = \frac{112.5}{\sin 41°}$$

$$= 171.5\,\text{mm}$$

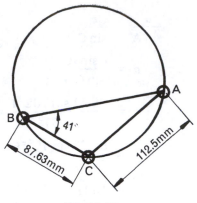

Fig. 13.37

THE COSINE RULE

The cosine rule is used in all cases where the sine rule cannot be
used. These are when given:

(1) two sides and the angle between them;

(2) three sides.

Whenever possible the sine rule is used because it results in a
calculation which is easier to perform. In solving a triangle it is
sometimes necessary to start with the cosine rule and then having
found one of the unknown elements to finish solving the triangle
using the sine rule.

The cosine rules states:

either

or

or

$$a^2 = b^2 + c^2 - 2bc \cos A$$
$$b^2 = a^2 + c^2 - 2ac \cos B$$
$$c^2 = a^2 + b^2 - 2ab \cos C$$

EXAMPLE 13.11

Solve the triangle ABC if $a = 70\,\text{mm}$,
$b = 40\,\text{mm}$ and $\angle C = 64°$.

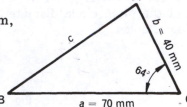

Fig. 13.38

Referring to Fig. 7.38, to find the side c we use

$$c^2 = a^2 + b^2 - 2ab \cos C$$
$$= 70^2 + 40^2 - 2 \times 70 \times 40 \times \cos 64° = 4044$$
$$\therefore \quad c = \sqrt{4044} = 63.6 \text{ mm}$$

We now use the sine rule to find $\angle A$:

$$\frac{a}{\sin A} = \frac{c}{\sin C}$$
$$\therefore \quad \sin A = \frac{a \sin C}{c} = \frac{70 \times \sin 64°}{63.6}$$

Thus $A = 81°36'$

and $B = 180° - 81°36' - 64° = 34°24'$

EXAMPLE 13.12

The mast AB of a job crane (Fig. 13.39) is 3 m long and the tie BC is 2.4 m long. The angle between AB and BC is 125°. Find the length of the job AC.

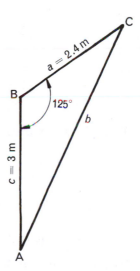

Fig. 13.39

Using the cosine rule:

$$b^2 = a^2 + c^2 - 2ac \cos B$$
$$= 2.4^2 + 3^2 - 2 \times 2.4 \times 3 \times \cos 125°$$

Now $\cos 125° = -\cos (180° - 125°) = -\cos 55° = -0.5736$

Hence $b^2 = 2.4^2 + 3^2 - 2 \times 2.4 \times 3 \times (-0.5736)$

$\therefore \qquad b = 4.80$

Therefore the jib of the crane is 4.80 m long.

EXAMPLE 13.13

The instantaneous values, i_1 and i_2, of two alternating currents are represented by the two sides of a triangle shown in Fig. 13.40. The resultant current i_R is represented by the third side. Calculate the magnitude of i_R and the angle ϕ between the current i_1 and i_R.

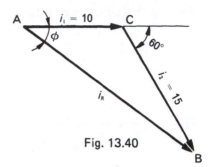

In $\triangle ABC$, Fig. 13.40, we have

$b = 10$, $a = 15$ and $\angle C = 120°$.

Fig. 13.40

Using the cosine rule gives

$$c^2 = a^2 + b^2 - 2ab \cos C$$

$$= 15^2 + 10^2 - 2 \times 15 \times 10 \times \cos 120°$$

$$= 225 + 100 - 300 \times (-\cos 60°)$$

$$\therefore \quad c = \sqrt{475} = 21.79 = i_R$$

To find $\angle A$ we use the sine rule,

$$\frac{a}{\sin A} = \frac{c}{\sin C}$$

$$\therefore \quad \sin A = \frac{a \sin C}{c} = \frac{15 \times \sin 120°}{21.79} = \frac{15 \times \sin 60°}{21.79} = 0.5962$$

$$\therefore \quad A = 39°56' = \phi$$

Hence the magnitude of i_R is 21.8 and the angle ϕ is 36°36'.

Exercise 13.4

1) The following are all exercises on the sine rule. Solve the following triangles ABC given:

(a) $A = 75°$	$B = 34°$	$a = 102\,\text{mm}$
(b) $C = 61°$	$B = 71°$	$b = 91\,\text{mm}$
(c) $A = 19°$	$C = 105°$	$c = 11.1\,\text{m}$
(d) $A = 116°$	$C = 18°$	$a = 170\,\text{mm}$

(e)	$A = 36°$	$B = 77°$	$b = 2.5\,m$
(f)	$A = 49°11'$	$B = 67°17'$	$c = 11.22\,mm$
(g)	$A = 17°15'$	$C = 27°7'$	$b = 221.5\,mm$
(h)	$A = 77°3'$	$C = 21°3'$	$a = 9.793\,m$
(i)	$B = 115°4'$	$C = 11°17'$	$c = 516.2\,mm$
(j)	$a = 17\,m$	$b = 15\,m$	$B = 39°$
(k)	$a = 7\,m$	$c = 11\,m$	$C = 22°7'$
(l)	$b = 92\,mm$	$c = 71\,mm$	$C = 39°8'$
(m)	$b = 15.13\,m$	$c = 11.62\,m$	$B = 85°17'$
(n)	$a = 23\,m$	$c = 18.2\,m$	$A = 49°19'$
(o)	$a = 9.217\,m$	$b = 7.152\,m$	$A = 105°4'$

2) Solve the following triangles ABC using the cosine rule:

(a)	$a = 9\,m$	$b = 11\,m$	$C = 60°$
(b)	$b = 10\,m$	$c = 14\,m$	$A = 56°$
(c)	$a = 8.16\,m$	$c = 7.14\,m$	$B = 37°18'$
(d)	$a = 5\,m$	$b = 8\,m$	$c = 7\,m$
(e)	$a = 312\,mm$	$b = 527.3\,mm$	$c = 700\,mm$
(f)	$a = 7.912\,m$	$b = 4.318\,m$	$c = 11.08\,m$

3) Three holes lie on a pitch circle and their chordal distances are 41.82 mm, 61.37 mm and 58.29 mm. Find their pitch circle diameter.

4) In Fig. 13.41 find the angle BCA given that BC is parallel to AD.

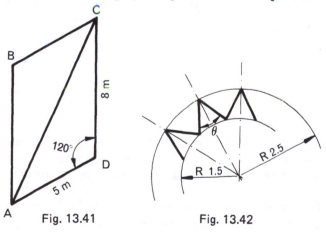

Fig. 13.41   Fig. 13.42

5) Calculate the angle θ in Fig. 13.42. There are 12 castellations and they are equally spaced.

6) Find the smallest angle in a triangle whose sides are 20, 25 and 30 m long.

7) In Fig. 13.43 find:

(a) the distance AB

(b) the angle ACB

8) Three holes are spaced in a plate detail as shown in Fig. 13.44. Calculate the centre distances from A to B and from A to C.

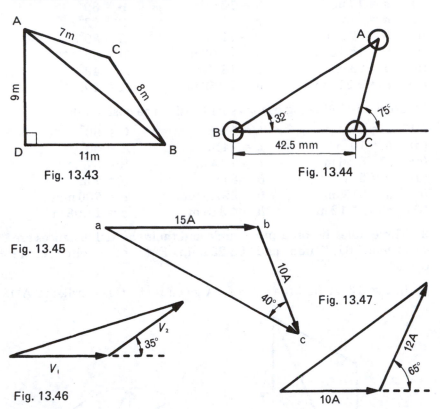

Fig. 13.43

Fig. 13.44

Fig. 13.45

Fig. 13.46

Fig. 13.47

9) In Fig. 13.45, *ab* and *bc* are phasors representing the alternating currents in two branches of a circuit. The line *ac* represents the resultant current. Find by calculation this resultant current.

10) Two phasors are shown in Fig. 13.46. If $V_1 = 8$ and $V_2 = 6$ calculate the value of their resultant and the angle it makes with V_1.

11) Calculate the resultant of the two phasors shown in Fig. 13.47.

12) The main chain lines of a small survey are shown in Fig. 13.48. Calculate the angle θ.

13) A part of a field survey consists of a triangle ABC in which the sides AB, BC and CA measure 36 m, 40 m and 47 m respectively. Find the size of each angle in the triangle.

14) Fig. 13.49 represents part of a roof truss. Calculate the length of the member BC.

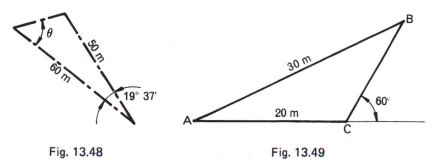

Fig. 13.48   Fig. 13.49

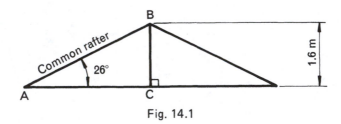

14. PRACTICAL APPLICATIONS OF TRIGONOMETRY

Outcome:

1. *Apply trigonometry to solutions of practical problems.*
2. *Calculate the area of any triangle using the formulae $\sqrt{s(s-a)(s-b)(s-c)}$ and $\frac{1}{2}ab\sin\theta$.*

3. *Solve problems on triangles and quadrilaterals involving the formulae for the areas of triangles.*

SOLVING PRACTICAL BUILDING PROBLEMS

Trigonometry is frequently used to solve problems which arise in practice. Some of these problems and their solutions are shown in the examples which follow.

EXAMPLE 14.1

The rise of a pitched roof is 1.6 metres and the roof makes an angle of 26° with the horizontal. Calculate the length of the common rafters.

Fig. 14.1

The conditions are shown in Fig. 14.1, the common rafter being AB. Now in △ABC,

$$\frac{AB}{BC} = \operatorname{cosec} 26°$$

$$AB = BC\operatorname{cosec} 26° = 1.6\operatorname{cosec} 26° = 3.65$$

Hence the common rafters are 3.65 metres long.

180

EXAMPLE 14.2

Fig. 14.2 shows a roof truss. Calculate the lengths of the members BC, BD and AC.

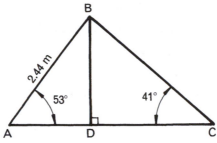

Fig. 14.2

In $\triangle ABD$,

$$\frac{BD}{AB} = \sin 53°$$

$\therefore$ $\quad BD = AB \sin 53° = 2.44 \sin 53° = 1.95$

In $\triangle BCD$,

$$\frac{BC}{BD} = \operatorname{cosec} 41°$$

$\therefore$ $\quad BC = BD \operatorname{cosec} 41° = 1.95 \operatorname{cosec} 41° = 2.97$

To find the length of AC we must first find the lengths of AD and DC.

In $\triangle ABD$, $\quad \dfrac{AD}{AB} = \cos 53°$

$\therefore$ $\quad AD = AB \cos 53° = 2.44 \cos 53° = 1.47$

In $\triangle BCD$, $\quad \dfrac{DC}{BC} = \cos 41°$

$\therefore$ $\quad DC = BC \cos 41° = 2.97 \cos 41° = 2.24$

Thus $\quad AC = AD + DC = 1.47 + 2.24 = 3.71$

Hence BD is 1.95 m. BC is 2.97 m and AC is 3.71 m.

ANGLE OF ELEVATION

If you look upwards at an object the angle formed between the horizontal and your line of sight is called the *angle of elevation* (Fig. 14.3).

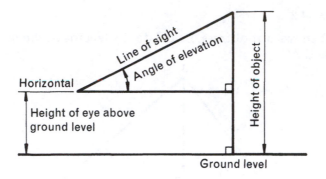

Fig. 14.3

EXAMPLE 14.3

To find the height of a tower a surveyor sets up his theodolite 100 m from the base of the tower. He finds the angle of elevation of the top of the tower to be 30°. If the instrument is 1.5 m from the ground, what is the height of the tower?

In Fig. 14.4,

$$\frac{BC}{AB} = \tan 30°$$

$$\therefore \ BC = AB \tan 30°$$

$$= 100 \tan 30° = 57.7$$

Hence, height of tower = 57.7 + 1.5 = 59.2 m.

Fig. 14.4

EXAMPLE 14.4

To find the height of a pylon, a surveyor sets up a theodolite some distance from the base of the pylon and finds that the angle of elevation to the top of the pylon to be 30°. He then moves 60 m nearer to the pylon and finds that the angle of elevation is 42°. Find the height of the pylon assuming that the ground is horizontal and that the theodolite stands 1.5 m above the ground.

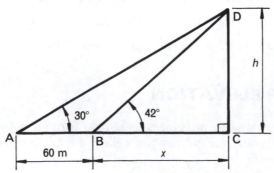

Fig. 14.5

Referring to Fig. 14.5, let $BC = x$ and $DC = h$.

In $\triangle ACD$,

$$\frac{DC}{AC} = \tan 30°$$

$\therefore \quad DC = AC \tan 30°$

or $\quad h = 0.5774(x + 60)$ [1]

In $\triangle BDC$,

$$\frac{DC}{BC} = \tan 42°$$

$\therefore \quad DC = BC \tan 42°$

or $\quad h = 0.9004x$ [2]

From equation [2], $x = \dfrac{h}{0.9004} = 1.1106h$

Substituting for x in equation [1] gives

$$h = 0.5774(1.1106h + 60) = 0.6413h + 34.64$$

$$h - 0.6413h = 34.64$$

or $\qquad 0.3587h = 34.64$

from which $\qquad h = \dfrac{34.64}{0.3587} = 96.6 \text{ m}$

Hence the height of the pylon is $96.6 + 1.5 = 98.1 \text{ m}$.

ANGLE OF DEPRESSION

If you look down at an object, the angle formed between the horizontal and your line of sight is called the *angle of depression* (Fig. 14.6).

EXAMPLE 14.5

From the top floor window of a house, 14 m above ground level, the angle of depression of an object in the street is $52°$. How far is the object from the house?

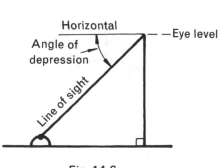

Fig. 14.6

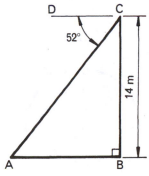

Fig. 14.7

The conditions are shown in Fig. 14.7. Since the angle of depression is $52°$, $\angle ACD = 52°$.

$$\angle ACB = 90° - 52° = 38°$$

In $\triangle ABC$, $\dfrac{AB}{CB} = \tan 38°$

$\therefore$ $AB = CB \tan 38° = 14 \tan 38° = 10.9$

Hence the object is 10.9 m from the house.

BEARINGS

The four cardinal directions are North, South, East and West (Fig. 14.8). The directions NE, NW, SE and SW are frequently used and are as shown in the diagram. A bearing of N20°E means an angle 20° measured from N towards E as shown in Fig. 14.9. Similarly a bearing of S40°E means an angle of 40° measured from S towards E (Fig. 14.10). A bearing of N50°W means an angle of 50° measured from N towards W (Fig. 14.11).

Bearings quoted in this way are always measured from N and S and never from E and W.

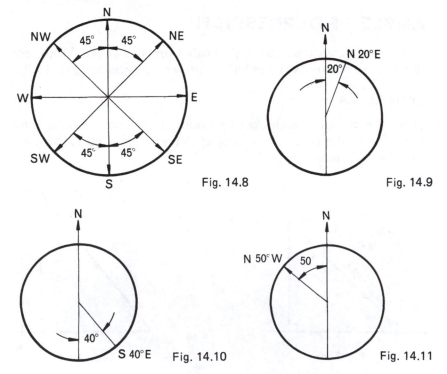

Fig. 14.8   Fig. 14.9

Fig. 14.10   Fig. 14.11

Another way of stating bearings is to measure the angle from N in a clockwise direction, N being taken as 0°. Three figures are always stated. For example 005° is written instead of 5° and 035° instead of 35° and so on. E will be 090°, S 180° and W 270°. Some typical bearings are shown in Fig. 14.12.

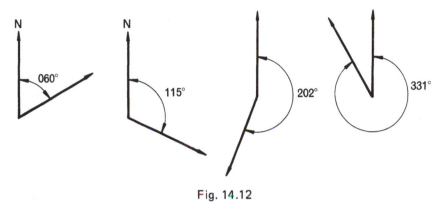

Fig. 14.12

EXAMPLE 14.6

In making a survey it is found that B is a point due east of a point A and a point C is 6 km due south of A. The distance BC is 7 km. Calculate the bearing of C from B.

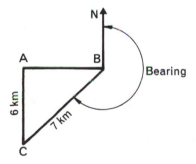

Fig. 14.13

In right-angled △ABC, Fig. 14.13 we have

$$\sin \angle B = \frac{AC}{BC} = \frac{6}{7}$$

∴ $\angle B = 59°$

The bearing of C from B = $270° - 59° = 211°$

EXAMPLE 14.7

Fig. 14.14 represents a survey of a plot of land. Calculate the distance PB and the bearing of B from P.

Draw the triangles PAC, ABD and BFP as shown in Fig. 14.15.

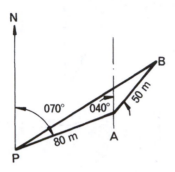

Fig. 14.14

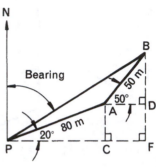

Fig. 14.15

In $\triangle$PAC, $\dfrac{PC}{AP} = \cos\angle APC$

$\therefore$ $PC = AP\cos\angle APC = 80\cos 20° = 75.18$

Also $\dfrac{AC}{AP} = \sin\angle APC$

$\therefore$ $AC = AP\sin\angle APC = 80\sin 20° = 27.36$

In $\triangle$ABD, $\dfrac{AD}{AB} = \cos\angle BAD$

$\therefore$ $AD = AB\cos\angle BAD = 50\cos 50° = 32.14$

Also $\dfrac{BD}{AB} = \sin\angle BAD$

$\therefore$ $BD = AB\sin\angle BAD = 50\sin 50° = 38.30$

In $\triangle$BPF, $PF = PC + AD = 75.18 + 32.14 = 107.32$

and $BF = AC + BD = 27.36 + 38.30 = 65.66$

$\therefore$ $\tan\angle BPF = \dfrac{BF}{PF} = \dfrac{65.66}{107.32}$ Hence $\angle BPF = 31°28'$

Thus the bearing of B from P is $90° - 31°28' = 58°32'$.

The distance PB may be found by using trigonometry or by using Pythagoras' theorem. Using trigonometry,

$$\frac{PB}{BF} = \text{cosec} \angle BPF$$

∴ $PB = BF \, \text{cosec} \angle BPF = 65.66 \, \text{cosec} \, 31°28' = 125.8$

Hence the distance PB is 125.8 metres.

Exercise 14.1

1) A straight stretch of road is 90 m long and rises at an angle of 20° to the horizontal. Find the vertical rise of this stretch of road.

2) The main rafters of a lean-to roof are each 3.96 m long. What is the span of the roof if the rise is 1.37 m?

3) A roof is pitched 31° to the horizontal and its rise is 2.62 m. Calculate the length of the main rafter.

4) A lean-to roof has a rise of 1.37 m and a span of 2.59 m. Find the angle that the roof makes with the horizontal.

5) A drain slopes downwards from a point A at an angle of 11° to the horizontal for 12 m to a point B. From B it slopes downwards at an angle of 14° from the horizontal for a further 13 m to a point C. From C it continues to D a distance of 25 m at an angle of 8° to the horizontal. Calculate the distance of D below A.

6) A road runs in a direction S30°E from a point A for a distance of 120 m to a point B. It then runs for a further distance of 140 m to a point C in a direction S35°E. Find: (a) how far C is south of A, (b) how far C is east of A, (c) the distance AC, (d) the bearing of C from A.

7) A tower is 25 m high. A man standing some distance from the tower finds the angle of elevation to the top of the tower to be 59°. How far is the man standing from the foot of the tower, if his eye level is 1.5 m above ground level?

8) A man whose eye level is 1.5 m above ground level is 15 m away from a tower 20 m tall. Determine the angle of elevation of the top of the tower from his eyes.

9) A man standing on top of a mountain 1200 m high observes the angle of depression of the top of a steeple to be 43°. If the height of the steeple is 50 m, how far is it from the mountain?

10) To find the height of a tower a surveyor stands some distance away from its base and he observes the angle of elevation to the top of the tower to be 45°. He then moves 80 m nearer to the tower and he then finds the angle of elevation to be 60°. Find the height of the tower. Assume the theodolite stands 1.5 m above ground level.

11) A tower is known to be 60 m high. A surveyor, using a theodolite, stands some distance away from the tower and measures the angle of elevation as 38°. How far away from the tower is he, if the theodolite stands 1.5 m above ground level? If the surveyor moves 80 m further away from the tower, what is now the angle of elevation of the tower?

12) A surveyor stands 100 m from the base of a tower on which an aerial stands. He measures the angles of elevation to the top and bottom of the aerial as 58° and 56°. Find the height of the aerial.

13) Triangle ABC (Fig. 14.16) represents the cross-section of a cutting. A bridge spanning the cutting is to be designed. Calculate the length of AC so that the design can be made.

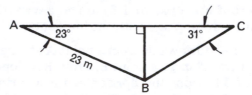

Fig. 14.16

14) Fig. 14.17 represents the framework for a footbridge AB = CD = EF = 2.4 m. Calculate the length of the members required to make this framework.

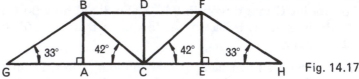

Fig. 14.17

15) To find the height of a feature a surveyor stands some distance away from its base and finds the angle of elevation to the top of the feature to be 49°. He then moves 60 m further away from the feature and finds the angle of elevation to be 38°. Calculate the height of the feature. Assume that the theodolite is 1.5 m above ground level.

16) A man standing on top of a building 50 m high is in line with two points A and B whose angles of depression are 17° and 21° respectively. Calculate the distance AB.

17) In surveying a plot of land it was found that A is 50 m due north of B. C is the point whose bearing from A is 150° and AC is 80 m. Calculate the distance BC and the bearing of C from B.

18) Three towns A, B and C lie on a straight road running east from A. B is 6 km from A and C is 22 km from A. Another town D lies to the north of this road and lies 10 km from both B and C. Calculate the distance of D from A and the bearing of D from A.

SOLVING PRACTICAL ENGINEERING PROBLEMS

Some of the problems which occur in Mechanical Engineering are discussed in the sections which follow. In all of them trigonometry is used.

Co-ordinate Hole Dimensions

In marking-out and in operating certain machine tools it is convenient to give the dimensions of holes relative to two axes which are at right angles to each other.

In graphical work the position of a point on a graph is specified by its co-ordinates (that is, the distances at which the point lies from the x- and y-axes respectively). Co-ordinate hole centres are specified in exactly the same way.

EXAMPLE 14.8

Three holes are to have their centres equally spaced on a 50.00 mm pitch circle diameter as shown in Fig. 14.18. Calculate the co-ordinate dimensions of the holes, relative to the axes Ox and Oy.

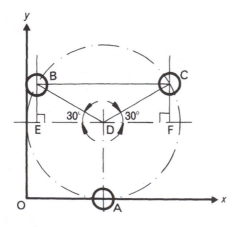

Hole A
x dimension $=$ 25.00 mm
y dimension $=$ 0

Fig. 14.18

To find the x and y dimensions for the holes B and C draw the $\triangle$DCF and the $\triangle$BED.

In $\triangle$BED, $\dfrac{ED}{BD} = \cos 30°$

$\therefore$ $ED = BD(\cos 30°) = 25(\cos 30°) = 21.65\,\text{mm}$

and $\dfrac{BE}{BD} = \sin 30°$

$\therefore$ $BE = BD(\sin 30°) = 25(\sin 30°) = 12.50\,\text{mm}$

Since $\triangle$BED is congruent with $\triangle$DCF,

$\qquad ED = DF = 21.65\,\text{mm} \quad \text{and} \quad BE = CF = 12.50\,\text{mm}$

Hole B

$\qquad x \text{ dimension} = 25.00 - ED = 25.00 - 21.65 = 3.35\,\text{mm}$

$\qquad y \text{ dimension} = 25.00 + BE = 25.00 + 12.50 = 37.50\,\text{mm}$

Hole C

$\qquad x \text{ dimension} = 25.00 + DF = 25.00 + 21.65 = 46.65\,\text{mm}$

$\qquad y \text{ dimension} = 25.00 + CF = 25.00 + 12.50 = 37.50\,\text{mm}$

Exercise 14.2

1) Fig. 14.19 shows 5 equally spaced holes on a 100 mm pitch circle diameter. Calculate their co-ordinate dimensions relative to the axes Ox and Oy.

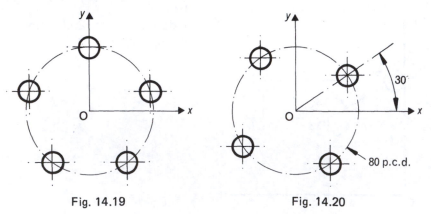

Fig. 14.19   Fig. 14.20

2) 4 holes are equally spaced as shown in Fig. 14.20. Find their co-ordinate dimensions, relative to the axes O*x* and O*y*.

3) Find the co-ordinate dimensions for the 3 holes shown in Fig. 14.21 relative to the axes O*x* and O*y*. The holes lie on a 75 mm pitch circle diameter.

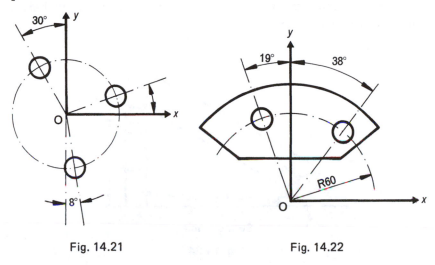

Fig. 14.21   Fig. 14.22

4) Find the co-ordinate hole centres for the two holes shown in Fig. 14.22.

5) Find the co-ordinate hole dimensions for the two holes shown in Fig. 14.23.

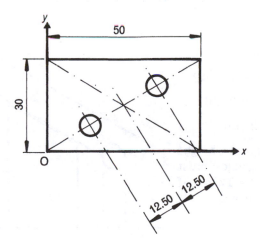

Fig. 14.23

The Sine Bar

A sine bar is used for measurements which require the angle to be determined closer than 5'. The method of using the instrument is shown in Fig. 14.24. The distance l is usually made 100 mm or 200 mm to facilitate calculations. h is the difference in height between the two rollers and $h = l \sin \theta$.

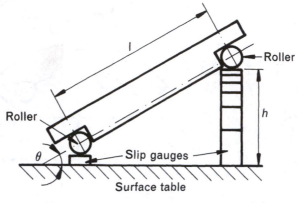

Fig. 14.24

EXAMPLE 14.9

The angle of 15°42' is to be checked on the metal block shown in Fig. 14.25. Find the difference in height between the two rollers which support the ends of the 200 mm sine bar.

Now

$$h = l \sin \theta$$

$$= 200 \times \sin 15°42'$$

$$= 200 \times 0.2706$$

$$= 54.12 \text{ mm}$$

The difference in height of the slip gauges must therefore be 54.12 mm if the angle is correct.

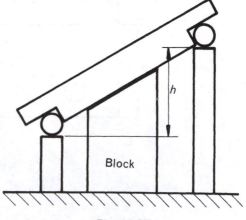

Fig. 14.25

Exercise 14.3

1) Calculate the setting of a 100 mm sine bar to measure an angle of 27°15′.

2) Find the setting of a 200 mm sine bar to check a taper piece which has a taper of 1 in 10 on diameter. The piece is mounted in a similar way to that shown in Fig. 14.27.

3) Calculate the setting of a 200 mm sine bar to check a taper of 1 in 8 on diameter.

4) Find the setting of a 100 mm sine bar to check the taper piece shown in Fig. 14.26. The taper piece is mounted in a similar way to the component shown in Fig. 14.27.

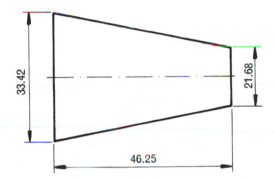

Fig. 14.26

5) Fig. 14.27 shows a component mounted on a 200 mm sine bar. Calculate the angle θ.

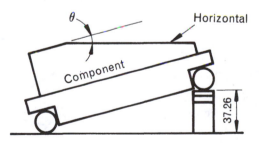

Fig. 14.27

TRIGONOMETRY AND THE CIRCLE

In this section we shall consider the application of trigonometry and geometry to fine measurement but we must first look at some important geometrical theorems.

(1) If two circles are tangential to each other then the straight line which passes through the centres of the circles also passes through the point of tangency.

Thus the line AB joining the centres of the circles also passes through C, the point of tangency (Fig. 14.28) and AB is perpendicular to DE.

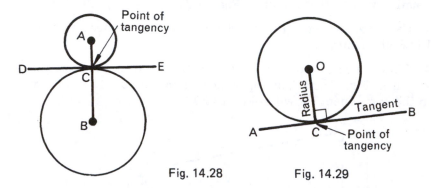

Fig. 14.28   Fig. 14.29

(2) If a line is tangential to a circle then it is at right angles to a radius drawn to the point of tangency.

Thus if AB is a tangent with C the point of tangency then the radius OC is at right angles to AB (Fig. 14.29).

(3) If from a point outside a circle tangents are drawn to the circle, then their lengths are equal.

Thus in Fig. 14.30 the lengths AC and BC are equal. It can also be proved that ∠ACB is bisected by CO, O being the centre of the circle.

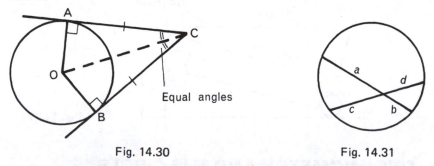

Fig. 14.30   Fig. 14.31

(4) If two chords intersect each other in a circle then the rectangle of the segments of the one equals the rectangle of the segments of the other.

Thus, in Fig. 14.31 $a \times b = c \times d.$

REFERENCE ROLLERS AND BALLS

Sets of rollers can be obtained which are guaranteed to be within 0.002 mm for both diameter and roundness. By using rollers and balls many problems in measurement can be solved, some examples of which are given below.

EXAMPLE 14.10

A taper angle of $7°$ is to be checked by means of an adjustable gauge. The gauge is to be set by means of two rollers, one of 20 mm diameter and the other of 25 mm diameter. Find the centre distance l between the rollers.

In Fig. 14.32, A and B are the centres of the rollers. E and D are the points where the rollers touch the top blade.

$$\angle AED = \angle BDE = 90°$$

(angles between a radius and a tangent).

Now draw AC parallel to ED,

In $\triangle CAB$,

$\angle CAB = 3°30'$ (half angle of taper)

$\angle ACB = 90°$,

$\quad BC = BD - AE = 12.5 - 10 = 2.50\,mm$

$\quad AB = l$

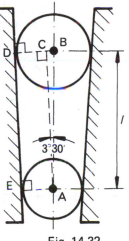

$\therefore \dfrac{l}{2.50} = \text{cosec } 3°30'$

$\quad l = 2.50 \times \text{cosec } 3°30' = 40.95\,mm$

Fig. 14.32

EXAMPLE 14.11

A taper piece has a taper of 1 in 8 on the diameter. Two pairs of rollers 15.00 mm in diameter are used to check the taper as shown in Fig. 14.33. The measurement over the top rollers is 55.87 mm. Find:

a) the measurement over the bottom rollers if the taper is correct,
b) the bottom diameter of the job.

The first step is to find the angle of the taper. Using Fig. 14.34 we see that $\tan\alpha = \dfrac{0.5}{8}$ so $\alpha = 3°34'$.

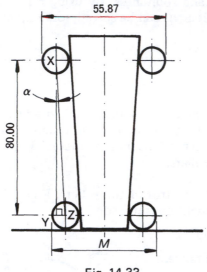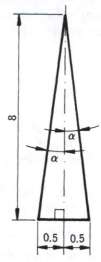

Fig. 14.33   Fig. 14.34

a) Referring to Fig. 14.33 we have in $\triangle XYZ$ that:

$$\angle\alpha = 3°34', \quad XY = 80.00\,\text{mm}$$

but

$$\frac{YZ}{XY} = \tan\alpha$$

$\therefore$

$$YZ = 80\times\tan 3°34' = 5.00$$

Thus

$$M = 55.87 - 2\times 5.00 = 45.87\,\text{mm}$$

b) Referring to Fig. 14.35, we need to find AB.

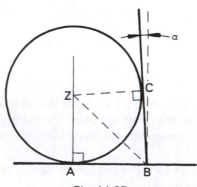

Fig. 14.35

$$\angle ABC = 90° - 3°34' = 86°26'$$

Since AB and BC are tangents, the line BZ bisects $\angle ABC$.
Hence $\qquad$ ABZ = 43°13′

In $\triangle ABZ$ $\qquad \dfrac{AB}{AZ} = \cot 43°13'$

$\therefore \qquad\qquad$ AB = AZ(cot 43°13′)

$$= 7.5(\cot 43°13') = 7.982 \text{ mm}$$

$\therefore$ Bottom diameter $= M - (2 \times AB) - (2 \times \text{Radius of roller})$

$$= 45.87 - 7.982 - 2 \times 7.50 = 14.91 \text{ mm}$$

EXAMPLE 14.12

The tapered hole shown was inspected by using two balls 25.00 and 20.00 mm diameter respectively. The measurements indicated in Fig. 14.36 were obtained. Find:

a) the included angle of taper 2α,
b) the top diameter d of the hole.

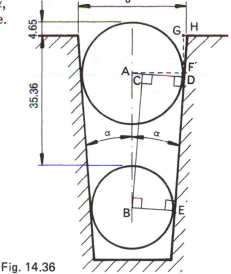

Fig. 14.36

a) In the Fig. 14.36, A and B are the centres of the balls and E and D are points where the balls just touch the sides of the hole.

$$\angle ADE = \angle BED = 90° \quad \text{(angles between radius and tangent)}$$

Draw BC parallel to DE; then in $\triangle ABC$

$$AC = 12.50 - 10.00 = 2.50 \text{ mm}$$

$$AB = 35.36 + 4.65 - 12.50 + 10.00 = 37.51 \text{ mm}$$

Now $\angle ACB = 90°$ and $\angle ABC = \alpha$

$\therefore$ $\sin\alpha = \dfrac{AC}{AB} = \dfrac{2.50}{37.51}$ $\therefore$ $\alpha = 3°49'$

$\therefore$ Included angle of taper $= 2\alpha = 2 \times 3°49' = 7°38'$.

b) To find d, draw AF horizontal and FG vertical.

Then $d = 2 \times (AF + GH)$

In $\triangle AFD$,

AD $= 12.50\,$mm, $\angle FAD = \alpha$, $\angle ADF = 90°$

$\therefore$ AF $= $ AD sec$\alpha = 12.50 \times \sec 3°49' = 12.53\,$mm

In $\triangle GFH$,

GF $= 12.50 - 4.65 = 7.85\,$mm, $\angle FGH = 90°$, $\angle GFH = \alpha$,

$\therefore$ GH $= $ GF tan$\alpha = 7.85 \times \tan 3°49' = 0.52\,$mm

Thus $d = 2(AF + GH) = 2 \times (12.53 + 0.52) = 26.10\,$mm

Exercise 14.4

1) A steel ball 40 mm in diameter is used to check the taper hole, a section of which is shown in Fig. 14.37. If the taper is correct, what is the dimension x?

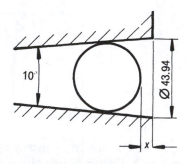

Fig. 14.37

2) A taper plug gauge is being checked by means of reference rollers and slip gauges. The set-up is as shown in Fig. 14.38. Find the included angle of the taper of the gauge and also the top and bottom diameters.

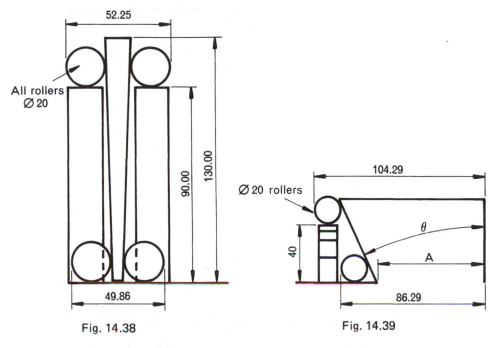

Fig. 14.38   Fig. 14.39

3) Fig. 14.39 shows a dovetail being checked by rollers and slip gauges. Find the angle θ and the dimension A.

4) Calculate the dimension M which is needed for checking the groove, a cross-section of which is shown in Fig. 14.40.

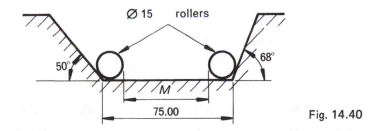

Fig. 14.40

5) A tapered hole has a maximum diameter of 32.00 mm and an included angle of 16°. A ball having a diameter of 20 mm is placed in the hole. Calculate the distance between the top of the hole and the top of the ball.

6) Fig. 14.41 shows the dimensions obtained in checking a tapered hole. Find the included angle of taper of the hole and the top diameter d.

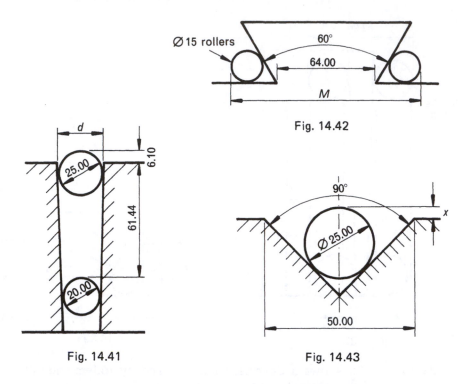

Fig. 14.42

Fig. 14.41

Fig. 14.43

7) Find the checking dimension M for the symmetrical dovetail slide shown in Fig. 14.42.

8) Fig. 14.43 shows a Vee block being checked by means of a reference roller. If the block is correct what is the dimension x?

LENGTHS OF BELTS

There are two distinct cases, open belts and crossed belts, as shown in Fig. 14.44.

With the open belt, pulleys revolve in the *same* direction.

With the crossed belt, pulleys revolve in *opposite* directions.

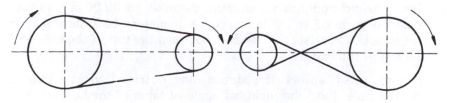

Fig. 14.44

EXAMPLE 14.13

Find the length of an open belt which passes over two pulleys of 200 mm and 300 mm diameter respectively. The distance between the pulley centres is 900 mm.

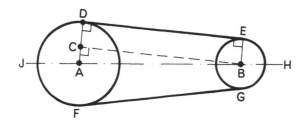

Fig. 14.45

Referring to Fig. 14.45, the total length of the belt is made up of the two straight lengths DE and FG and the arcs DJF and EHG.

$$\angle ADE = \angle BED = 90° \quad \text{(angle between radius and tangent)}$$

Draw CB parallel to DE. Then in $\triangle ABC$,

$$AC = AD - EB = 150 - 100 = 50\,\text{mm}$$

$$AB = 900\,\text{mm (given)}, \quad \angle ACB = 90°$$

$$\therefore \quad \cos C\hat{A}B = \frac{AC}{AB} = \frac{50}{900} = 0.0556 \quad \therefore \quad \angle C\hat{A}B = 86°49'$$

Now $\quad BC = AB \sin C\hat{A}B = 900 \times \sin 86°49' = 898.6\,\text{mm}$

Also $\qquad\qquad \angle EBH = \angle CAB = 86°49'$

Hence the arc EHG subtends an angle of $2 \times 86°49' = 173°38'$ at the centre.

$$\text{Length of arc EHG} = 2\pi \times 100 \times \frac{173°38'}{360°} = 303.0\,\text{mm}$$

Now $\quad \angle DAJ = 180° - \angle CAB = 180° - 86°49' = 93°11'$

The arc DJF therefore subtends an angle of $2 \times 93°11' = 186°22'$ at the centre.

$$\text{Length of arc DJF} = \frac{2\pi \times 150 \times 186°22'}{360°} = 488.0\,\text{mm}$$

$\therefore \quad$ Total length of belt $= 2 \times BC + \text{Arc EHG} + \text{Arc DJF}$

$$= 2 \times 898.6 + 303.0 + 488.0 = 2588\,\text{mm}$$

EXAMPLE 14.14

Two pulleys 200 mm and 300 mm in diameter respectively are placed 1200 mm apart. They are connected by a closed belt. Find the length of the belt required.

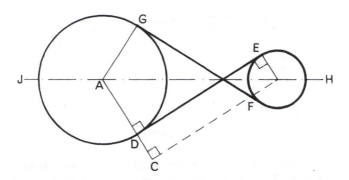

Fig. 14.46

The length of the belt is made up of two straight lengths ED and FG, arc GJD and arc EHF, as shown in Fig. 14.46.

$\angle ADE = \angle DEB = 90°$ (angle between radius and tangent)

Draw CB parallel to DE; then in $\triangle ABC$

$AC = 100 + 150 = 250$ mm, $AB = 1200$ mm, $\angle ACB = 90°$

$\therefore \ \cos C\hat{A}B = \dfrac{AC}{AB} = \dfrac{250}{1200} = 0.2083$ $\therefore \ \angle CAB = 77°59'$

Thus $BC = AB \times \sin C\hat{A}B = 1200 \times \sin 77°59' = 1174$ mm

Also $\angle EBA = \angle CAB = 77°59'$

$\therefore$ $\angle EBH = 180° - 77°59' = 102°1'$

The arc EHF therefore subtends an angle $2 \times 102°1'$ at the centre.

$\therefore$ Length of arc EHF $= 2\pi \times 100 \times \dfrac{204°2'}{360°} = 356$ mm

Similarly

Length of arc GJD $= 2\pi \times 150 \times \dfrac{204°2'}{360°} = 534$ mm

$\therefore$ Total length of belt $= 2 \times 1174 + 356 + 534 = 3238$ mm

SCREW THREAD MEASUREMENT

When accurate screw thread measurement is required the method of 2 or 3 wire measurement is used. The 3-wire method is used when checking with a hand micrometer and the 2-wire method is used with bench micrometers. The methods are fundamentally the same and allow the pitch or effective diameters to be measured (Fig. 14.47).

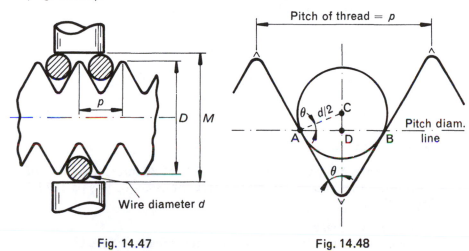

Fig. 14.47   Fig. 14.48

For the most accurate results each type and size of thread requires wires of a certain size. The best wire size is one which just touches the thread flanks at the pitch diameter as shown in Fig. 14.48.

At the pitch line diameter the distance between the flanks of the thread is equal to half the pitch.

Hence in Fig. 14.48

$$AB = \frac{p}{2} \quad \text{and} \quad AD = \frac{p}{4}$$

In $\triangle ADC$, $$\frac{AC}{AD} = \sec \theta$$

$\therefore$ $$AC = AD \times \sec \theta$$

$\therefore$ $$\frac{d}{2} = \frac{p}{4} \sec \theta \quad \therefore d = \frac{p}{2} \sec \theta$$

For a metric thread, $\theta = 30°$ and hence the best wire size is

$$d = \frac{p}{2} \sec 30° = 0.5774p$$

Formula for Checking the Form of a Metric Thread

Although the best wire size should be used, in practice the wire used will vary a little from this best size. In order to determine the distance over the wires for a specific thread (M in Fig. 14.47) the following formula is used

$$M = D - \frac{5p}{6} \cot \theta + d(\operatorname{cosec} \theta + 1)$$

For a metric thread, the formula becomes

$$M = D - \frac{5p}{6} \cot 30° + d(\operatorname{cosec} 30° + 1)$$

$$= D - 14434p + d(2 + 1) = D - 1.4434p + 3d$$

EXAMPLE 14.15

A metric thread having a major diameter of 20 mm and a pitch of 2.5 mm is to be checked using the best wire size for this particular thread. Find:

a) the best wire size,
b) the measurement over the wires if the thread is correct.

a) The best wire size is

$$d = 0.5774p = 0.5774 \times 2.5 = 1.4435 \text{ mm}$$

b) The measurement over the wires is

$$M = D - 1.4434p + 3d$$

$$= 20 - 1.4434 \times 2.5 + 3 \times 1.4435 = 20.722 \text{ mm}$$

Exercise 14.5

1) A belt passes over a pulley 1200 mm in diameter. The angle of contact between the pulley and the belt is 230°. Find the length of belt in contact with the pulley.

2) An open belt passes over two pulleys 900 mm and 600 mm in diameter respectively. If the centres of the pulleys are 1500 mm apart, find the length of the belt required.

3) Two pulleys of diameters 1400 mm and 900 mm respectively, with centres 4.5 m apart, are connected by an open belt. Find its length.

4) An open belt connects two pulleys of diameters 120 mm and 300 mm with centres 300 mm apart. Calculate the length of the belt.

5) A crossed belt passes over two pulleys each of 450 mm diameter. If their centres are 600 mm apart, calculate the length of the belt.

6) If, in Question 2, a crossed belt is used, what will be its length?

7) A crossed belt passes over two pulleys 900 mm and 1500 mm in diameter respectively, which have their centres 6 m apart. Find its length.

8) A crossed belt passes over two pulleys, one of 280 mm diameter and the other of 380 mm diameter. The angle between the straight parts of the belt is 90°. Find the length of the belt.

9) A metric thread having a major diameter of 52 mm and a pitch of 5 mm is to be checked by the 3-wire method. Determine the best size of wire and, using this best wire size, determine the measurement over the wires if the thread is correct.

10) A metric thread having a major diameter of 30 mm and a pitch of 2 mm is to be checked using wires whose diameters are 1.14 mm. Calculate the measurement over the wires that will be obtained if the thread is correct.

AREA OF A TRIANGLE

Three formulae are commonly used for finding the areas of triangles:

(1) If given the base and the altitude (i.e. vertical height).

(2) If any given two sides and the included angle.

(3) If given the three sides.

Case (1) Given the base and the altitude

In Fig. 14.49,

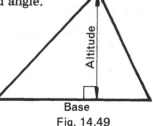

Area of triangle $= \frac{1}{2} \times$ **Base** $\times$ **Altitude**

Base

Fig. 14.49

EXAMPLE 14.16

Find the areas of the triangles shown in Fig. 14.50.

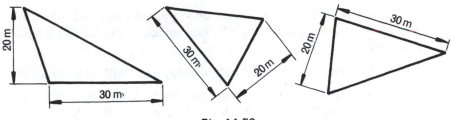

Fig. 14.50

In each case the 'base' is taken as the side of given length and the 'altitude' is measured perpendicular to this side.

Hence

Triangular area $= \frac{1}{2} \times$ Base $\times$ Altitude

$= \frac{1}{2} \times 30 \times 20 = 300 \, \text{m}^2$ in each case

EXAMPLE 14.17

A trapezium is shown in Fig. 14.51 in which AB is parallel to CD. Find its area.

If we join AD then the trapezium is divided into two triangles, the 'bases' and 'altitudes' of which are known.

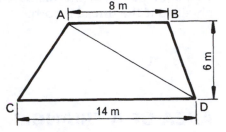

Fig. 14.51

$\therefore$ Area of trapezium $=$ Area of $\triangle$ABD $+$ Area of $\triangle$ADC

$= (\frac{1}{2} \times 8 \times 6) + (\frac{1}{2} \times 14 \times 6) = 66 \, \text{m}^2$

Case (2) If given any two sides and the included angle

In Fig. 14.52,

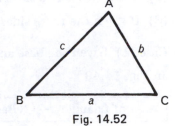

	Area of triangle $= \frac{1}{2}bc \sin A$
or	Area of triangle $= \frac{1}{2}ac \sin B$
or	Area of triangle $= \frac{1}{2}ab \sin C$

Fig. 14.52

EXAMPLE 14.18

Find the area of the triangle shown in Fig. 14.53.

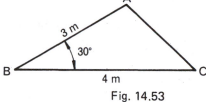

$$\text{Area} = \tfrac{1}{2} \times a \times c \times \sin B$$
$$= \tfrac{1}{2} \times 4 \times 3 \times \sin 30°$$
$$= 3 \, \text{m}^2$$

Fig. 14.53

EXAMPLE 14.19

Find the area of the triangle shown in Fig. 14.54.

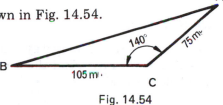

$$\text{Area} = \tfrac{1}{2} ab \sin C$$
$$= \tfrac{1}{2} \times 105 \times 75 \times \sin 140°$$

Fig. 14.54

We find the value of $\sin 140°$ by using the method shown in Fig. 14.55 from which it may be seen that

$$\sin 140° = \sin(180° - 140°)$$
$$= \sin 40°$$
$$\therefore \quad \text{Area} = \tfrac{1}{2} \times 105 \times 75 \times \sin 40°$$
$$= 2530 \, \text{m}^2$$

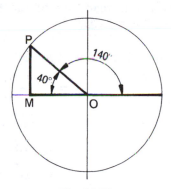

Fig. 14.55

Case (3) If given the three sides

In Fig. 14.56,

Area of triangle $= \sqrt{s(s-a)(s-b)(s-c)}$
where $\quad s = \dfrac{a+b+c}{2}$

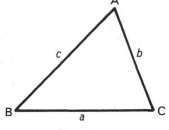

Fig. 14.56

EXAMPLE 14.20

A triangle has sides of lengths 3 m, 5 m and 6 m. What is its area?

Since we are given the lengths of 3 sides we use

$$\text{Area} = \sqrt{s(s-a)(s-b)(s-c)}$$

Now, $s = \dfrac{3+5+6}{2} = 7$

$\therefore$ Area $= \sqrt{7\times(7-3)\times(7-5)\times(7-6)} = \sqrt{56} = 7.48\,\text{m}^2$

EXAMPLE 14.21

The chain lines of a survey are shown in Fig. 14.57. Find the area of the plot of land.

The plot is the quadrilateral ABCD which is made up of the triangles ABC and ACD.

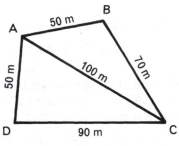

Fig. 14.57

To find the area of $\triangle$ABC,

$$s = \frac{50+70+100}{2} = 110$$

$\therefore$ Area of $\triangle$ABC $= \sqrt{s(s-a)(s-b)(s-c)}$

$$= \sqrt{110\times(110-50)\times(110-70)\times(110-100)}$$

$$= \sqrt{110\times60\times40\times10} = 1625\,\text{m}^2$$

To find the area of $\triangle$ACD,

$$s = \frac{50+90+100}{2} = 120$$

$\therefore$ Area of $\triangle$ACD $= \sqrt{s(s-a)(s-b)(s-c)}$

$$= \sqrt{120\times(120-50)\times(120-90)\times(120-100)}$$

$$= \sqrt{120\times70\times30\times20} = 2245\,\text{m}^2$$

$\therefore$ Area of quadrilateral $=$ Area of $\triangle$ABC $+$ Area of $\triangle$ACD

$$= 1625+2245 = 3870\,\text{m}^2$$

Exercise 14.6

1) Find the area of a triangle whose base is 75 mm and whose altitude is 59 mm.

2) Find the area of an isosceles triangle whose equal sides are 82 mm and whose base is 95 mm.

3) A plate in the shape of an equilateral triangle has a mass of 12.25 kg. If the material has a mass of 3.7 kg/m², find the dimensions of the plate in mm.

4) Obtain the area of a triangle whose sides are 39.3 m and 41.5 m if the angle between them is 41°30′.

5) Find the area of the playground shown in Fig. 14.58.

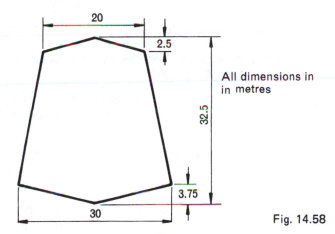

Fig. 14.58

6) Calculate the area of a triangle ABC if:
(a) $a = 4$ m, $b = 5$ m and $\angle C = 49°$,
(b) $a = 3$ m, $c = 6$ m and $\angle B = 63°44′$.

7) A triangle has sides 4 m, 7 m and 9 m long. What is its area?

8) A triangle has sides 37 mm, 52 mm and 63 mm long. What is its area?

9) Find the areas of the quadrilaterals shown in Fig. 14.59.

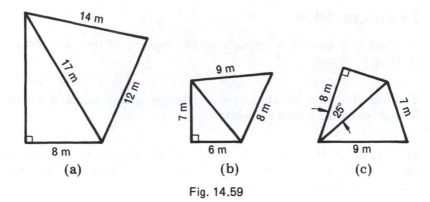

Fig. 14.59

10) Find the area of the triangle shown in Fig. 14.60.

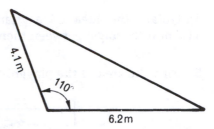

Fig. 14.60

11) What is the area of a parallelogram whose base is 7 m long and whose vertical height is 4 m?

12) Obtain the area of a parallelogram if two adjacent sides measure 112.5 mm and 105 mm and the angle between them is 49°.

13) Determine the length of the side of a square whose area is equal to that of a parallelogram with a 3 m base and a vertical height of 1.5 m.

14) Find the area of a trapezium whose parallel sides are 75 mm and 82 mm long respectively and whose vertical height is 39 mm.

15) Find the area of a regular hexagon,

(a) which is 4 m wide across flats,

(b) which has sides 5 m long.

16) Find the area of a regular octagon,

(a) which is 2 m wide across flats,

(b) which has sides 2 m long.

17) The parallel sides of a trapezium are 12 m and 16 m long. If its area is 220 m² what is its altitude?

15. THREE-DIMENSIONAL (SOLID) TRIGONOMETRY

Outcome:

1. Define the angle between a line and a plane.
2. Define the angle between two relevant planes in a given three-dimensional problem.
3. Solve three-dimensional triangulation problems

capable of being specified within a rectangular prism.

4. Relate lengths and areas on an inclined plane to corresponding lengths and areas on plan.

LOCATION OF A POINT IN SPACE

The location of a point in space requires co-ordinates referred to the origin O, and three mutually perpendicular axes Ox, Oy and Oz (Fig. 15.1).

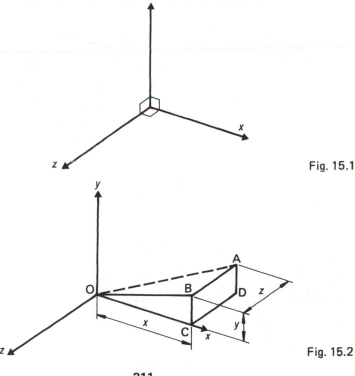

Fig. 15.1

Fig. 15.2

Consider the point A (Fig. 15.2). Let the co-ordinates measured parallel to the axes Ox, Oy and Oz be x, y and z respectively. To calculate the true lengths of the line OA we use the theorem of Pythagoras.

In the right-angled triangle OBC we have

$$OB^2 = OC^2 + BC^2 = x^2 + y^2$$

and in the right-angled triangle OAB we have

$$OA^2 = OB^2 + AB^2 = x^2 + y^2 + z^2$$

$$\therefore \qquad OA = \sqrt{x^2 + y^2 + z^2}$$

EXAMPLE 15.1

Fig. 15.3 shows part of a hipped roof. Find the length of the hip rafter AB and the common rafter BD.

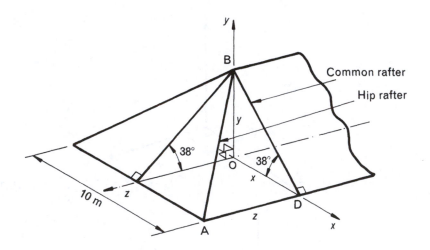

Fig. 15.3

In $\triangle BOD$, $\quad OD = 5 \quad$ and $\quad \angle BDO = 38°$

thus $\qquad \dfrac{OB}{OD} = \tan \angle BDO$

$\therefore \qquad OB = OD \tan \angle BDO = 5 \tan 38° = 3.906$

We now have the values $x = 5$ and $y = 3.906$

Thus $BD = \sqrt{x^2 + y^2} = \sqrt{5^2 + 3.906^2} = \sqrt{40.257} = 6.345$

Hence the length of the common rafter is 6.345 m.

For the line AB we have $x = 5$, $y = 3.906$ and $z = 5$

Thus

$$AB = \sqrt{x^2 + y^2 + z^2} = \sqrt{5^2 + 3.906^2 + 5^2} = \sqrt{65.257} = 8.08 \, \text{m}$$

Hence the length of the hip rafter is 8.08 m.

THE ANGLE BETWEEN A LINE AND A PLANE

In Fig. 15.4 the line AP intersects the xz plane at A. To find the angle between AP and the plane, draw PM perpendicular to the plane and then join AM.

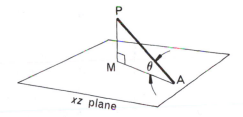

The angle between the line and the plane is $\angle$PAM.

Fig. 15.4

THE ANGLE BETWEEN TWO PLANES

Two planes which are not parallel intersect in a straight line. Examples of this are the floor and a wall of a room, and two walls of a room. To find the angle between two planes draw a line in each plane which is perpendicular to the common line of intersection. The angle between the two lines is the same as the angle between the two planes.

Three planes usually intersect at a point as, for instance, two walls and the floor of a room.

Problems with solid figures are solved by choosing suitable right-angled triangles in different planes. It is essential to make a clear three-dimensional drawing in order to find these triangles. The examples which follow show the methods that should be adopted.

EXAMPLE 15.2

Fig. 15.5 shows a cuboid. Find the angle between the diagonal AG and the plane EFGH.

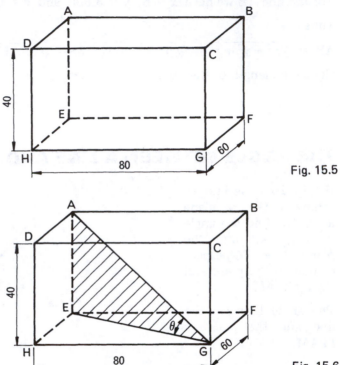

Fig. 15.5

Fig. 15.6

In order to find the required angle (θ in Fig. 15.6) we must use the right-angled triangle AEG, GE being the diagonal of the base rectangle.

In $\triangle$EFG, EF = 80 mm, GF = 60 mm and EFG = 90°. Using Pythagoras' theorem gives

$$EG^2 = EF^2 + GF^2 = 80^2 + 60^2 = 10\,000$$

$$EG = \sqrt{10\,000} = 100\,\text{mm}$$

In $\triangle$AEG, $\tan \theta = \dfrac{AE}{EG} = \dfrac{40}{100} = 0.4$

$$\theta = 21.8°$$

Hence the angle between the diagonal AG and the plane EFGH is 21.8°.

EXAMPLE 15.3

Fig. 15.7 shows a pyramid with a square base. The base has sides 60 mm long and the edges of the pyramid, VA, VB, VC and VD are each 100 mm long. Find the altitude of the pyramid.

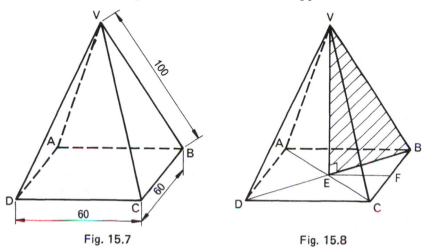

Fig. 15.7   Fig. 15.8

The right-angled triangle VBE (Fig. 15.8) allows the altitude VE to be found, but first we must find BE from the right-angled triangle BEF.

In $\triangle$BEF, BF = EF = 30 mm and $\angle$BFE = 90°.

Using Pythagoras' theorem gives

$$BE^2 = BF^2 + EF^2 = 30^2 + 30^2 = 900 + 900 = 1800$$

$$BE = \sqrt{1800} = 42.43 \text{ mm}$$

In $\triangle$VBE, BE = 42.43 mm, VB = 100 mm and $\angle$VEB = 90°.

Using Pythagoras' theorem gives

$$VE^2 = VB^2 - BE^2 = 100^2 - 42.43^2$$
$$= 10000 - 1800 = 8200$$
$$VE = \sqrt{8200} = 90.6 \text{ mm}$$

PROJECTED AREAS

In Fig. 15.9, the edges AD and BC of the rectangle ABCD are inclined at an angle θ to the horizontal plane. The rectangle *abcd* represents the plane of ABCD.

$$\text{Area of rectangle ABCD} = \text{AB} \times \text{BC}$$

But $\dfrac{bc}{BC} = \cos\theta$ or $BC = \dfrac{bc}{\cos\theta}$,   and $AB = ab$

∴   Area of rectangle ABCD $= ab \times \dfrac{bc}{\cos\theta}$

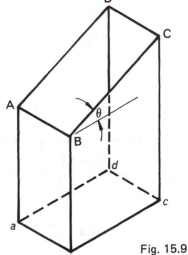

$$= \dfrac{ab \times bc}{\cos\theta}$$

$$= \dfrac{\text{Area of rectangle } abcd}{\cos\theta}$$

$$= \dfrac{\text{Projected plan area of ABCD}}{\cos\theta}$$

Although we have used a rectangle the formula applies to any area inclined to the horizontal at an angle θ. Thus:

Fig. 15.9

Area of inclined surface $= \dfrac{\text{Projected plan area of the surface}}{\cos\theta}$

EXAMPLE 15.4

A vertical chimney stack passes through a roof which is inclined at 41° to the horizontal (Fig. 15.10). The diameter of the stack is 150 mm. Find the area at the place where the stack passes through the roof.

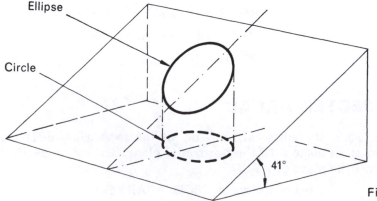

Fig. 15.10

$$\text{Projected plan area of stack} = \pi \times 75^2 = 17700 \, \text{mm}^2$$

$$\text{Area of stack at roof surface} = \frac{17700}{\cos 41°} = 23400 \, \text{mm}^2$$

EXAMPLE 15.5

The plan of a lean-to shed is a rectangle 3 m by 4 m. The roof of the shed is inclined at $22°$ to the horizontal. Calculate the roof area.

$$\text{Roof area} = \frac{\text{Plan area}}{\cos \theta} = \frac{3 \times 4}{\cos 22°} = 12.9 \, \text{m}^2$$

EXAMPLE 15.6

The plan of a steeple is a regular octagon of side 4 m. The steeple is 20 m high. Find the area of the roof to be covered.

The first step is to find the plan area. In Fig. 15.11, the octagon is divided into eight equal triangles. AOB is one of these triangles.

In $\triangle$AOB,

$$\angle AOB = \frac{360°}{8} = 45°$$

Since $\triangle$AOB is isosceles,

$$\angle AON = \frac{45°}{2} = 22°30'$$

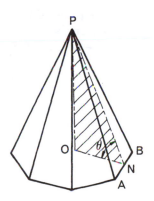

Now

$$\frac{ON}{AN} = \cot 22°30'$$

$$\therefore \quad ON = AN \cot 22°30'$$

$$= 2 \cot 22°30'$$

$$= 4.828 \, \text{m}$$

Area $\triangle$AOB $= \frac{1}{2} \times 4 \times 4.828$

$$= 9.656 \, \text{m}^2$$

Area of octagon $= 8 \times \text{Area} \, \triangle\text{AOB}$

$$= 8 \times 9.656$$

$$= 77.25 \, \text{m}^2$$

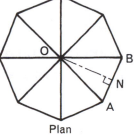

Plan

Fig. 15.11

To find the angle of slope of the roof the triangle PON is used (Fig. 15.11).

$$\tan \theta = \frac{OP}{ON} = \frac{20}{4.828} \qquad \therefore \ \theta = 76.42°$$

Thus Roof area $= \dfrac{\text{Plan area}}{\cos \theta} = \dfrac{77.25}{\cos 76.42°} = 329 \, \text{m}^2$

HEIGHTS OF INACCESSIBLE OBJECTS

The height of a hill, tower, etc. cannot usually be measured directly. However if the angle of elevation to the top of the feature is measured from two points whose distance apart is known, the height of the feature can be determined. When the two points are in line with the feature we proceed as shown on page 169. We now deal with the case of the two points not being in line.

EXAMPLE 15.7

From a point due south of a small hill, the angle of elevation of its summit is 23°. After moving 100 m due east the angle is 22°. Find the height of the hill.

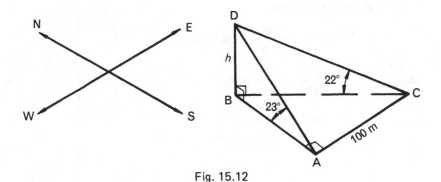

Fig. 15.12

In Fig. 15.12, BD represents the hill. If we let BD $= h$, then:

In $\triangle ABD$, $\angle DBA = 90°$ and $\angle DAB = 23°$

$\therefore \qquad\qquad AB = h \cot 23° = 2.356h$

In $\triangle BCD$, $\angle DBC = 90°$ and $\angle BCD = 22°$

$\therefore \qquad\qquad BC = h \cot 22° = 2.475h$

In $\triangle ABC$, $\angle BAC = 90°$.

Using Pythagoras' theorem $BC^2 = AB^2 + AC^2$

then $\qquad\qquad (2.475h)^2 = (2.356h)^2 + 100^2$

or $\qquad\qquad 6.126h^2 = 5.551h^2 + 10000$

from which $\qquad\qquad h = 132$

Hence the height of the hill is 132 metres.

Exercise 15.1

1) Fig. 15.13 shows a cuboid.

(a) Sketch the rectangle EFGH.
(b) Calculate the diagonal FH of rectangle EFGH.
(c) Sketch the rectangle FHDB adding known dimensions.
(d) Calculate the diagonal BH of rectangle FHDB.

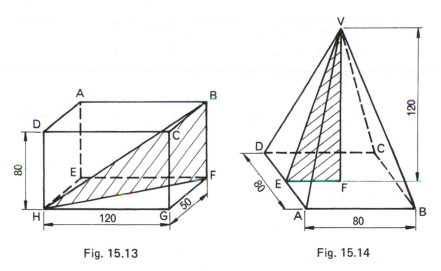

Fig. 15.13   Fig. 15.14

2) Fig. 15.14 shows a pyramid on a square base of side 80 mm. The altitude of the pyramid is 120 mm.

(a) Calculate EF.
(b) Draw the triangle VEF adding known dimensions.
(c) Find the angle VEF.
(d) Calculate the slant height VE.
(e) Calculate the area of $\triangle VAD$.
(f) Calculate the complete surface area of the pyramid.

3) Fig. 15.15 shows a pyramid on a rectangular base. Calculate the length VA.

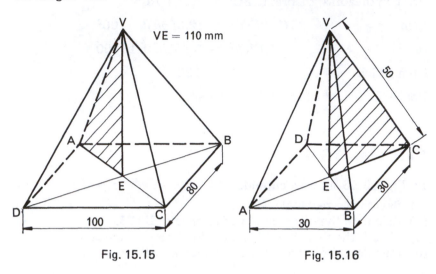

Fig. 15.15   Fig. 15.16

4) Fig. 15.16 shows a pyramid on a square base with

$$VA \ = \ VB \ = \ VC \ = \ VD \ = \ 50\,mm$$

Calculate the altitude of the pyramid.

5) A hipped roof makes an angle of 42° with the horizontal and its span is 12 m. Calculate the length of a common rafter and the length of a hip rafter.

6) In Fig. 15.17, ABCD represents part of a hill-side. A line of greatest slope AB is inclined at 36° to the horizontal AE and runs due north of A. The line AF bears 050° (N50°E) and C is 2500 m east of B. The lines of BE and CF are vertical. Calculate:

(a) the height of C above A,
(b) the angle between AB and AC.

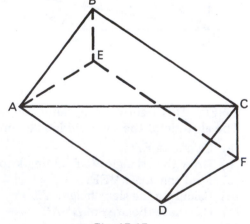

Fig. 15.17

7) A lean-to shed has a base which is 6 m long by 3 m wide. The roof makes an angle of 24° to the horizontal. Calculate the area of roof to be covered.

8) A roof is in the shape of an octagonal pyramid. In plan the edges of the octagon are each 4.2 m long. If the rise of the roof is 12.8 m calculate:

(a) the length of a hip rafter,
(b) the area of the roof.

9) Fig. 15.18 shows the plan of a hipped roof. Calculate the total area of the roof if it is inclined at 40° to the horizontal.

10) Fig. 15.19 is a sketch of a roof feature whose horizontal base is the rectangle ABCD 37 m by 23 m. The vertex V is 12 m directly above A. Calculate:

(a) the length VC,
(b) the total area of the inclined surfaces VBC and VCD.

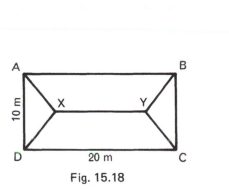

Fig. 15.18

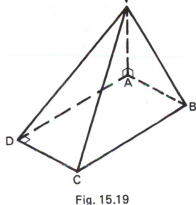

Fig. 15.19

11) To obtain the height of a tower a surveyor sets up his theodolite some distance to the north of the base of the tower and finds that the angle of elevation to the top of the tower is 31°. He then moves a distance of 60 m to the west of the first point and finds the angle of elevation to the top of the tower to be 22°. If the instrument stands 1.5 m above ground level calculate the height of the tower.

12) A point on top of a hill has an angle of elevation from a point P, due east of it, of 29°. By travelling 180 m due north of P the angle of elevation is found to be 26°. Calculate the height of the hill.

13) Fig. 15.20 shows the plan and elevation of a north light roof. Calculate the area of the roof surfaces.

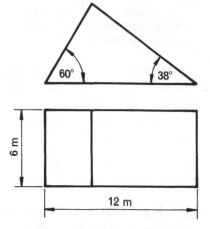

Fig. 15.20

14) A spire is covered with sheet copper. Its plan is a regular hexagon of side 2 m. Each sloping face is inclined at 76° to the horizontal. Calculate the area of the copper covering.

15) Part of a hillside is inclined uniformly at an angle of 39° to the horizontal. A straight road on the hillside makes an angle of 58° with the line of greatest slope. Determine the angle that the road makes with the horizontal.

16. AREAS AND VOLUMES

UNITS OF AREA

The standard abbreviation for units of area are:

$$\text{Square metres} = \text{m}^2$$

$$\text{Square millimetres} = \text{mm}^2$$

Conversion of square units of area are:

$$1\,\text{m}^2 = (1000\,\text{mm})^2 = (1000 \times 1000)\,\text{mm}^2 = 10^6\,\text{mm}^2$$

For large areas the hectare is used such that:

$$1\,\text{hectare (ha)} = 10000\,\text{m}^2$$

AREAS AND PERIMETERS

Rectangle

Area $= l \times b$
Perimeter $= 2l + 2b$

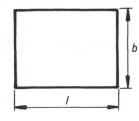

EXAMPLE 16.1

Find the area of the section shown in Fig. 16.1.

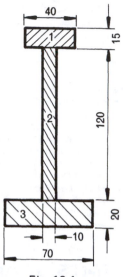

The section can be split up into three rectangles as shown. The total area can be found by calculating the areas of the three rectangles separately and then adding these together. Thus,

Area of rectangle 1 $= 15 \times 40$

$= 600 \text{ mm}^2$

Area of rectangle 2 $= 10 \times 120$

$= 1200 \text{ mm}^2$

Area of rectangle 3 $= 20 \times 70$

$= 1400 \text{ mm}^2$

Total area of section $= 600 + 1200$

$+ 1400 = 3200 \text{ mm}^2$

Fig. 16.1

Parallelogram

$$\boxed{\text{Area} = b \times h}$$

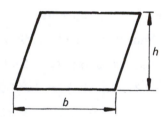

EXAMPLE 16.2

Find the area of the parallelogram shown in Fig. 16.2.

The first step is to find the vertical height h.

In $\triangle BCE$,

$h = BC \times \sin 60° = 3 \times 0.866 = 2.598$

$\begin{pmatrix} \text{Area of} \\ \text{parallelogram} \end{pmatrix} = \text{Base} \times \text{Vertical height}$

$= 5 \times 2.598$

$= 13.0 \text{ m}^2$

Fig. 16.2

Triangle

$$\boxed{\text{Area} = \tfrac{1}{2} \times b \times h}$$

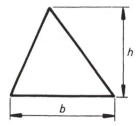

EXAMPLE 16.3

The inner shape of a pattern in a large hall is a regular octagon (8-sided polygon) which is 5 m across flats (Fig. 16.3). Find its area.

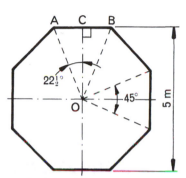

Fig. 16.3

The angle subtended at the centre by a side of the octagon $= \dfrac{360°}{8} = 45°$.

Now triangle AOB is isosceles, since $OA = OB$.

$\therefore \qquad \angle AOC = \dfrac{45°}{2} = 22°30'$

But $\qquad OC = \dfrac{5}{2} = 2.5 \, m$

Also $\qquad \dfrac{AC}{OC} = \tan 22°30'$

$\therefore \qquad AC = OC \times \tan 22°30' = 2.5 \times 0.4142 = 1.035 \, m$

Thus $\quad$ Area of $\triangle AOB = AC \times OC = 1.035 \times 2.5 = 2.588 \, m^2$

$\therefore \qquad$ Area of octagon $= 2.588 \times 8 = 20.7 \, m^2$

Trapezium (or Trapezoid)

$$\boxed{\text{Area} = \tfrac{1}{2} \times h \times (a + b)}$$

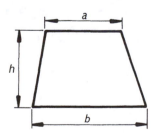

EXAMPLE 16.4

Fig. 16.4 shows the cross-section of a retaining wall. Calculate its cross-sectional area.

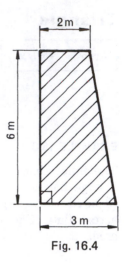

Fig. 16.4

Since the section is a trapezium:

$$\text{Area} = \tfrac{1}{2} \times h \times (a + b)$$
$$= \tfrac{1}{2} \times 6 \times (2 + 3)$$
$$= \tfrac{1}{2} \times 6 \times 5$$
$$= 15\,\text{m}^3$$

THE SURVEYOR'S FIELD BOOK

Fig. 16.5 shows the measurements made when surveying a plot of land. The offsets CK, GF, BH and JE are measured perpendicularly to the diagonal AD and the distances AG, AH, AK and AD are also measured. Hence the vertices A, B, C, D, E and F are fixed.

In recording the information in the surveyor's field book, distances from one end of the base line AD are written down in order in the centre column and the offsets are recorded to the right or left of the centre column as shown opposite.

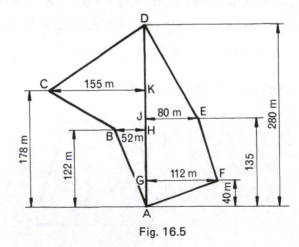

Fig. 16.5

	280 (D)	
155 (C)	178 (K)	
	135 (J)	80 (E)
52 (B)	122 (H)	
	40 (G)	112 (F)
	From A	
	in metres	

The area of the plot is now found by considering the areas of the trapeziums and triangles formed by the base line AD and the offsets, CK, BH, GF and JE. Thus,

$$\text{Area of triangle AGF} = \tfrac{1}{2} \times 40 \times 112 = 2240 \text{ m}^2$$

$$\text{Area of trapezium GFEJ} = \tfrac{1}{2} \times 95 \times (80 + 112) = 9120 \text{ m}^2$$

$$\text{Area of triangle JED} = \tfrac{1}{2} \times 145 \times 80 = 5800 \text{ m}^2$$

$$\text{Area of triangle DCK} = \tfrac{1}{2} \times 102 \times 155 = 7905 \text{ m}^2$$

$$\text{Area of trapezium CKHB} = \tfrac{1}{2} \times 56 \times (155 + 52) = 5796 \text{ m}^2$$

$$\text{Area of triangle ABH} = \tfrac{1}{2} \times 122 \times 52 = 3172 \text{ m}^2$$

$$\text{Hence the total area of the surveyed plot} = 34033 \text{ m}^2$$

Circle

$$\text{Area} = \pi r^2 = \frac{\pi d^2}{4}$$

$$\text{Circumference} = 2\pi r = \pi d$$

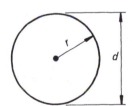

EXAMPLE 16.5

A pipe has an outside diameter of 32.5 mm and an inside diameter of 25 mm. Calculate the cross-sectional area of the shaft (Fig. 16.6).

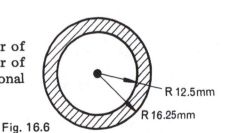

R 12.5mm

R 16.25mm

Fig. 16.6

$$\text{Area of cross-section} = \text{Area of outside circle} - \text{Area of inside circle}$$

$$= \pi \times 16.25^2 - \pi \times 12.5^2 = 338 \text{ mm}^2$$

Sector of a Circle

$$\text{Length of arc} = 2\pi r \times \frac{\theta^\circ}{360}$$

$$\text{Area of sector} = \pi r^2 \times \frac{\theta^\circ}{360}$$

EXAMPLE 16.6

Calculate a) the length of arc of a circle whose radius is 8 m and which subtends an angle of 56° at the centre, and b) the area of the sector so formed.

a)   Length of arc $= 2\pi r \times \dfrac{\theta^\circ}{360} = 2 \times \pi \times 8 \times \dfrac{56}{360} = 7.82\,\text{m}$

b)   Area of sector $= \pi r^2 \times \dfrac{\theta^\circ}{360} = \pi \times 8^2 \times \dfrac{56}{360} = 31.3\,\text{m}^2$

EXAMPLE 16.7

In a circle of radius 40 mm a chord is drawn which subtends an angle of 120° at the centre. What is the area of the minor segment?

Area of sector MCNO

$$= \frac{\pi r^2 \theta^\circ}{360} = \frac{\pi \times 40^2 \times 120}{360} = 1676\,\text{mm}^2$$

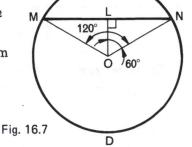

In the $\triangle\text{MON}$, $\text{MO} = \text{NO} = 40\,\text{mm}$ and the included angle $\text{MON} = 120^\circ$.

Hence $\angle\text{LON} = 60^\circ$

Now $\dfrac{\text{OL}}{\text{ON}} = \cos 60^\circ$

Fig. 16.7

$\therefore$ $\text{OL} = \text{ON} \times \cos 60^\circ = 40 \times 0.5000 = 20\,\text{mm}$

Also $\dfrac{\text{LN}}{\text{ON}} = \sin 60^\circ$

$\therefore$ $\text{LN} = \text{ON} \times \sin 60^\circ = 40 \times 0.8660 = 34.64\,\text{mm}$

Area of $\triangle\text{MON} = \frac{1}{2} \times \text{OL} \times \text{MN} = \frac{1}{2} \times 20 \times 69.28 = 693\,\text{mm}^2$

$\left(\begin{array}{c}\text{Area of minor} \\ \text{segment MCNL}\end{array}\right) = 1676 - 693 = 983\,\text{mm}^2$

Exercise 16.1

1) The area of a metal plate is $220\,mm^2$. If its width is 25 mm, find its length.

2) A sheet metal plate has a length of 147.5 mm and a width of 86.5 mm. Find its area in m^2.

3) Find the areas of the sections shown in Fig. 16.8.

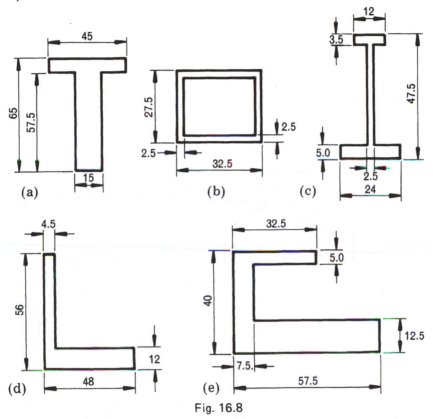

Fig. 16.8

4) What is the area of a parallelogram whose base is 70 mm long and whose vertical height is 40 mm?

5) Obtain the area of a parallelogram if two adjacent sides measure 112.5 mm and 105 mm and the angle between them is 49°.

6) Determine the length of the side of a square whose area is equal to that of a parallelogram with a 3 m base and a vertical height of 1.5 m.

7) Find the area of a trapezium whose parallel sides are 75 mm and 82 mm long respectively and whose vertical height is 39 mm.

8) Find the area of a regular hexagon,

(a) which is 40 mm wide across flats,

(b) which has sides 50 mm long.

9) Find the area of a regular octagon,

(a) which is 2 mm wide across flats,

(b) which has sides 2 mm long.

10) The parallel sides of a trapezium are 120 mm and 160 mm long. If its area is 22 000 mm² what is its altitude?

11) If the area of cross-section of a circular shaft is 700 mm², find its diameter.

12) Find the areas of the shaded portions of each of the diagrams of Fig. 16.9.

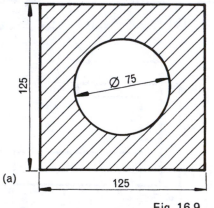

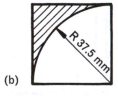

(a)   125   Ø 75   125   (b)   R 37.5 mm

Fig. 16.9

13) A hollow shaft has a cross-sectional area of 868 mm². If its inside diameter is 7.5 mm, calculate its outside diameter.

14) Find the area of the blank shown in Fig. 16.10.

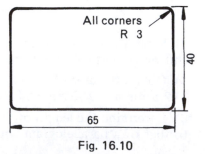

All corners
R 3

40

65

Fig. 16.10

15) How many revolutions will a wheel make in travelling 2 km if its diameter is 700 mm?

16) A bay window, semi-circular in plan, is to be covered with lead. Its radius is 2.4 m. Calculate:

(a) the area of lead required,

(b) the length of skirting board required to go round the bay.

17) A rectangular piece of insulating material is required to wrap round a pipe which is 560 mm diameter. Allowing 150 mm for overlap, calculate the width of material required.

18) The following figures are taken from a surveyor's field book. Roughly sketch the survey and find the area of the plot.

(a)

	To E	
	150	
	90	26.4 to D
To F 32.0	78	
	76	13.6 to C
	16	52.0 to B
	From A	

(b)

	To Y	
	50	
	40	15 to C
	32	18 to D
To B 6	31	
To A 6	24	25 to E
	19	15 to F
	From X	

19) The centre panel of a large ceiling consists of a 2.2 m diameter circle circumscribed by a regular hexagon. Find:

(a) the area of the circle,
(b) the area of the hexagon,
(c) the area between the circle and the hexagon.

20) The floor of a summer house is an octagon which could be inscribed inside a circle whose diameter is 7 m. Find the area of the floor.

UNITS OF VOLUME OR CAPACITY

The capacity of a container is the volume that it will contain. It is often measured in the same units as volume, that is cubic metres.

Sometimes however, as in the case of liquid measure, the litre* (abbreviation ℓ) unit is used such that

$$1\,m^3 = 1000\,\ell \quad \text{to four figure accuracy}$$

Small capacities are often measured in millilitres (mℓ) and

$$1000\,\text{millilitres (m}\ell) = 1\,\text{litre}\,(\ell) \quad \text{to four figure accuracy}$$

but there are $1\,000\,000\,mm^3$ in 1 litre

$$\therefore \qquad 1\,m\ell = 1000\,mm^3 \quad \text{to four figure accuracy}$$

VOLUMES AND SURFACE AREAS

Any Solid Having a Uniform Cross-section and Parallel End Faces

Volume = Cross-sectional area × Length of solid
Surface area = Longitudinal surface + Ends i.e. (Perimeter of cross-section × Length of solid) + (Total area of ends)

A *prism* is the name often given this type of solid if the cross-section is triangular or polygonal.

EXAMPLE 16.8

A piece of timber has the cross-section shown in Fig. 16.11. If its length is 300 mm, find its volume and total surface area.

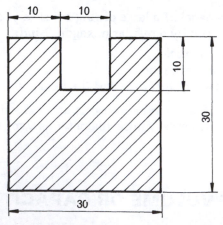

Fig. 16.11

*A litre is 1.000 028 cubic decimetres, but the term 'litre' is *not* to be used for precise measurements.

Area of cross-section $= (30 \times 30) - (10 \times 10) = 800\,\text{mm}^2$

Volume $=$ (Area of cross-section) $\times$ (Length)

$\qquad = 800 \times 300 = 240\,000\,\text{mm}^3$

Perimeter of cross-section $= (3 \times 30) + (5 \times 10) = 140\,\text{mm}$

Total surface area $=$ (Perimeter of cross-section $\times$ Length)
$\qquad\qquad\qquad + $ (Area of ends)

$\qquad\qquad = (140 \times 300) + (2 \times 800) = 43\,600\,\text{mm}^2$

EXAMPLE 16.9

A steel section has the cross-section shown in Fig. 16.12. If it is 9 m long, calculate its volume and total surface area.

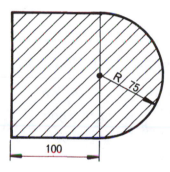

Fig. 16.12

To find the volume

$\quad$ Area of cross-section $= \frac{1}{2} \times \pi \times 75^2 + 100 \times 150 = 23\,840\,\text{mm}^2$

$$= \frac{23\,840}{(1000)^2} = 0.023\,84\,\text{m}^2$$

$\therefore\qquad$ Volume of solid $= 0.023\,84 \times 9 = 0.215\,\text{m}^3$

To find the surface area

$\quad$ Perimeter of cross-section $= \pi \times 75 + 2 \times 100 + 150$

$$= 585.5\,\text{mm} = \frac{585.5}{1000} = 0.5855\,\text{m}$$

$\qquad$ Lateral surface area $= 0.5855 \times 9 = 5.270\,\text{m}^2$

$\qquad$ Surface area of ends $= 2 \times 0.024 = 0.048\,\text{m}^2$

$\therefore\qquad$ Total surface area $= 5.27 + 0.05 = 5.32\,\text{m}^2$

Cylinder

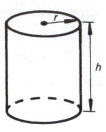

Volume $= \pi r^2 h$
Surface area $= 2\pi rh + 2\pi r^2 = 2\pi r(h + r)$

EXAMPLE 16.10

A cylindrical can holds 18 litres of petrol. Find the depth of the petrol if the can has a diameter of 600 mm.

Now   18 litres $= 18 \times 10^6$ mm^3

and if the depth of the petrol is h mm

then   Volume of petrol $= \pi (\text{Radius})^2 \times h$

$\therefore$ $18 \times 10^6 = \pi \times 300^2 \times h$

$\therefore$ $h = \dfrac{18\,000\,000}{\pi \times 90\,000} = 63.7$ mm

EXAMPLE 16.11

A metal bar of length 200 mm and diameter 75 mm is melted down and cast into washers 2.5 mm thick with an internal diameter of 12.5 mm and external diameter 25 mm. Calculate the number of washers obtained assuming no loss of metal.

$$\text{Volume of original bar of metal} = \pi \times 37.5^2 \times 200$$

$$= 883\,500 \text{ mm}^3$$

$$\text{Volume of one washer} = \pi \times (12.5^2 - 6.25^2) \times 2.5$$

$$= \pi \times 117.2 \times 2.5$$

$$= 920.4 \text{ mm}^3$$

$$\text{Number of washers obtained} = \frac{883\,500}{920.4} = 960$$

Cone

Volume $= \frac{1}{3}\pi r^2 h$ (h is the vertical height)
Curved surface area $= \pi r l$ (l is the slant length)

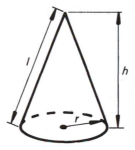

EXAMPLE 16.12

A hopper is in the form of an inverted cone. It has a maximum internal diameter of 2.4 m and an internal height of 2.1 m.

a) If the hopper is to be lined with lead, calculate the area of lead required.

b) Determine the capacity of the hopper before lining.

a) The slant height may be found by using Pythagoras' theorem on the triangle shown in Fig. 16.13.

$$l^2 = r^2 + h^2 = 1.2^2 + 2.1^2 = 5.85$$

$$l = \sqrt{5.85} = 2.42 \, \text{m}$$

Surface area $= \pi r l = \pi \times 1.2 \times 2.42 = 9.12 \, \text{m}^2$

Hence the area of lead required is $9.12 \, \text{m}^2$

b) Volume $= \frac{1}{3}\pi r^2 h = \frac{1}{3} \times \pi \times 1.2^2 \times 2.1 = 3.17 \, \text{m}^3$

Hence the capacity of the hopper is $3.17 \, \text{m}^3$

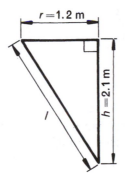

Fig. 16.13

Frustum of a Cone

A *frustum* is the portion of a cone or pyramid between the base and a horizontal slice which removes the pointed portion.

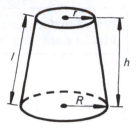

Volume $= \frac{1}{3}\pi h(R^2 + Rr + r^2)$ (h is the vertical height)
Curved surface area $= \pi l(R + r)$ Total surface area $= \pi l(R + r) + \pi R^2 + \pi r^2$ (l is the slant height)

EXAMPLE 16.13

A concrete column is shaped like a frustum of a cone. It is 2.8 m high. The radius at the top is 0.6 m and the base radius is 0.9 m. Calculate the volume of concrete in the column.

$$\text{Volume} = \frac{1}{3}\pi h(R^2 + Rr + r^2)$$
$$= \frac{1}{3} \times \pi \times 2.8 \times (0.9^2 + 0.9 \times 0.6 + 0.6^2)$$
$$= \frac{1}{3} \times \pi \times 2.8 \times 1.71 = 5.01\,\text{m}^3$$

EXAMPLE 16.14

The bowl shown in Fig. 16.14 is made from sheet steel and has an open top. Calculate the total cost of painting the vessel (inside and outside) at a cost of 1 p per 10000 mm^2.

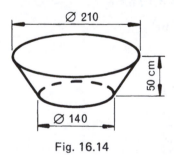

Fig. 16.14

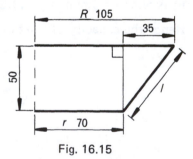

Fig. 16.15

Fig. 16.15 shows a half section of the bowl. Using Pythagoras' theorem on the right-angled triangle

$$l^2 = 50^2 + 35^2$$

∴ $$l = 61.0\,\text{mm}$$

Now the required total surface area, i.e. inside and outside

$$= 2\,\{\text{Curved surface area}\} + 2(\text{Base area})$$
$$= 2\,\{\pi l(R + r)\} + 2(\pi r^2)$$
$$= 2\,\{\pi(61)(105 + 70)\} + 2(\pi 70^2) = 97\,800\ \text{mm}^2$$

At 1p per 10 000 mm^2 total cost $= \dfrac{97\,800}{10\,000} = 9.78\text{p}$

Sphere

Volume $= \frac{4}{3}\pi r^3$

Surface area $= 4\pi r^2$

EXAMPLE 16.15

A concrete dome is in the shape of a hemisphere. Its internal and external diameters are 4.8 m and 5 m respectively.

a) Calculate the volume of concrete used in its construction.
b) If the inside is to be painted, calculate the area to be covered.

a) If R is the outside radius, r the inside radius and V the volume, then

$$V = \tfrac{1}{2}(\tfrac{4}{3}\pi R^3 - \tfrac{4}{3}\pi r^3)$$
$$= \tfrac{1}{2} \times \tfrac{4}{3} \times \pi \times (R^3 - r^3)$$
$$= \tfrac{2}{3} \times \pi \times (2.5^3 - 2.4^3) = 3.772\ \text{m}^3$$

b)   Area of inside of dome $= \tfrac{1}{2} \times 4\pi r^2$
$$= \tfrac{1}{2} \times 4 \times \pi \times 2.4^2 = 36.1911\ \text{m}^2$$

Spherical Cap

$$\begin{pmatrix}\text{Total}\\ \text{surface area}\end{pmatrix} = \begin{pmatrix}\text{Curved}\\ \text{surface area}\end{pmatrix} + \begin{pmatrix}\text{Flat}\\ \text{base area}\end{pmatrix}$$
$$= 2\pi Rh + \pi r^2$$
$$\text{or } \pi(r^2 + h^2) + \pi r^2$$

Volume $= \dfrac{\pi h^2}{3}(3R - h)$ or $\dfrac{\pi h}{6}(3r^2 + h^2)$

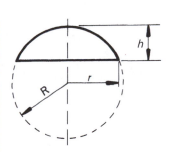

EXAMPLE 16.16

Calculate the volume and outer curved surface area of a spherical dome having a span of 12 m and a rise of 4 m.

Using the formula

$$V = \frac{\pi h}{6}(3r^2 + h^2)$$

when $r = 6$ and $h = 4$, $V = \frac{\pi \times 4}{6} \times (3 \times 6^2 + 4^2)$

$$= 2.09 \times (108 + 16) = 260 \text{ m}^3$$

Using the formula $A = \pi(r^2 + h^2)$

when $r = 6$ and $h = 4$, $A = \pi(6^2 + 4^2) = 163 \text{ m}^2$

Pyramid

> Volume $= \frac{1}{3}Ah$
>
> Surface area $=$ Sum of the areas of the triangles forming the sides plus the area of the base

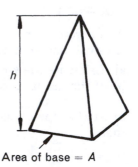

Area of base $= A$

EXAMPLE 16.17

Find the volume and total surface area of a symmetrical pyramid whose base is a rectangle $7 \text{ m} \times 4 \text{ m}$ and whose height is 10 m.

Base area $= 7 \times 4 = 28 \text{ m}^2$

Height $= 10 \text{ m}$

Volume $= \frac{1}{3}Ah = \frac{1}{3} \times 28 \times 10$

$$= 93.3 \text{ m}^3$$

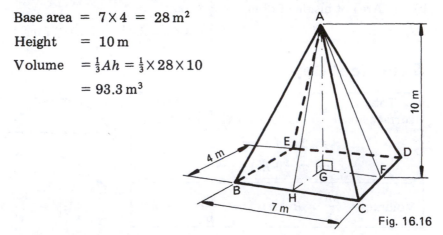

Fig. 16.16

From Fig. 16.16 the surface area consists of two sets of equal triangles (that is △ABC and △ADE, and also △ABE and △ACD) together with the base BCDE. To find the area of △ABC we must find the slant height AH. From the apex, A, drop a perpendicular AG on to the base and draw GH perpendicular to BC. H is then the mid-point of BC.

In △AHG, $\angle AGH = 90°$ and, by Pythagoras' theorem,

$$AH^2 = AG^2 + HG^2 = 10^2 + 2^2 = 104$$

∴
$$AH = \sqrt{104} = 10.2\,m$$

∴ Area of △ABC $= \frac{1}{2} \times$ Base $\times$ Height

$$= \frac{1}{2} \times 7 \times 10.20 = 35.7\,m^2$$

Similarly, to find the area of △ACD we must find the slant height AF. Draw GF, F being the mid-point of CD. Then in △AGF, $\angle AGF = 90°$ and by Pythagoras' theorem,

$$AF^2 = AG^2 + GF^2 = 10^2 + 3.5^2 = 112.3$$

∴
$$AF = \sqrt{112.3} = 10.6\,m$$

∴ Area of △ACD $= \frac{1}{2} \times$ Base $\times$ Height

$$= \frac{1}{2} \times 4 \times 10.6 = 21.2\,m^2$$

∴ Total surface area $= (2 \times 35.7) + (2 \times 21.2) + (7 \times 4)$

$$= 142\,m^2$$

Frustum of a Pyramid

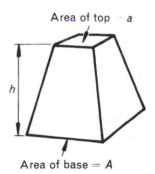

Area of top = a

Area of base = A

> Volume $= \frac{1}{3}h(A + \sqrt{Aa} + a)$
>
> Surface area = Sum of the areas of the trapeziums forming the sides plus the areas of the top and base of the frustum

EXAMPLE 16.18

A casting has a length of 2 m and its cross-section is a regular hexagon. The casting tapers uniformly along its length, the hexagon having a side of 200 mm at one end and 100 mm at the other. Calculate the volume of the casting.

In Fig. 16.17 the area of the hexagon of 200 mm side

$$= 6 \times \text{Area} \triangle ABO$$

In $\triangle AOC$, $\angle AOC = 30°$, $AC = 100 \text{ mm}$

and $\tan \angle AOC = \dfrac{AC}{OC}$

$$\therefore \quad OC = \frac{AC}{\tan AOC} = \frac{100}{\tan 30°}$$

$$= 173 \text{ mm}$$

$\therefore \quad$ Area of $\triangle ABO = \frac{1}{2} \times 200 \times 173$

$\therefore \quad$ Area of hexagon $= 6 \times \frac{1}{2} \times 200 \times 173 = 103\,800 \text{ mm}^2$

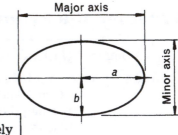

Plan view
of large end

Fig. 16.17

The area of the hexagon of 100 mm side can be found in the same way. It is $25\,950 \text{ mm}^2$.

The casting is a frustum of a pyramid with $A = 103\,800$, $a = 25\,950$ and $h = 2000$

$\therefore \quad$ Volume $= \frac{1}{3}h(A + \sqrt{Aa} + a)$

$$= \frac{1}{3} \times 2000 \times (103\,800 + \sqrt{103\,800 \times 25\,950} + 25\,950)$$

$$= 1.21 \times 10^8 \text{ mm}^3$$

The Ellipse

Area $= \pi ab$ exactly

Perimeter $= \pi(a + b)$ approximately

EXAMPLE 16.19

Find the area and the approximate perimeter of an ellipse with a major axis of 200 mm and a minor axis of 150 mm.

Area $= \pi ab = \pi \times 100 \times 75 = 23\,600 \text{ mm}^2$

Perimeter $= \pi(a + b) = \pi \times (100 + 75) = 550 \text{ mm}$ approximately

EXAMPLE 16.20

Part of an air ducting system calls for an ellipse whose area is $20\,000\,\text{mm}^2$ to be placed on a sheet metal cylinder whose diameter is $150\,\text{mm}$ (Fig. 16.18). Find the angle θ at which the cylinder must be cut.

We have $\pi a b = 20\,000$

and since $b = 75\,\text{mm}$

then $a = \dfrac{20\,000}{\pi \times 75} = 84.9\,\text{mm}$

From Fig. 10.24,

$$\cos\theta = \frac{150}{2 \times 84.9} = 0.883$$

$$\theta = 28°$$

Fig. 16.18

Exercise 16.2

1) A beam has a cross-sectional area of $18\,000\,\text{mm}^2$ and it is $6\,\text{m}$ long. Calculate, in cubic metres, the volume of material in the beam.

2) A block of lead $0.15\,\text{m}$ by $0.1\,\text{m}$ by $0.075\,\text{m}$ is hammered out to make a square sheet $10\,\text{mm}$ thick. What are the dimensions of the square?

3) Calculate the volume of metal in a pipe which has a bore of $50\,\text{mm}$, a thickness of $8\,\text{mm}$ and a length of $6\,\text{m}$.

4) The vertical section of a cutting for a road is a trapezium $100\,\text{m}$ wide at the top and $30\,\text{m}$ wide at the bottom. If it is $15\,\text{m}$ deep and $40\,\text{m}$ long, calculate its volume.

5) The cross-section of a retaining wall is a trapezium as shown in Fig. 16.19. Its length is $8\,\text{m}$. Calculate the volume of the wall.

6) A hot-water cylinder whose length is $1.12\,\text{m}$ is to hold 200 litres. Find the diameter of the cylinder in millimetres.

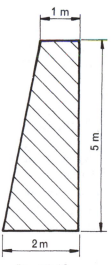

Fig. 16.19

7) Calculate the heating surface (in square metres) of a steam pipe whose external diameter is 60 mm and whose length is 8 m.

8) A small cone has a maximum diameter of 70 mm and a vertical height of 100 mm. Calculate is volume and its total surface area.

9) A cement silo is in the form of a frustum of a cone. It is 3 m high, 2.4 m diameter at the top and 1.2 m diameter at the bottom. Calculate the volume of cement that it will hold.

10) A metal bucket is 400 mm deep. It is 300 mm diameter at the top and 200 mm diameter at the bottom. Calculate the number of litres of water that the bucket will hold assuming that it is a frustum of a cone.

11) It is required to replace two pipes with bores of 28 mm and 70 mm respectively with a single pipe which has the same area of flow. Find the bore of this single pipe.

12) A pyramid has a square base of side 2 m and a height of 4 m. Calculate its volume and its total surface area.

13) A column is a regular octagon (8-sided polygon) in cross-section. It is 460 mm across flats at the base and it tapers uniformly to 300 mm across flats at the top. If it is 3.6 m high calculate the volume of material required to make it.

14) A bucket used on a crane is in the form of a frustum of a pyramid. Its base is a square of 600 mm side and its top is a square of 750 mm side. It has a depth of 800 mm.
(a) Calculate, in cubic metres, the volume of cement that it will hold.
(b) Twenty of these buckets of cement are emptied into a cylindrical cavity 4 m diameter. Calculate the depth to which the cavity will be filled.

15) A small turret roof is in the form of a pyramid. The base is a regular pentagon (5-sided polygon) with each side 3 m long. Its height is 4 m.
(a) Calculate the volume enclosed by the roof.
(b) Determine the total area of the inclined surfaces.

16) The ball of a float valve is a sphere 200 mm diameter. Calculate the volume and surface area of the ball.

17) A hemispherical dome has a diameter of 7 m.

(a) Calculate the volume that the dome encloses.

(b) If the outside of the dome is to be painted, calculate the area, in square metres, which must be covered.

18) A hemispherical dome has a radius of 4 m. It has its top 2 m cut off to provide for a horizontal laylight.

(a) Calculate the volume enclosed by the dome below the laylight.

(b) Find the area of internal plastering required to cover the curved surface below the laylight.

19) A dome is in the form of a cap of a sphere. The base radius of the cap is 6 m and the height of the dome is 5 m. Calculate the air space contained in the dome and the inner curved surface area.

20) The ball of a float valve is a sphere 200 mm diameter. It is immersed to a depth of 80 mm in water. How many litres of water does it displace?

21) Find the area and approximate perimeter of an ellipse which has a major axis of 40 mm and a minor axis of 30 mm.

22) An air duct is 1.20 m in diameter. At one point it is transformed into an ellipse of equal area. If the minor axis is 0.8 m, what is the length of the major axis?

23) A cylindrical vessel, made of sheet metal, is 200 mm in diameter. It is cut at an angle of 40° as shown in Fig. 16.20. Find the area and the perimeter of the ellipse so formed.

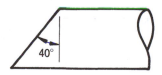

Fig. 16.20

24) A swimming bath is 50 m long by 12 m wide. It is 3 m deep at the deep end and 1.5 m deep at the shallow end. Find the capacity of the bath in litres.

25) A roof is in the form of a frustum of a rectangular pyramid. At the base the roof is 4 m long and 3 m wide whilst at the top it is 2 m long and 1.5 m wide. If it is 2 m high calculate the volume enclosed by the roof and the total surface area of the inclined faces.

26) A tapering column is 7 m tall and its cross-section is a regular octagon. The octagon is 3 m across flats at the base and 2 m across flats at the top.
(a) Calculate the volume of the column.
(b) If the material used in the construction of the column has a density of 2200 kg/m³, calculate the mass of the column.

27) The base of a ventilating turret is in the form of the frustum of a pyramid with a height of 1.3 m. The bottom of the turret is a square of side 2 m and the top is a square of side 1 m. Calculate the volume of the turret and the total area of the inclined surfaces.

28) The length of a conical nozzle is 600 mm long. The diameters at the ends are 150 mm and 220 mm respectively. Calculate the volume of water, in litres, that the nozzle will hold.

29) A lime kiln has the dimensions shown in Fig. 16.21. Calculate the volume enclosed by the kiln.

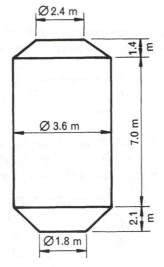

Fig. 16.21

30) A tub holding 58 litres of water when full is shaped like the frustum of a cone. If the radii at the ends of the tub are 400 mm and 300 mm, calculate the height of the tub.

31) A cupola is a cap of a 1.5 mm radius sphere. If it is 2 m high, calculate its volume and curved surface area.

32) A dome is in the form of the cap of a sphere. The radius at the base of the cap is 7 m and its height is 5 m. Calculate the area for plastering the inside of the dome.

NUMERICAL METHODS FOR CALCULATING IRREGULAR AREAS AND VOLUMES

Areas

An irregular area is one whose boundary does not follow a definite pattern, e.g. the cross-section of a river.

In these cases practical measurements are made and the results plotted to give a graphical display.

Various numerical methods may then be used to find the area.

Mid-ordinate Rule

Suppose we wish to find the area shown in Fig. 16.22. Let us divide the area into a number of vertical strips, each of equal width b.

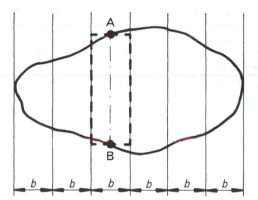

Fig. 16.22

Consider the 3rd strip, whose centre line is shown cutting the curved boundaries of the area at A and B respectively. Through A and B horizontal lines are drawn and these help to make the dotted rectangle shown. The rectangle has approximately the same area as that of the original 3rd strip, and this area will be $b \times$ AB.

AB is called the mid-ordinate of the 3rd strip, as it is mid-way between the vertical sides of the strip.

To find the *whole area*, the areas of the other strips are found in a similar manner and then all are added together for the final result.

∴   Area = Width of strip × Sum of the mid-ordinates

A useful practical tip to avoid measuring each separate mid-ordinate is to use a strip of paper and mark off along its edge successive mid-ordinate lengths, as shown in Fig. 16.23. The total area will then be found by measuring the whole length marked out (in the case shown this is HR) and multiplying by the strip width b.

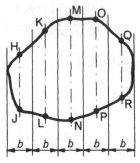

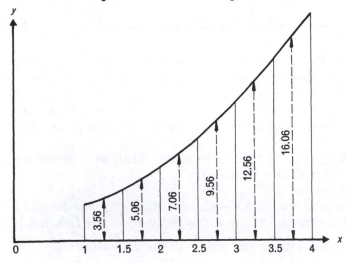

Fig. 16.23

EXAMPLE 16.21

Find the area under the curve $y = x^2 + 2$ between $x = 1$ and $x = 4$

The curve is sketched in Fig. 16.24. Taking 6 strips we may calculate the mid-ordinates.

x	1.25	1.75	2.25	2.75	3.25	3.75
y	3.56	5.06	7.06	9.56	12.56	16.06

Since the width of the strips $= \frac{1}{2}$, the mid-ordinate rule gives

$$\text{Area} = \tfrac{1}{2} \times (3.56 + 5.06 + 7.06 + 9.56 + 12.56 + 16.06)$$
$$= \tfrac{1}{2} \times 53.86 = 26.93 \text{ square units}$$

Fig. 16.24

It so happens that in this example it is possible to calculate an exact answer. How this is done need not concern us at this stage, but by comparing the exact answer with that obtained by the mid-ordinate rule we can see the size of the error.

$$\text{Exact answer} = 27 \text{ square units}$$

$$\text{Approximate answer (using the mid-ordinate rule)}$$

$$= 26.93 \text{ square units}$$

$$\therefore \qquad \text{Error} = 0.07 \text{ square units}$$

$$\therefore \qquad \text{Percentage error} = \frac{0.07}{27} \times 100 = 0.26\%$$

From the above it is clear that the mid-ordinate rule gives a good approximation to the correct answer.

Trapezoidal Rule

Consider the area having boundary ABCD shown in Fig. 16.25.

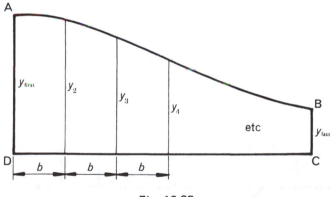

Fig. 16.25

The area is divided into a number of vertical strips of equal width b.

Each vertical strip is assumed to be a trapezium. Hence the third strip, for example, will have an area $= b \times \frac{1}{2}(y_3 + y_4)$.

But

Area ABCD = Sum of all the vertical strips

$$= b \times \tfrac{1}{2}(y_{\text{first}} + y_2) + b \times \tfrac{1}{2}(y_2 + y_3) + b \times \tfrac{1}{2}(y_3 + y_4) + \dots$$

$$= b[\tfrac{1}{2}y_{\text{first}} + \tfrac{1}{2}y_2 + \tfrac{1}{2}y_2 + \tfrac{1}{2}y_3 + \dots + \tfrac{1}{2}y_{\text{last}}]$$

$$= b[\tfrac{1}{2}(y_{\text{first}} + y_{\text{last}}) + y_2 + y_3 + y_4 \dots]$$

= Width of strips × [½(Sum of the first and last ordinates) + (Sum of the remaining ordinates)]

The accuracy of the trapezoidal rule is similar to that of the mid-ordinate rule. A comparison may be made by solving Example 16.21 using the trapezoidal rule as in Example 16.22.

EXAMPLE 16.22

Find the area under the curve $y = x^2 + 2$ between $x = 1$ and $x = 4$

The curve is sketched in Fig. 16.26, the lengths of the ordinates having been calculated.

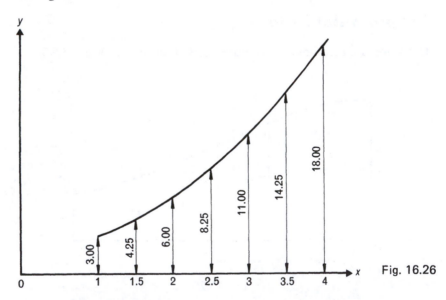

Fig. 16.26

Since the width of the strips $= \tfrac{1}{2}$, the trapezoidal rule gives

$$\text{Area} = \tfrac{1}{2} \times [\tfrac{1}{2}(3 + 18) + 4.25 + 6 + 8.25 + 11 + 14.25]$$

$$= \tfrac{1}{2} \times [10.5 + 43.75]$$

$$= 27.13 \text{ square units}$$

The exact answer is 27 square units and therefore

$$\text{Percentage error} = \frac{27.13 - 27}{27} \times 100 = 0.48\%$$

EXAMPLE 16.23

The table gives the values of a force required to pull a trolley when measured at various distances from a fixed point in the direction of the force.

F (N)	51	49	45	37	26	15	10
s (m)	0	1	2	3	4	5	6

Calculate the total work done by this force.

The force–distance graph is plotted as shown in Fig. 16.27. The required work done is given by the shaded area under the curve. This may be found by dividing the area into strips and using the trapezoidal rule.

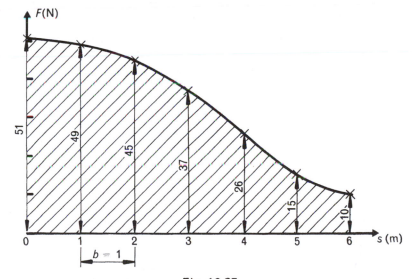

Fig. 16.27

Therefore

$$\text{Area} = 1[\tfrac{1}{2}(51 + 10) + (49 + 45 + 37 + 26 + 15)] = 203$$

Thus

$$\text{Work done} = 203 \, \text{Nm} = 203 \, \text{J} \quad (\text{since joule} = \text{newton} \times \text{metre})$$

Simpson's Rule

The required area is divided into an *even* number of vertical strips of equal width b.

Then Simpson's rule gives

$$\text{Area} = \frac{b}{3}\,[(\text{Sum of the first and last ordinates})$$
$$+ 2(\text{Sum of the remaining odd ordinates})$$
$$+ 4(\text{Sum of the even ordinates})]$$

This rule usually gives a more accurate result than either the mid-ordinate or trapezoidal rules, but is slightly more complicated to use.

Note.   There must be an *even* number of strips.

EXAMPLE 16.24

One boundary of a plot of land is a straight line 60 m long. The lengths of perpendicular offsets at 10 m intervals from this line to the curved boundary are as follows:

Distance (m)	0	10	20	30	40	50	60
Length of offset (m)	0	16	28	36	40	42	43

Draw a plan of the plot of land and find its area.

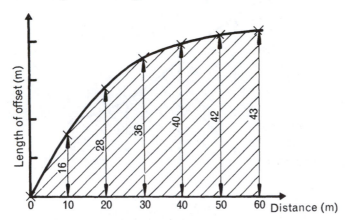

Fig. 16.28

The area under the graph gives the area of the plot and it is shown shaded in Fig. 16.28.

Simpson's rule gives

$$\text{Area} = \tfrac{10}{3}[(0 + 43) + 2(28 + 40) + 4(16 + 36 + 42)] = 1850$$

Therefore area of the plot of land is $1850\,\text{m}^2$.

Volumes of Irregular Solids

All the methods explained in this chapter for finding irregular areas may be applied to finding volumes. The following example shows a typical problem solved by the use of Simpson's rule and the trapezoidal rule.

EXAMPLE 16.25

The diameters in metres of a felled tree trunk at one metre intervals along its length are as follows:

$$1.00, \ 0.90, \ 0.81, \ 0.74, \ 0.68, \ 0.64 \ \text{and} \ 0.61$$

Assuming that the cross-sections of the trunk are circular, estimate the volume of timber.

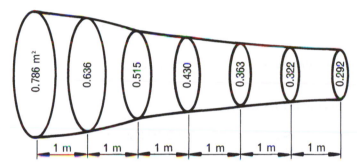

Fig. 16.29

The areas corresponding to each diameter are as shown in Fig. 16.29 and are calculated using the formula

$$\text{Area} = \frac{\pi}{4} (\text{Diameter})^2$$

The graph of area against length could be plotted and the cross-sectional areas would be the lengths of ordinates. The area under this curve would then represent volume.

Using Simpson's rule we have

$$
\begin{aligned}
\text{Volume} \ = \ & \frac{b}{3} \, [(\text{Sum of the first and last areas}) \\
& + 2(\text{Sum of the remaining odd areas}) \\
& + 4(\text{Sum of the even areas})] \\
= \ & \tfrac{1}{3}[(0.786 + 0.292) + 2(0.515 + 0.363) \\
& + 4(0.636 + 0.430 + 0.322)] \\
= \ & 2.79 \, \text{m}^3 \quad \text{the units being the result of multiplying} \\
& \qquad \text{m}^2 \text{ (area) by m (length)}
\end{aligned}
$$

An alternative layout using a table to show the calculations is often used:

Area number	Area	Simpson's multiplier	Product
1	0.786	1	0.786
2	0.636	4	2.544
3	0.515	2	1.030
4	0.430	4	1.720
5	0.363	2	1.726
6	0.322	4	1.288
7	0.292	1	0.292
		$\therefore$ Total product = 8.386	

Hence

$$\text{Volume} = \tfrac{1}{3}(8.386) = 2.795 \text{ m}^3$$

Using the trapezoidal rule we have

$$\text{Volume} = b[\tfrac{1}{2}(\text{Sum of the first and last areas}) + (\text{Sum of the remaining areas})]$$

$$= 1[\tfrac{1}{2}(0.786 + 0.292) + 0.636 + 0.515 + 0.430 + 0.363 + 0.322)]$$

$$= 2.805 \text{ m}^3$$

These results are reasonably close and we could safely assume that the volume of timber is 2.8 cubic metres.

Prismoidal Rule for Calculating Volumes

The prismoidal formula is a general formula by which the volume of any prism, pyramid or frustum of a pyramid may be found. It is

$$V = \frac{h}{6}(A_1 + 4A_m + A_2)$$

where

A_1 = Area of one end of object, A_2 = Area of other end

A_m = Area of the section mid-way between the two end surfaces

h = Distance between areas A_1 and A_2

It will be noticed that the prismoidal rule is really Simpson's rule for two strips.

EXAMPLE 16.26

A cutting is 180 m long. The cross-sectional areas at the ends of the cutting are 22 m² and 25 m² whilst the cross-sectional area at the middle of the cutting is 28 m². Find the volume of earth excavated.

We have $h = 180$, $A_1 = 22$, $A_m = 28$ and $A_2 = 25$

Volume of earth excavated $= \dfrac{180}{6}(22 + 4 \times 28 + 25) = 4770\,\text{m}^3$

Exercise 16.3

It is suggested that the following questions are solved using at least two of the methods covered in the preceding text, i.e., using the trapezoidal rule, using the mid-ordinate rule, or by using Simpson's rule.

1) The table below gives corresponding values of x and y. Plot the graph and by using the mid-ordinate rule find the area under the graph.

x	1.5	1.7	1.9	2.1	2.3	2.5	2.7	2.9	3.1
y	800	730	622	528	438	366	306	262	214

2) The table below gives corresponding values of two quantities A and x. Draw the graph and hence find the area under it. (Plot x horizontally.)

A	53.2	35	22.2	21.8	24.2	23.6	18.7	0
x	0	1	2	3	4	5	6	7

3) Plot the curve given by the following values of x and y and hence find the area included by the curve and the axes of x and y.

x	1	2	3	4	5
y	1	0.25	0.11	0.063	0.040

4) Plot the curve of $y = 2x^2 + 7$ between $x = 2$ and $x = 5$ and find the area under this curve.

5) Plot the graph of $y = 2x^3 - 5$ between $x = 0$ and $x = 3$ and find the area under the curve.

6) A series of soundings taken across a section of a river channel are given in Fig. 16.30. Find an approximate value for the cross-sectional area of the river at this section.

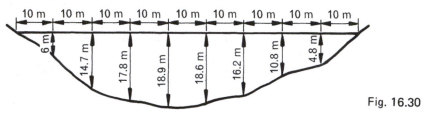

Fig. 16.30

7) The cross-sectional areas of a tree trunk are given in the table below. Find its volume.

Distance from one end (m)	0	1	2	3	4	5	6
Area (m²)	5.1	4.1	3.4	2.7	2.2	1.8	1.3

8) The width of a river at a certain section is 60 m. Soundings of the depth of the river taken at this section are recorded as follows:

Distance from left bank (m)	0	5	10	15	20	25	30	35	40	45	50	55	60
Sounding depth (m)	4	8	9	19	30	35	30	24	20	16	10	8	0

Plot the above information and drawing a fair curve through the points, calculate the cross-sectional area of the river at this section in m². If the volume of water flowing past this point per second is $10\,000\,\text{m}^3$, calculate the speed, in m/s, at which the river is flowing.

Hint: Volume of flow per second = (Cross-sectional area)
$\times$ (Velocity of flow)

9) Observation by surveyors show that the cross-sectional areas at 100 m intervals of a cutting are as shown in Fig. 16.31. Find the volume of soil required to fill the cutting.

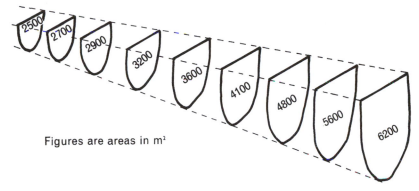

Figures are areas in m²

Fig. 16.31

10) A drain 15 m long is to be laid on level ground. The bottom of the trench is to be 0.5 m wide over the entire length. The depths at the ends of the trench are 0.7 m and 0.9 m whilst half way along the trench the depth is 0.85 m. The width of the top of the trench at the shallow end is 0.62 m, at the middle section it is 0.66 m and at the deep end it is 0.59 m. Calculate the amount of earth excavated.

11) A cutting for a road is 150 m long. The cross-sectional areas at the two ends and at the middle are $24\,m^2$, $31\,m^2$ and $29\,m^2$. Calculate the amount of earth excavated.

12) Part of a ventilation system consists of an equal tapered elliptical duct. It is 300 mm by 240 mm at one end and 220 mm by 160 mm at the other end. If it is 600 mm long, use the prismoidal rule to calculate the volume of the duct.

17. VOLUMES AND SURFACE AREAS

Outcome:

1. *Use the theorem of Pappus to calculate the volume of a solid.*

2. *Use the theorem of Pappus to calculate the surface area of a solid.*

CENTROIDS OF AREAS

The centroid of an area is at the point which corresponds to the centre of gravity of a lamina of the same shape as the area. A thin flat sheet of metal of uniform thickness is an example of a lamina. For calculation purposes the centroid of an area is the point at which the total area may be considered to be situated.

The positions of centroids of the areas met most frequently in mensuration problems are given in Fig. 17.1.

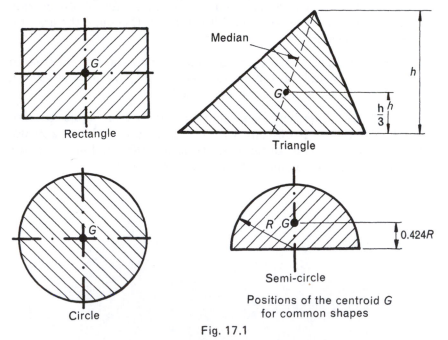

Rectangle

Triangle

Median

Circle

Semi-circle

Positions of the centroid G for common shapes

Fig. 17.1

THEOREMS OF PAPPUS (or GULDINUS)

Theorem 1

If a plane area rotates about a line in its plane (which does not cut the area) then the volume generated is given by the equation

Volume = Area × Length of path of its centroid

EXAMPLE 17.1

Find the volume of the circular ring of concrete shown in Fig. 17.2.

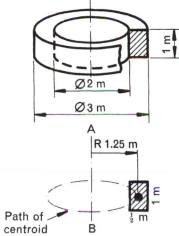

Fig. 17.2

Consider that the volume of the ring is made by one revolution of the cross-sectional area shown about the axis AB. Pappus' theorem 1 states

Volume = Area × Length of path of centroid of area

Hence, for one revolution of the area about AB, we have

$$\text{Volume} = (1 \times \tfrac{1}{2}) \times 2\pi\,1.25 = 1.25\theta = 3.93\,\text{m}^3$$

This can be checked by considering the ring as shown in Fig. 17.3.

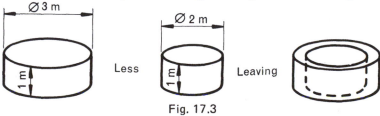

Fig. 17.3

Vol. of ring = (Vol. of 3 m diam. disc) − (Vol. of 2 m diam. disc)

$$= \left(\frac{\pi 3^2}{4}\right) \times 1 - \left(\frac{\pi 2^2}{4}\right) \times 1 = 3.93\,\text{m}^3$$

COMPOSITE AREAS

A composite area is an area which comprises two or more simple shapes as for example in Fig. 17.4 where the cross-sectional area comprises a rectangle and a triangle.

When using the theorem of Pappus with a composite area rotate each simple shape individually and then add together the separate volumes generated.

EXAMPLE 17.2

Calculate the volume of concrete required to make the retaining wall shown in Fig. 17.4.

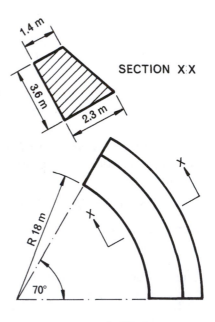

Fig. 17.4

Fig. 17.5 shows the given sectional area as comprising a rectangle and a triangle and the distances of their centroids from the axis of rotation.

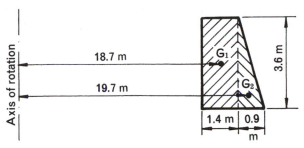

Fig. 17.5

Now

$$\text{Volume of wall} = \begin{bmatrix}\text{Volume generated} \\ \text{by the revolution,} \\ \text{through } 70°, \text{ of the} \\ \text{rectangular area}\end{bmatrix} + \begin{bmatrix}\text{Volume generated} \\ \text{by the revolution,} \\ \text{through } 70°, \text{ of the} \\ \text{triangular area}\end{bmatrix}$$

But theorem 1 of Pappus gives:

$$\text{Volume} = (\text{Area}) \times (\text{Length of path of centroid})$$

and if this is used to find each of the generated volumes then we have

$$\text{Volume of wall} = \left[\left(\begin{matrix} \text{Rectangular} \\ \text{area} \end{matrix} \right) \times \left(\begin{matrix} \text{Length of path} \\ \text{of centroid} \end{matrix} \right) \right]$$

$$+ \left[\left(\begin{matrix} \text{Triangular} \\ \text{area} \end{matrix} \right) \times \left(\begin{matrix} \text{Length of path} \\ \text{of centroid} \end{matrix} \right) \right]$$

$$= \left[\left(3.6 \times 1.4 \right) \times \left(\frac{70}{360} \times 2 \times \pi \times 18.7 \right) \right]$$

$$+ \left[\left(\frac{1}{2} \times 0.9 \times 3.6 \right) \times \left(\frac{70}{360} \times 2 \times \pi \times 19.7 \right) \right]$$

$$= 115 + 39 = 154 \text{ m}^3$$

EXAMPLE 17.3

A uniform solid circular cylinder of diameter 60 mm and height 15 mm is to be made into a pulley wheel by cutting a groove round the curved surface. The cross-section of the groove is to be a semi-circle of diameter 10 mm. Find the volume of the pulley wheel.

The distance of the centroid of the semi-circular cross-sectional area of the groove from its diameter is

$0.424 \times \dfrac{10}{2} = 2.12\,\text{mm}$, so the dis-

tance of the centroid from the axis of the cylinder

$$= \dfrac{60}{2} - 2.12 = 27.88\,\text{mm}$$

(see Fig. 17.6).

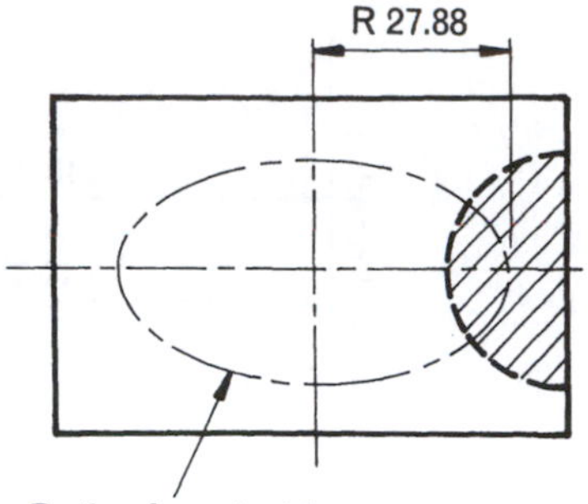

R 27.88

Path of centroid

Fig. 17.6

By Pappus' theorem 1:

Volume of cut-away portion = Area of groove
X Length of path of centroid

and so for one revolution of the area about the cylinder axis,

Volume of cut-away portion = $\frac{1}{2}\pi 5^2 \times 2\pi(27.88) = 6880\,\text{mm}^3$

but

Volume of pulley = Volume of cylinder − Volume of cut-away

$$= \dfrac{\pi}{4}(60)^2(15) - 6880 = 35500\,\text{mm}^3$$

Theorem 2

If an arc rotates about a line in its plane (which does not cut the arc) then the surface area generated is given by the equation

Area = Length of arc X Length of path of the arc centroid

EXAMPLE 17.4

Find the surface area of the circular ring of concrete shown in Fig. 17.7.

Consider that the surface area of the ring is made by one revolution of a wire, bent to the shape of the perimeter of the rectangular cross-section around the polar axis of the ring (Fig. 17.7).

Pappus' theorem 2 states that

Area swept = Length of arc X Path of centroid

Therefore

Ring surface area = $(1 + 1 + \frac{1}{2} + \frac{1}{2}) \times (2 \times 1.25 \times \pi) = 23.5\,\text{m}^2$

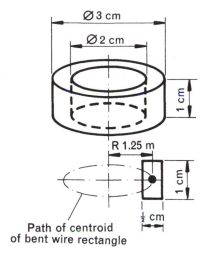

Ø 3 cm

Ø 2 cm

1 cm

R 1.25 m

1 cm

½ cm

Path of centroid
of bent wire rectangle

Fig. 17.7

Exercise 17.1

1) A reinforced concrete ring 4 m internal diameter whose cross-section is a square of 0.4 m side is to be cast. Calculate the volume of concrete required. If the ring is to be painted, calculate the area to be covered.

2) The ramp (Fig. 17.8) is to surround a monument. Calculate the volume of material needed to make the ramp.

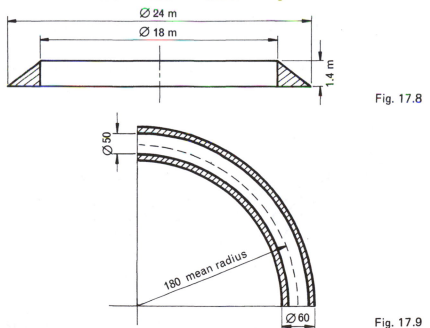

Ø 24 m

Ø 18 m

1.4 m

Fig. 17.8

Ø 50

180 mean radius

Ø 60

Fig. 17.9

3) The right-angled bend shown in Fig. 17.9 is to be formed on a copper pipe. Find the volume of the bend.

4) Part of the ventilation system consists of the ducting shown in Fig. 17.10. Calculate the total outer curved surface area of the ducting which is required for estimating the cost of decorating it.

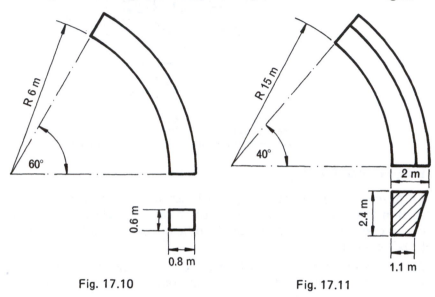

Fig. 17.10   Fig. 17.11

5) Fig. 17.11 shows a retaining wall. Calculate the volume of concrete required to make it.

6) A concrete channel (Fig. 17.12) is to be made. What volume of concrete is required?

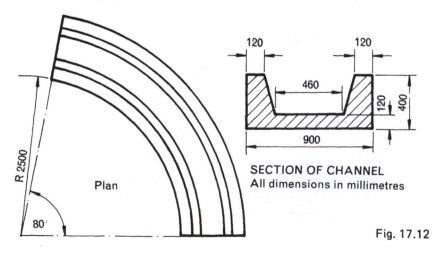

Plan

SECTION OF CHANNEL
All dimensions in millimetres

Fig. 17.12

7) A metal casting (Fig. 17.13) is to be made. Find the volume of metal required.

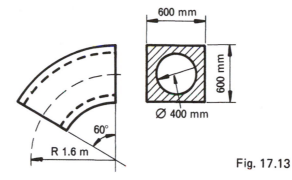

Fig. 17.13

8) A triangular ring (Fig. 17.14) is required as part of a design.
(a) What is the total surface area of the ring?
(b) Calculate the volume of the ring.

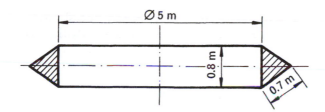

Fig. 17.14

9) Find the volume of the turned component shown in Fig. 17.15:
(a) using a theorem of Pappus,
(b) by considering the shape as the frustum of a cone.

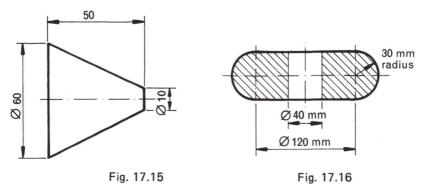

Fig. 17.15   Fig. 17.16

10) Find the volume of the ring shown in Fig. 17.16.

18.

GRAPHS OF LINEAR EQUATIONS

THE LAW OF A STRAIGHT LINE

In Fig. 18.1, the point B is any point on the line shown and has co-ordinates x and y. Point A is where the line cuts the y-axis and has co-ordinates $x = 0$ and $y = c$.

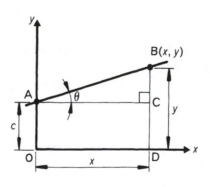

Fig. 18.1

In $\triangle ABC$ $\qquad \dfrac{BC}{AC} = \tan \theta$

$\therefore \qquad BC = (\tan \theta).AC$

264

but also $\qquad\qquad y \;=\; BC + CD$

$\qquad\qquad\qquad\quad = (\tan\theta).AC + CD$

$\therefore \qquad\qquad \boxed{y \;=\; mx + c}$

This is called the *standard equation*, or *law*, of a straight line

where $\qquad\qquad m \;=\; \tan\theta$

and $\qquad\qquad c \;=\; \text{Distance CD} \;=\; \text{Distance OA}$

m is called the *gradient of the line*.

c is called the *intercept on the y-axis*. Care must be taken as this only applies if the origin (i.e. the point $(0,0)$) is at the intersection of the axes.

In mathematics the gradient of a line is defined as the tangent of the angle that the line makes with the horizontal, and is denoted by the letter m.

Hence in Fig. 4.1 the gradient $= m = \tan\theta = \dfrac{BC}{AC}$ *

Fig. 18.2 shows the difference between positive and negative gradients.

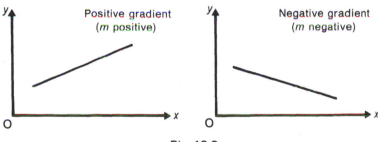

Fig. 18.2

Summarising:

The standard equation, or law, of a straight line is $y = mx + c$

where m is the gradient

and c is the intercept on the y-axis

*Care should be taken not to confuse this with the gradient given on maps, railways, etc. which is the sine of the angle (not the tangent) — e.g. a railway slope of 1 in 100 is one unit vertically for every 100 units measured along the slope.

LINEAR EQUATIONS

Equations in which the highest power of the variables is the first, are called equations of the *first degree*. Thus, $y = 3x + 5$, $y = 2 - 7x$ and $y = 0.3x - 0.5$ are all equations of the first degree. All equations of this type give graphs which are straight lines and hence they are often called *linear equations*.

EXAMPLES OF LINEAR EQUATIONS AND THEIR GRAPHS

You should remember that in the standard equation of a straight line $y = mx + c$ the value of the constant c represents the intercept on the y-axis *only* when the origin (i.e. the point $(0, 0)$) is at the intersection of the axes.

Two Special Cases

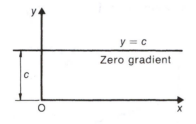

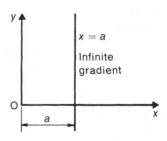

Linear equations which are not stated in standard straight line form must be rearranged if they are to be compared with the standard equation $y = mx + c$.

EXAMPLE 18.1

a) The equation $y = x + 1$

    may be written as $y = 1x + 1$

    Now comparing with $y = mx + c$

    the gradient $m = 1$

    and the intercept $c = 1$

b) The equation $y = 2x - 1.5$

may be compared with $y = mx + c$

the gradient $m = 2$

and the intercept $c = -1.5$

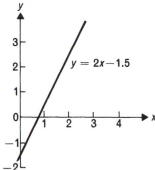

c) The equation $2y + x = 4$

must be rearranged: $y = -\dfrac{x}{2} + 2$

Now comparing with $y = mx + c$

the gradient $m = -\dfrac{1}{2}$

and the intercept $c = 2$

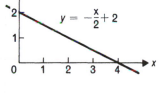

d) The equation $3y = x$

must be rearranged: $y = \dfrac{x}{3} + 0$

Now comparing with $y = mx + c$

the gradient $m = \dfrac{1}{3}$

and the intercept $c = 0$

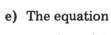

e) The equation $y = -2$

may be written as $y = 0x - 2$

Now comparing with $y = mx + c$

the gradient $m = 0$

and the intercept $c = -2$

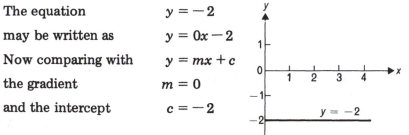

OBTAINING THE STRAIGHT LINE LAW OF A GRAPH

Two methods are used:

1. Origin at the Intersection of the Axes

When it is convenient to arrange the origin, i.e. the point $(0, 0)$, at the intersection of the axes the values of gradient m and intercept c may be found directly from the graph as shown in Example 18.2.

EXAMPLE 18.2

Find the law of the straight line shown in Fig. 18.3.

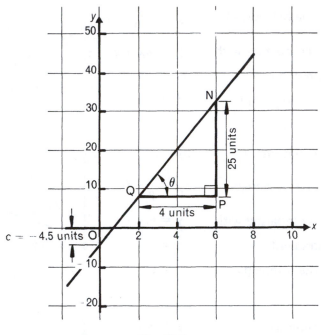

Fig. 18.3

To find gradient m. Take any two points Q and N on the line and construct the right-angled triangle QPN. This triangle should be of reasonable size, since a small triangle will probably give an inaccurate result. Note that if we can measure to an accuracy of 1 mm using an ordinary rule, then this error in a length of 20 mm is much more significant than the same error in a length of 50 mm.

The lengths of NP and QP are then found using the scales of the x and y axes. Direct lengths of these lines as would be obtained using an ordinary rule, e.g. both in centimetres, must *not* be used — the scales of the axes must be taken into account.

$$\therefore \qquad \text{Gradient } m = \tan \theta = \frac{\text{NP}}{\text{QP}} = \frac{25}{4} = 6.25$$

To find intercept c.   This is measured again using the scale of the y-axis.

$$\therefore \text{ intercept} \qquad\qquad c = -4.5$$

The law of the straight line

The standard equation is

$$y = mx + c$$

$\therefore$ the required equation is

$$y = 6.25x + (-4.5)$$

i.e.
$$y = 6.25x - 4.5$$

2.   Origin not at the Intersection of the Axes

This method is applicable for all problems — it may be used, therefore, when the origin is at the intersection of the axes.

If a point lies on line then the co-ordinates of that point satisfy the equation of the line, e.g. the point $(2, 7)$ lies on the line $y = 2x + 3$ because if $x = 2$ is substituted in the equation, $y = 2 \times 2 + 3 = 7$ which is the correct value of y. Two points, which lie on the given straight line, are chosen and their co-ordinates are substituted in the standard equation $y = mx + c$. The two equations which result are then solved simultaneously to find the values of m and c.

EXAMPLE 18.3

Determine the law of the straight line shown in Fig. 18.4.

Choose two convenient points P and Q and find their co-ordinates. Again these points should not be close together, but as far apart as conveniently possible. Their co-ordinates are as shown in Fig. 18.4.

Let the equation of the line be $y = mx + c$.

Now P(22, 19.8) lies on the line   ∴ $19.8 = m(22) + c$

and Q(28, 16.4) lies on the line   ∴ $16.4 = m(28) + c$

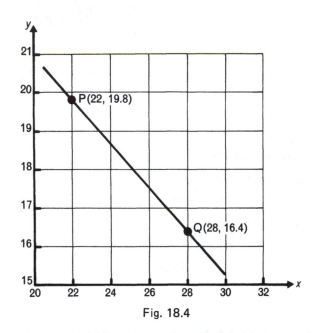

Fig. 18.4

To solve these two equations simultaneously we must first elimi-nate one of the unknowns. In this case c will disappear if the second equation is subtracted from the first, giving

$$19.8 - 16.4 = m(22 - 28)$$

∴ $$m = \frac{3.4}{-6} = -0.567$$

To find c the value of $m = -0.567$ may be substituted into either of the original equations. Choosing the first equation we get

$$19.8 = -0.567(22) + c$$

∴ $$c = 19.8 + 0.567(22) = 32.3$$

Hence the required law of the straight line is

$$y = -0.567x + 32.3$$

GRAPHS OF EXPERIMENTAL DATA

Best Fit Straight Line

Readings which are obtained as a result of an experiment will usually contain errors owing to inaccurate measurement and other experimental errors. If the points, when plotted, show a trend towards a straight line or a smooth curve, this is usually accepted and the best fit straight line or curve drawn. In this case the line will not pass through some of the points and an attempt must be made to ensure an even spread of these points above and below the line or the curve.

Alternatively the method of least squares may be used — shown on p. 348.

One of the most important applications of the straight line law is the determination of a law connecting two quantities when values have been obtained from an experiment, as Example 18.4 illustrates.

EXAMPLE 18.4

During a test to find how the power of a lathe varied with the depth of cut results were obtained as shown in the table. The speed and feed of the lathe were kept constant during the test.

Depth of cut, d (mm)	0.51 1.02 1.52 2.03 2.54 3.0
Power, P (W)	0.89 1.04 1.14 1.32 1.43 1.55

Show that the law connecting d and P is of the form $P = ad + b$ and find the law. Hence find the value of d when P is 1.2 watts.

The standard equation of a straight line is $y = mx + c$. It often happens that the variables are *not* x and y. In this example d is used instead of x and is plotted on the horizontal axis, and P is used instead of y and is plotted on the vertical axis.

Similarly the gradient $= a$ instead of m, and the intercept on the y-axis $= b$ instead of c.

On plotting the points (Fig. 18.5) it will be noticed that they deviate slightly from a straight line. Since the data are experimental we must expect errors in observation and measurement and hence a slight deviation from a straight line must be expected.

The points, therefore, approximately follow a straight line and we can say that the equation connecting P and d is of the form $P = ad + b$.

Because the origin is *not* at the intersection of the axes, to find the values of constants a and b we must choose two points *which lie on the line*. These two points must be as far apart as possible in order to obtain maximum accuracy.

In Fig. 18.5 the points P(0.90, 1.00) and Q(2.76, 1.50) have been chosen.

The point P(0.90, 1.00) lies on the line $\therefore$ $1.00 = a(0.90) + b$

and point Q(2.76, 1.50) lies on the line $\therefore$ $1.50 = a(2.76) + b$

Now subtracting the first equation from the second we get

$$1.50 - 1.00 = a(2.76 - 0.90)$$

$$\therefore \qquad a = \frac{0.50}{1.86} = 0.27$$

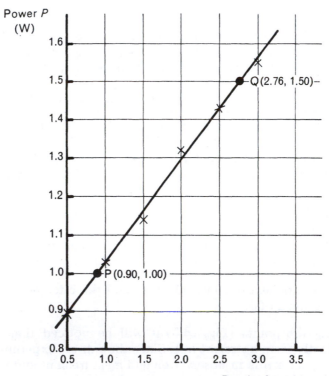

Power P
(W)

Depth of cut (d mm)   Fig. 18.5

Now substituting the value $a = 0.27$ into the first equation we get

$$1.00 = 0.27(0.90) + b$$

$\therefore$ $$b = 1.00 - 0.27(0.90) = 0.76$$

Hence the required law of the line is $P = 0.27d + 0.76$

To find d when $P = 1.2\,\text{W}$. The value $P = 1.2$ is substituted into the equation giving

$$1.2 = 0.27d + 0.76$$

$$d = \frac{1.2 - 0.76}{0.27} = 1.63\,\text{mm}$$

(since all values of d are mm when values of P are watts).

This value of d may be verified by checking the corresponding value of d corresponding to $P = 1.2$ on the straight line in Fig.18.5.

Any inaccuracies may be due to rounding off calculations to two significant figures, e.g. the value of m is $\dfrac{0.5}{1.86} = 0.269$ if three significant figures are considered. Bearing in mind, however, the experimental errors etc. the rounding off as shown seems reasonable. This question of accuracy is always open to debate, the most dangerous error being to give the calculated results to a far greater accuracy than the original given data.

EXAMPLE 18.5

Hooke's law states that for an elastic material, up to the limit of proportionality, the stress, σ, is directly proportional to the strain, ϵ, it produces. In equation form this is $\sigma = E\epsilon$ where the constant E is called the modulus of elasticity of the material.

Find the value of E using the following results obtained in an experiment

σ MN/m^2	125	110	95	80
ϵ no units	0.000 600	0.000 522	0.000 466	0.000 382

σ MN/m^2	63	54	38
ϵ no units	0.000 367	0.000 269	0.000 168

In order to compare the equation $\sigma = E\epsilon$ with the standard straight line equation $y = mx + c$ we must plot σ on the vertical axis, and ϵ on the horizontal axis.

Inspection of the values shows that it is convenient to arrange for the origin, i.e. the point $(0, 0)$ to be at the intersection of the axes. The graph is shown in Fig. 18.6.

Since the values are the result of an experiment it is unlikely that the points plotted will lie exactly on a straight line. Having arranged for the origin to be at the intersection of the axes we can see from the given equation that the intercept on the y-axis is zero. This means that the graph passes through the origin and this helps us in drawing the 'best' straight line to 'fit' the points. Judgement is needed here—for instance the point $(0.000\,367, 63)$ is obviously an incorrect result as it lies well away from the line of the other points, and so we should ignore this point.

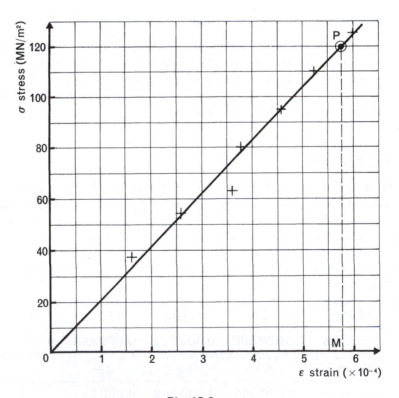

Fig. 18.6

We see also that the gradient $m = E$ on our graph. Thus we may find the value of E by calculating the gradient of the straight line. From the suitable right-angled triangle POM

$$\text{the gradient is given by } \frac{PM}{OM} = \frac{120}{0.000\,58} = 207\,000$$

The units of the ratio $\dfrac{PM}{OM}$ will be those of PM, i.e. MN/m^2, since OM represents strain which has no units (this is because strain is the ratio of two lengths).

Hence the value of the modulus of elasticity of the material $E = 207\,000 \text{ MN/m}^2 = 207 \text{ GN/m}^2$.

Exercise 18.1

1) Draw the straight line which passes through the points $(4, 7)$ and $(-2, 1)$. Hence find the gradient of the line and its intercept on the y-axis.

2) The following equations represent straight lines. Sketch them and find in each case the gradient of the line and the intercept on the y-axis.

(a) $y = x + 3$ (b) $y = -3x + 4$

(c) $y = -3.1x - 1.7$ (d) $y = 4.3x - 2.5$.

3) A straight line passes through the points $(-2, -3)$ and $(3, 7)$. *Without* drawing the line find the values of m and c in the equation $y = mx + c$.

4) The width of keyways for various shaft diameters are given in the table below.

Diameter of shaft D (mm)	10	20	30	40	50	60	70	80
Width of keyway W (mm)	3.75	6.25	8.75	11.25	13.75	16.25	18.75	21.25

Show that D and W are connected by a law of the type $W = aD + b$ and find the values of a and b.

5) During an experiment to find the coefficient of friction between two metallic surfaces the following results were obtained.

Load W (N)	10	20	30	40	50	60	70
Friction force F (N)	1.5	4.3	7.6	10.4	13.5	15.6	18.8

Show that F and W are connected by a law of the type $F = aW + b$ and find the values of a and b.

6) In a test on a certain lifting machine it is found that an effort of 50 N will lift a load of 324 N and that an effort of 70 N will lift a load of 415 N. Assuming that the graph of effort plotted against load is a straight line find the probable load that will be lifted by an effort of 95 N.

7) The following results were obtained from an experiment on a set of pulleys. W is the load raised and E is the effort applied. Plot these results and obtain the law connecting E and W.

W (N)	15	20	25	30	35	40	45
E (N)	2.3	2.7	3.2	3.8	4.3	4.7	5.3

8) During a test with a thermocouple pyrometer the e.m.f. (E millivolts) was measured against the temperature at the hot junction ($t°C$) and the following results were obtained:

t	200	300	400	500	600	700	800	900	1000
E	6	9.1	12.0	14.8	18.2	21.0	24.1	26.8	30.2

The law connecting t and E is supposed to be $E = at + b$. Test if this is so and find suitable values for a and b.

9) The resistance (R ohms) of a field winding is measured at various temperatures ($t°C$) and the results are recorded in the table below:

t (°C)	21	26	33	38	47	54	59	66	75
R (ohms)	109	111	114	116	120	123	125	128	132

If the law connecting R and t is of the form $R = a + bt$ find suitable values of a and b.

10) The rate of a spring, λ, is defined as force per unit extension. Hence for a load, F N producing an extension x mm the law is $F = \lambda x$. Find the value of λ in units of N/m using the following values obtained from an experiment:

F (N)	20	40	60	80	100	120	140
x (mm)	37	79	111	156	197	229	270

11) A circuit contains a resistor having a fixed resistance of R ohms. The current, I amperes and the potential difference, V volts are related by the expression $V = IR$. Find the value of R given the following results obtained from an experiment:

V (volts)	3	7	11	13	17	20	24	29
I (amperes)	0.066	0.125	0.209	0.270	0.324	0.418	0.495	0.571

19. GRAPHS OF NON-LINEAR EQUATIONS

Outcome:

1. *Draw up suitable tables of values and plot the curves of the types:*

   $$y = ax^2 + bx + c, \quad y = \frac{a}{x}, \quad y = x^{1/2},$$

   $$x^2 + y^2 = a^2, \quad \frac{x^2}{a^2} + \frac{y^2}{b^2} = 1, \quad \frac{x^2}{a^2} - \frac{y^2}{b^2} = 1,$$

   $$xy = c^2.$$

2. *Recognise the effects on the above curves by changes in the constants* a, b, *and* c.

3. *Reduce non-linear physical laws, such as*

   $$y = ax^2 + b \quad or \quad y = \frac{a}{x} + b, \quad and \quad y = a.x^n$$

   (using logarithms), to a straight line graph form.

4. *Plot the corresponding straight line graph to verify the law, determine the values of the constants* a *and* b, *and find intermediate values.*

GRAPHS OF FAMILIAR EQUATIONS

As technicians it is important that we can recognise the shapes and layout of curves related to their equations. In this section we will draw graphs of the more common equations.

Graph of $y = ax^2 + bx + c$: Parabola

The important part of the curve is usually the portion in the vicinity of the vertex of the parabola (Fig. 19.1).

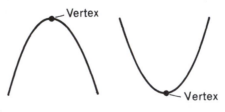

Fig. 19.1

The shape and layout will depend on the values of the constants a, b and c and we will examine the effect of each constant in turn.

278

Constant *a*

Consider the equation $y = ax^2$. The table of values given below is for $a = 4$, $a = 2$ and $a = -1$:

x	-3	-2	-1	0	1	2	3
$y = 4x^2$	36	16	4	0	4	16	36
$y = 2x^2$	18	8	2	0	2	8	18
$y = -x^2$	-9	-4	-1	0	-1	-4	-9

Fig. 19.2 shows the graphs of $y = ax^2$ when $a = 4$, $a = 2$ and $a = -1$

We can see that if the value of a is positive the curve is shaped $\smile$, and the greater the value of a, the 'steeper' the curve rises.

Negative values of a give a curve shaped $\frown$.

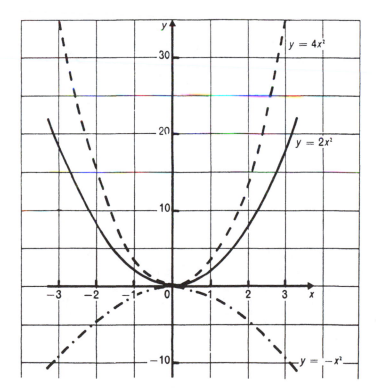

Fig. 19.2

Constant b

Consider the equation $y = x^2 + bx$. The table of values given below is for $b = 2$ and $b = -3$:

x	-3	-2	-1	0	1	2	3	4
$y = x^2 + 2x$	3	0	-1	0	3	8	15	24
$y = x^2 - 3x$	18	10	4	0	-2	-2	0	4

Fig. 19.3 shows the graphs of $y = x^2 + bx$ when $b = 2$ and $b = -3$. The effect of a positive value of b is to move the vertex to the left of the vertical y-axis, whilst a negative value of b moves the vertex to the right of the vertical axis.

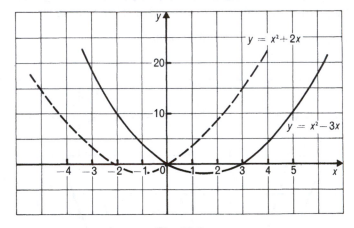

Fig. 19.3

Constant c

Consider the equation $y = x^2 - 2x + c$. The table of values given below is for $c = 10$, $c = 5$, $c = 0$ and $c = -5$

x	-3	-2	-1	0	1	2	3	4
$y = x^2 - 2x + 10$	25	18	13	10	9	10	13	18
$y = x^2 - 2x + 5$	20	13	8	5	4	5	8	13
$y = x^2 - 2x$	15	8	3	0	-1	0	3	8
$y = x^2 - 2x - 5$	10	3	-2	-5	-6	-5	-2	3

Fig. 19.4 shows the graphs of $y = x^2 - 2x + c$ when $c = 10$, $c = 5$, $c = 0$ and $c = -5$. As we can see the effect is to move the vertex up or down according to the magnitude of c.

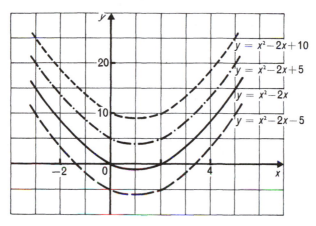

Fig. 19.4

The graph of $y = ax^{\frac{1}{2}}$　Parabola
or $y = a\sqrt{x}$:　(with horizontal axis)

The table of values given below is for $a = 4$ which is the graph of $y = 4\sqrt{x}$:

x	All negative values	0	1	2	3	4	5	6
$y = 4\sqrt{x}$	No real values	0	± 4	± 5.66	± 6.93	± 8	± 8.94	± 9.80

The graph of $y = 4\sqrt{x}$ is shown in Fig. 19.5. We can see that as the value of a is increased so the 'overall depth' of the figure increases.

If the given equation is rearranged we may obtain

$$x = \frac{y^2}{a^2}$$

or $\qquad\qquad x = (\text{A constant}) \times y^2$

which is the equation of a parabola with a horizontal axis of symmetry.

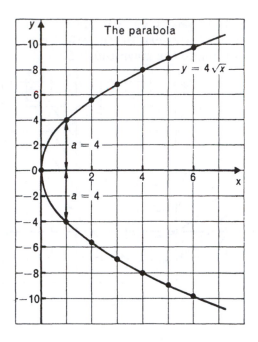

Fig. 19.5

The graph of $y = \dfrac{a}{x}$: **Reciprocal curve (rectangular hyperbola)**

The table of values given below is for $a = 4$ which is the graph of $y = \dfrac{4}{x}$:

x	-10	-8	-6	-4	-2	-1	-0.5	0
$y = \dfrac{4}{x}$	-0.4	-0.5	-0.67	-1	-2	-4	-8	
	Similar numerical results will occur for positive values							

The graph of $y = \dfrac{4}{x}$ is shown in Fig. 19.6. We can see that as the value of a decreases the curves are brought nearer to the axes, and vice-versa.

We should also note that for extreme values the curves will become nearer and nearer to the axes, but will never actually 'touch' them. At these extreme values the axes are said to be 'asymptotic' to the curves. In this case the axes are also called 'asymptotes'.

We shall see later on page 285 that the curves comprise what is called a 'rectangular hyperbola'.

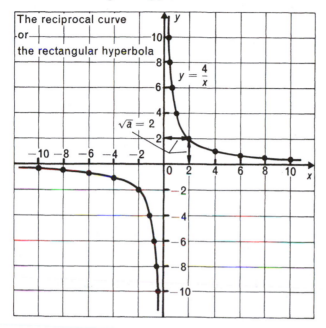

Fig. 19.6

The graph of $x^2 + y^2 = a^2$:   Circle

The graph of $x^2 + y^2 = 4$, which is when $a = 2$, is shown in Fig. 19.7.

The table of values has been omitted here but the reader may find it useful to rearrange the equation and plot the curve.

We can see that the radius of the circle is given by the value of constant a.

The circle

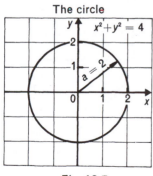

Fig. 19.7

The ellipse

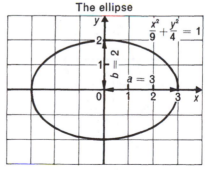

Fig. 19.8

The graph of $\dfrac{x^2}{a^2} + \dfrac{y^2}{b^2} = 1$:   Ellipse

The graph of $\dfrac{x^2}{9} + \dfrac{y^2}{4} = 1$, which is when $a = 3$ and $b = 2$, is shown in Fig. 19.8.

The line along which the greatest dimension (given by $2a$) across the ellipse is measured is called the major axis. In this case it lies along the x-axis.

The line along which the least dimension (given by $2b$) across the ellipse is measured is called the minor axis. In this case it lies along the y-axis.

The graph of $\dfrac{x^2}{a^2} + \dfrac{y^2}{b^2} = 1$:   Hyperbola

The graph of $\dfrac{x^2}{9} - \dfrac{y^2}{4} = 1$, which is when $a = 3$ and $b = 2$, is shown in Fig. 19.9.

The ratio $\pm\dfrac{b}{a}$, in this case $\pm\dfrac{2}{3}$, gives the slopes of two straight lines called asymptotes. At extreme values these lines will become nearer and nearer to the curves but will never actually 'touch' them, and are said to be *asymptotic* to the curves.

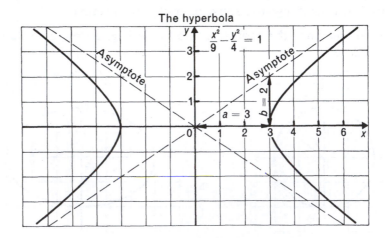

Fig. 19.9

The graph of $xy = c^2$:   Rectangular hyperbola

The graph of $xy = 9$, which is when $c = 3$, is shown in Fig. 19.10.

We can see that the curve is similar to the reciprocal curve discussed earlier. It is also similar to the ordinary hyperbola, the description 'rectangular' referring to the fact that the asymptotes are also the rectangular axes.

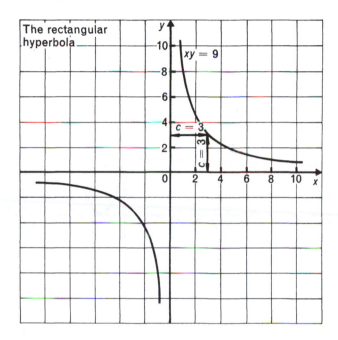

Fig. 19.10

Exercise 19.1

State which answer or answers are correct in Questions 1-9. In every diagram the origin is at the intersection of the axes.

1) The graph of $y = x^2$ is:

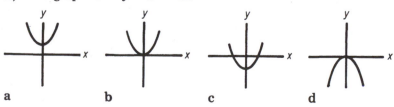

a   b   c   d

2) The graph of $y = x^2 + 2$ is:

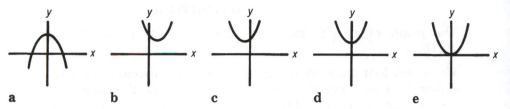

a   b   c   d   e

3) The graph of $y = -2x^2$ is:

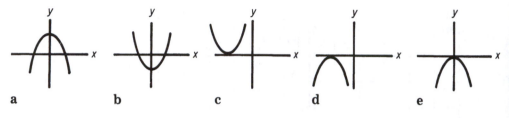

a   b   c   d   e

4) The graph of $y = 4 - x^2$ is:

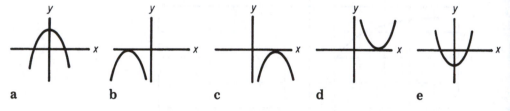

a   b   c   d   e

5) The graph of $y = x^2 - 3x + 2$ is:

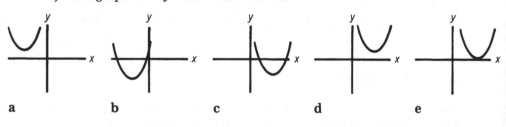

a   b   c   d   e

6) The graph of $y = x^2 + 4x + 4$ is:

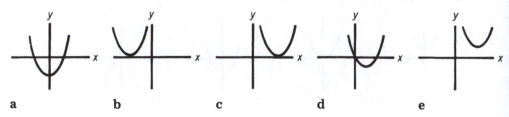

a   b   c   d   e

7) The graph of $\dfrac{x^2}{16} + \dfrac{y^2}{9} = 1$ is:

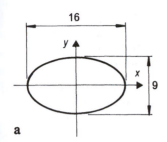

a

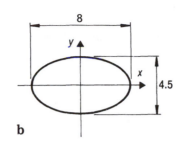

b

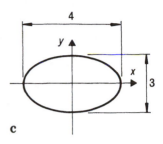

c

8) The graph of $\dfrac{x^2}{9} - \dfrac{y^2}{4} = 1$ is:

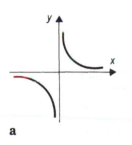

a

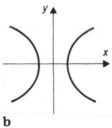

b

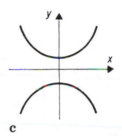

c

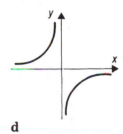

d

9) The graph of $xy = 9$ is:

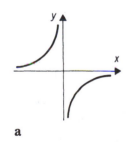

a

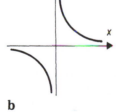

b

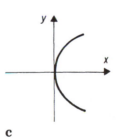

c

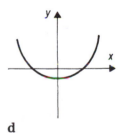

d

10) What is the equation of the circle shown in the diagram?

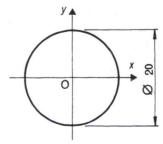

11) The profile of the cross-section of a headlamp reflector is parabolic in shape, having an equation $y = 0.00617x^2$. Plot the profile for values of x from 0 to 90 mm.

12) An elliptical template has an equation: $\dfrac{x^2}{3600} + \dfrac{y^2}{1600} = 1$ where x and y have mm units. Plot the shape of the template.

13) The pressure, p MN/m^2, and volume, V m^3, of a fixed mass of air at a constant temperature are connected by the equation: $pV = 0.16$. Plot the graph of pressure against volume, as the volume changes from 0 to 1 m^3.

NON-LINEAR LAWS WHICH CAN BE REDUCED TO THE LINEAR FORM

Many non-linear equations can be reduced to the linear form by making a suitable substitution.

Common forms of non-linear equations are (a and b constants):

$y = \dfrac{a}{x} + b$	$y = \dfrac{a}{x^2} + b$	$y = ax^2 + b$	$y = a\sqrt{x} + b$

$y = \dfrac{a}{\sqrt{x}} + b$

The form $y = \dfrac{a}{x} + b$

Let $z = \dfrac{1}{x}$ so that the equation becomes $y = az + b$. If we now plot values of y against the corresponding values of z we will get a straight line since $y = az + b$ is of the standard linear form. In effect y has been plotted against $\dfrac{1}{x}$. The following example illustrates this method.

EXAMPLE 19.1

An experiment connected with the flow of water over a rectangular weir gave the following results:

C	0.503	0.454	0.438	0.430	0.425	0.421
H	0.1	0.2	0.3	0.4	0.5	0.6

The relation between C and H is thought to be of the form $C = \dfrac{a}{H} + b$. Test if this is so and find the values of the constants a and b.

In the suggested equation C is the sum of two terms, the first of which varies as $\dfrac{1}{H}$. If the equation $C = \dfrac{a}{H} + b$ is correct then when we plot C against $\dfrac{1}{H}$ we should obtain a straight line. To do this we draw up the following table:

C	0.503	0.454	0.438	0.430	0.425	0.421
$\dfrac{1}{H}$	10.00	5.00	3.33	2.50	2.00	1.67

The graph obtained is shown in Fig. 19.11. It is a straight line and hence the given values follow a law of the form $C = \dfrac{a}{H} + b$.

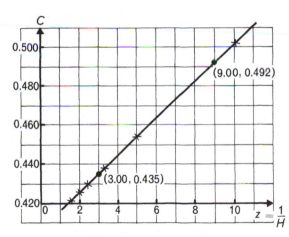

Fig. 19.11

To find the values of a and b we choose two points which lie on the straight line.

The point $(3.00, 0.435)$ lies on the line.

$$\therefore \qquad\qquad 0.435 = 3.00a + b \qquad\qquad [1]$$

The point $(9.00, 0.492)$ also lies on the line.

$$\therefore \qquad\qquad 0.492 = 9.00a + b \qquad\qquad [2]$$

Subtracting equation [1] from equation [2] gives

$$0.492 - 0.435 = a(9.00 - 3.00)$$

$$\therefore \qquad a = 0.0095$$

Substituting this value for a in equation [1] gives

$$0.435 = 3.00 \times 0.0095 + b$$

$$\therefore \qquad b = 0.435 - 0.0285 = 0.407$$

Hence the values of a and b are 0.0095 and 0.407 respectively.

The form $y = ax^2 + b$

Let $z = x^2$ and as previously if we plot values of y against z (in effect x^2) we will get a straight line since $y = az + b$ is of the standard form. The following example illustrates this method.

EXAMPLE 19.2

The fusing current I amperes for wires of various diameters d mm is as shown below:

d (mm)	5	10	15	20	25
I (amperes)	6.25	10	16.25	25	36.25

It is suggested that the law $I = ad^2 + b$ is true for the range of values given, a and b being constants. By plotting a suitable graph show that this law holds and from the graph find the constants a and b. Using the values of these constants in the equation $I = ad^2 + b$ find the diameter of the wire required for a fusing current of 12 amperes.

By putting $z = d^2$ the equation $I = ad^2 + b$ becomes $I = az + b$ which is the standard form of a straight line. Hence by plotting I against d^2 we should get a straight line if the law is true. To try this we draw up a table showing corresponding values of I and d^2.

d	5	10	15	20	25
$z = d^2$	25	100	225	400	625
I	6.25	10	16.25	25	36.25

From the graph (Fig. 19.12) we see that the points do lie on a straight line and hence the values obey a law of the form $I = ad^2 + b$.

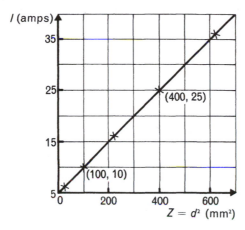

Fig. 19.12

To find the values of a and b choose two points which lie on the line and find their co-ordinates.

The point $(400, 25)$ lies on the line

$$\therefore \qquad 25 = 400a + b \qquad\qquad [1]$$

The point $(100, 10)$ lies on the line

$$\therefore \qquad 10 = 100a + b \qquad\qquad [2]$$

Subtracting equation [2] from equation [1] gives

$$15 = 300a$$

$$\therefore \qquad a = 0.05$$

Substituting $a = 0.05$ in equation [2] gives

$$10 = 100 \times 0.05 + b$$

$$\therefore \qquad b = 5$$

Therefore the law is

$$I = 0.05d^2 + 5$$

When $I = 12$

$$12 = 0.05d^2 + 5$$

$$\therefore \qquad d = \sqrt{140} = 11.8\,\text{mm}$$

Consider $y = \dfrac{a}{x^2} + b$

Let $z = \dfrac{1}{x^2}$ so that the equation becomes $y = az + b$. If we now plot values of y against corresponding values of z we will get a straight line since $y = az + b$ is of the standard linear form. In effect y has been plotted against $\dfrac{1}{x^2}$.

Consider $y = a\sqrt{x} + b$

Let $z = \sqrt{x}$ and as previously, if we plot values of y against z (in effect $\sqrt{x}$) we will obtain a straight line since $y = az + b$ is of the standard linear form.

The form $y = \dfrac{a}{\sqrt{x}} + b$

This may also be written equivalently as $y = \dfrac{a}{x^{1/2}} + b$ or $y = ax^{-1/2} + b$.

Let $z = \dfrac{1}{\sqrt{x}}$ and as previously, if we plot values of y against z $\left(\text{in effect } \dfrac{1}{\sqrt{x}}\right)$ we will obtain a straight line since $y = az + b$ is the standard linear form.

Exercise 19.2

1) The following readings were taken during a test:

R (ohms)	85	73.3	64	58.8	55.8
I (amperes)	2	3	5	8	12

R and I are thought to be connected by an equation of the form $R = \dfrac{a}{I} + b$.

Verify that this is so by plotting R (y-axis) against $\dfrac{1}{I}$ (x-axis) and hence find values for a and b.

2) In the theory of the moisture content of thermal insulation efficiency of porous materials the following table gives values of μ, the diffusion constant of the material, and k_m, the thermal conductivity of damp insulation material:

μ	1.3	2.7	3.8	5.4	7.2	10.0
k_m	0.0336	0.0245	0.0221	0.0203	0.0192	0.0183

Find the equation connecting μ and k_m if it is of the form $k_m = a + \dfrac{b}{\mu}$ where a and b are constants.

3) The accompanying table gives the corresponding values of the pressure, p, of mercury and the volume, v, of a given mass of gas at constant temperature.

p	90	100	130	150	170	190
v	16.66	13.64	11.54	9.95	8.82	7.89

By plotting p against the reciprocal of v obtain some relation between p and v. Evaluate any constants used in your method.

4) The approximate number of a type of bacteria, B, is checked regularly and recorded in the table below.

Bacteria, $B\,(\times 10^3)$	5	28.5	41.0	113.0	253.6	450.0
Time, t (hours)	1.0	2.5	3.0	5.0	7.5	10.0

It is thought that the growth is related according to the law $B = mt^2 + c$ where m and c are constants. By plotting a suitable graph verify this to be true and evaluate m and c.

5) In an experiment, the resistance R of copper wire of various diameters d mm was measured and the following readings were obtained.

d (mm)	0.1	0.2	0.3	0.4	0.5
R (ohms)	20	5	2.2	1.3	0.8

Show that $R = \dfrac{k}{d^2}$ and find a suitable value for k.

6) The following table gives the thickness T mm of a brass flange brazed to a copper pipe of internal diameter D mm:

T mm	15.5	17.8	19.5	20.9	22.2	23.3
D mm	50	100	150	200	250	300

Show that T and D are connected by an equation of the form $T = a\sqrt{D} + b$, find the values of constants a and b, and find the thickness of the flange for a 70 mm diameter pipe.

7) The table shows how the coefficient of friction, μ, between a belt and a pulley varies with the speed, v m/s, of the belt. By plotting a graph show that $\mu = m\sqrt{v} + c$ and find the values of constants m and c.

μ	0.26	0.29	0.32	0.35	0.38
v	2.22	5.00	8.89	13.89	20.00

8) Using the table below show that the values are in agreement with the law $y = \dfrac{m}{\sqrt{x}} + c$. Hence evaluate the constants m and c.

x	0.2	0.8	1.2	1.8	2.5	4.4
y	1.62	1.51	1.49	1.47	1.46	1.44

9) On a test bed a projectile, starting from rest over a constant measured distance, is subjected to different accelerations. The table below records these accelerations, A, and the times, t, taken to complete the distance.

A (ms^{-2})	5	10	15	20	25	30
t (s)	5.00	3.37	2.65	2.20	1.90	1.70

If the law connecting A and t is thought to be of the form $t = aA^{-1/2} + b$ plot a suitable straight line graph, and find the constants a and b.

REDUCING TO THE LOGARITHMIC FORM

Consider the following relationship, in which z and t are the
variables whilst a and n are
constants.

$$z = a.t^n$$

Taking logs we get:

$$\log z = \log(t^n .a)$$
$$= \log t^n + \log a$$
$$\therefore \log z = n.(\log t) + \log a$$

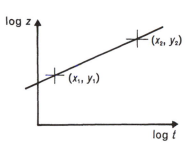

Fig. 19.13

Comparing this with the equation $y = mx + c$ of a straight line,
we see that if we plot $\log z$ on the y-axis and $\log t$ on the x-axis
(Fig. 19.13) then the result will be a straight line.

Two convenient points, (x_1, y_1) and (x_2, y_2), which lie on the line
should be chosen. The co-ordinates are now substituted in the
straight line equation:

The point (x_1, y_1) lies on the line — thus $y_1 = nx_1 + \log a$

and point (x_2, y_2) lies on the line — thus $y_2 = nx_2 + \log a$

These two equations may be solved simultaneously to find the
values of n and a.

EXAMPLE 19.3

The law connecting two quantities z and t is of the form $z = a.t^n$
Find the values of the constants a and n given the following pairs
of values:

z	3.170	4.603	7.499	10.50	15.17
t	7.980	9.863	13.03	15.81	19.50

By taking logs and rearranging (see text) we have

$$\log z = n.\log t + \log a \qquad [1]$$

For the numerical part of the solution we may use common
logarithms (logs to the base 10) or natural logarithms (logs to the
base e).

The solution given uses common logarithms. The reader may find it instructive to work through this example using natural logarithms and verify that the same results are obtained.

From the given values, using logarithms to the base 10:

$\log_{10} z$	0.5011	0.6631	0.8750	1.0212	1.1810
$\log_{10} t$	0.9020	0.9940	1.1149	1.1990	1.2900

Since it is not convenient to show the origin (point $0, 0$) we shall use the two-point method of finding the constants (Fig. 19.14).

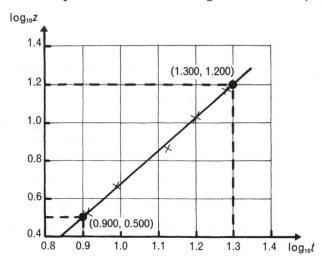

Fig. 19.14

Point $(0.900, 0.500)$ lies on the line, and substituting in equation [1] gives

$$0.500 = n(0.900) + \log_{10} a \qquad [2]$$

Point $(1.300, 1.200)$ lies on the line, and substituting in equation [1] gives

$$1.200 = n(1.300) + \log_{10} a \qquad [3]$$

Subtracting equation [2] from equation [3],

$$0.7 = 0.4n \qquad \therefore \ n = 1.75$$

Substituting in the equation [2] gives

$$0.500 = 1.75(0.900) + \log_{10} a$$

$$\therefore \qquad \log_{10} a = -1.075$$

$$\therefore \qquad a = 0.084$$

Exercise 19.3

1) The following values of x and y follow a law of the type $y = ax^n$. By plotting $\log y$ (vertically) against $\log x$ (horizontally) find values for a and n.

x	1	2	3	4	5
y	3	12	27	48	75

2) The following results were obtained in an experiment to find the relationships between the luminosity I of a metal filament lamp and the voltage V.

V	40	60	80	100	120
I	5.1	26.0	82	200	414

The law is thought to be of the type $I = aV^n$ Test this by plotting $\log I$ (vertically) against $\log V$ (horizontally) and find suitable values for a and n.

3) The relationship between power P (watts), the e.m.f. E (volts) and the resistance R (ohms) is thought to be of the form $P = \dfrac{E^n}{R}$. In an experiment in which R was kept constant the following results were obtained:

E	5	10	15	20	25	30
P	2.5	10	22.5	40	62.5	90

Verify the law and find the values of the constants n and R.

4) The following results were obtained in an experiment to find the relationship between the luminosity I of a metal filament lamp and the voltage V.

V	60	80	100	120	140
I	11	20.5	89	186	319

Allowing for the fact that an error was made in one of the readings show that the law between I and V is of the form $I = aV^n$ and find the probable correct value of the reading. Find the value of n.

5) Two quantities t and m are connected by a law of the type $t = a.m^b$ and the co-ordinates of two points which satisfy the equation are $(8, 6.8)$ and $(20, 26.9)$. Find the law.

6) The intensity of radiation, R, from certain radioactive materials at a particular time t is thought to follow the law $R = kt^n$. In an experiment to test this the following values were obtained:

R	58	43.5	26.5	14.5	10
t	1.5	2	3	5	7

Show that the assumption was correct and evaluate k and n.

20. GRAPHICAL SOLUTION OF EQUATIONS

Outcome:

1. Solve a pair of two simultaneous linear equations graphically.
2. Determine the roots of a quadratic equation by the intersection of the graph with the x-axis.
3. Solve simultaneously linear and quadratic equations by the intersection of their graphs.
4. Plot the graph of a cubic equation with specified interval and range.
5. Solve a cubic equation using 4.

GRAPHICAL SOLUTION OF TWO SIMULTANEOUS LINEAR EQUATIONS

Since the solutions we require have to satisfy both the given equations, they will be given by the values of x and y where the graphs of the equations intersect.

The two examples which follow are first solved graphically and then by an alternative theoretical method.

EXAMPLE 20.1

Solve graphically:

$$y - 2x = 2 \qquad [1]$$
$$3y + x = 20 \qquad [2]$$

Equation [1] may be written as

$$y = 2x + 2$$

and equation [2] may be written as

$$y = -\frac{x}{3} + \frac{20}{3}$$

We will first draw up a table of values. Only three values are necessary in each case since both equations are linear and will have straight line graphs.

x	-3	0	3
$y = 2x + 2$	-4	2	8
$y = -\dfrac{x}{3} + \dfrac{20}{3}$	7.7	6.7	5.7

The two graphs are now plotted on the same axes, as shown in Fig. 20.1.

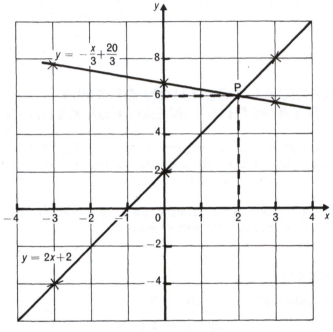

Fig. 20.1

The solutions of the equations will be given by the co-ordinates of the point where the two lines cross (that is, point P in Fig. 20.1). The co-ordinates of P are $x = 2$ and $y = 6$. Hence the solutions are:

$$x = 2 \quad \text{and} \quad y = 6$$

An Alternative Solution by the Elimination of One Unknown

Multiplying equation [2] by 2 we have

$$6y + 2x = 40 \qquad\qquad [3]$$

Adding equations [1] and [3] we get

$$y - 2x + 6y + 2x = 2 + 40$$

$$\therefore \qquad\qquad y = 6$$

Substituting $y = 6$ into equation [1] we have

$$6 - 2x = 2$$

from which $\qquad\qquad x = 2$

Thus the solutions are $x = 2$ and $y = 6$ which confirms the graphical results.

EXAMPLE 20.2

Two operators are producing the same assemblies. Their total output per week is 220 assemblies. If the ratio of their individual outputs is $6:5$, find the number of assemblies per week that each operator produces.

Let the number of assemblies produced by the faster operator be x, and the number produced by the slower operator be y.

$$x + y = 220 \qquad\qquad [1]$$

and $$\frac{x}{y} = \frac{6}{5} \qquad\qquad [2]$$

We now have two simultaneous equations for x and y.

Rearranging equation [1] gives

$$y = -x + 220$$

and rearranging equation [2] gives

$$y = \frac{5}{6}x$$

We will first draw up a table of values. Only three values are necessary in each case since both equations are linear and will have straight line graphs.

x	0	100	200
$y = -x + 220$	220	120	20
$y = \dfrac{5}{6}x$	0	83	167

The two graphs are now plotted on the same axes, as shown in Fig. 20.2.

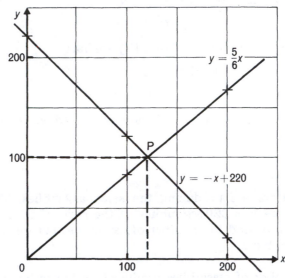

Fig. 20.2

The solutions of the equations will be given by the co-ordinates of the point where the two lines cross (that is, point P in Fig. 20.2). The co-ordinates of P are $x = 120$ and $y = 100$.

Hence the operators produce 120 and 100 assemblies each respectively per week.

An Alternative Solution Using Substitution

Rearranging equation [2] gives $y = \dfrac{5}{6}x$

We will now substitute this expression for y into equation [1] and obtain

$$x + \frac{5}{6}x = 220$$

from which $x = 120$

Now substituting $x = 120$ into equation [1] we have

$$120 + y = 220$$

from which $y = 100$

Thus the individual outputs per week of each operator are 120 and 100 assemblies per week. This confirms the result obtained by the graphical method.

Exercise 20.1

Solve the following equations:

1) $3x + 2y = 7$
   $x + y = 3$

2) $4x - 3y = 1$
   $x + 3y = 19$

3) $2x - 3y = 5$
   $x - 2y = 2$

4) A motorist travels x km at 40 km/h and y km at 50 km/h. The total time taken is $2\frac{1}{2}$ hours. If the time taken to travel $6x$ km at 30 km/h and $4y$ km at 50 km/h is 14 hours find x and y.

5) An alloy containing 8 cm^3 of copper and 7 cm^3 of tin has a mass of 122.3 g. A second alloy containing 9 cm^3 of copper and 7 cm^3 of tin has a mass of 131.2 g. Find the densities of copper and tin respectively in g/cm^3.

6) The resistance R ohms of a wire at a temperature of $t\,^\circ$C is given by the formula $R = R_0(1 + \alpha t)$ where R_0 is the resistance at 0 °C and α is a constant. The resistance is 35 ohms at a temperature of 80 °C and 42.5 ohms at a temperature of 140 °C. Find R_0 and α. Hence find the resistance when the temperature is 50 °C.

7) If 100 m of wire and 8 plugs cost £12.40 and 150 m of wire and 10 plugs cost £18. Find the cost of 1 m of wire and the cost of a plug.

8) A heating installation for one house consists of 5 radiators and 4 convector heaters and the cost, including labour, is £2080. In a second house 6 radiators and 7 convector heaters are used and the cost, including labour, is £3076. In each house the installation costs are £400. Find the cost of a radiator and the cost of a convector heater.

9) If 100 m of tubing and 8 elbow fittings cost £100.00, and 150 m of tubing and 10 elbow fittings cost £147.50, find the cost of 1 m of tubing and the cost of an elbow fitting.

10) For a builder's winch it is found that the effort E newtons and the load W newtons are connected by the equation $E = aW + b$. An effort of 90 N lifts a load of 100 N whilst an effort of 130 N lifts a load of 200 N. Find the values of a and b and hence determine the effort required to lift a load of 300 N.

11) A penalty clause states that a contractor will forfeit a certain sum of money for each day that he is late in completing a contract (i.e. the contractor gets paid the value of the original contract less any sum forfeit). If he is 6 days late he receives £5000 and if he is 14 days late he receives £3000. Find the amount of the daily forfeit and determine the value of the original contract.

12) The total cost of equipping two laboratories, A and B, is £30 000. If laboratory B costs £2000 more than laboratory A find the cost of the equipment for each of them.

13) x kg of a chemical and $2y$ kg of a second chemical cost £90. If the mass ratio of the chemicals is inverted the cost is only £60. Assuming that each chemical costs £5 per kg find the masses x and y.

14) The sum of the ages of two installations is 46 months. The modern version is 10 months younger than the original. Calculate their present ages.

15) A company's annual net profit of £8800 is divided amongst the two partners in the ratio $x : 2y$. If the first shareholder receives £2000 more than the other, find the values of x and y, and hence the respective shares of the profit.

16) The cost of a bacteria culture b is £0.60 and of a bacteria culture B is £0.90. A combination of several samples of each culture costs £9. Inflation raises the cost of each sample by 15 p and the cost of the combination to £10.65. Find the number of samples of each bacteria cultured.

GRAPHICAL SOLUTION OF QUADRATIC EQUATIONS

An equation in which the highest power of the unknown is two, and containing no higher powers of the unknown, is called a quadratic equation. It is also known as an equation of the *second degree*. Thus

$$x^2 - 9 = 0 \qquad 2.5x^2 - 3.1x - 2 = 0$$
$$x^2 + 2x - 8 = 0 \qquad 2x^2 - 5x = 0$$

are all examples of quadratic equations.

Quadratic equations may be solved by plotting graphs. This method is explained fully by the examples which follow.

EXAMPLE 20.3

Plot the graph of $y = 3x^2 + 10x - 8$ between $x = -6$ and $x = +4$. Hence solve the equation $3x^2 + 10x - 8 = 0$.

A table can be drawn up as follows giving values of y for the chosen values of x.

x	-6	-5	-4	-3	-2	-1
$3x^2$	108	75	48	27	12	3
$10x$	-60	-50	-40	-30	-20	-10
-8	-8	-8	-8	-8	-8	-8
y	40	17	0	-11	-16	-15

x	0	1	2	3	4	
$3x^2$	0	3	12	27	48	
$10x$	0	10	20	30	40	
-8	-8	-8	-8	-8	-8	
y	-8	5	24	49	80	

The graph of $y = 3x^2 + 10x - 8$ is shown plotted in Fig. 20.3.

To solve the equation $3x^2 + 10x - 8 = 0$ we have to find the value of x when $y = 0$, that is, the value of x where the graph cuts the x-axis.

These are the points A and B in Fig. 20.3.

Hence the solutions of $3x^2 + 10x - 8 = 0$ are

$$x = -4 \quad \text{and} \quad x = 0.7$$

The accuracy of the results obtained by this method will depend on the scales chosen. The value $x = -4$ is exact as this value was taken when drawing up the table of values, and gave $y = 0$. The value $x = 0.7$ is as accurate as may be read from the scale chosen for the x-axis.

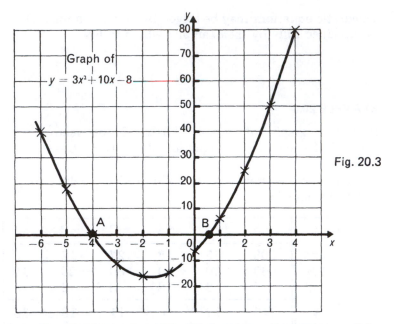

Fig. 20.3

The next example shows how to obtain a more accurate result by plotting a portion of the graph with larger scales.

EXAMPLE 20.4

Find the roots of the equation $x^2 - 1.3 = 0$ by drawing a suitable graph.

Using the same method as in Example 20.3 we need to plot a graph of $y = x^2 - 1.3$ and find where it cuts the x-axis.

We have not been given a range of values of x between which the curve should be plotted and so we must make our own choice.

A good method is to try first a range from $x = -4$ to $x = +4$. If only five values of y are calculated for values of x of $-4, -2, 0, +1$ and $+2$, we shall not have wasted much time if these values are not required—in any case we shall learn from this trial and be able to make a better choice at the next attempt.

The first table of values is as follows:

x	-4	-2	0	2	4
x^2	16	4	0	4	16
-1.3	-1.3	-1.3	-1.3	-1.3	-1.3
y	14.7	2.7	-1.3	2.7	14.7

The graph of these values is shown plotted in Fig. 20.4.

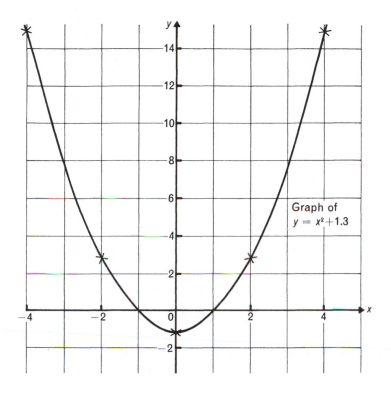

Graph of
$y = x^2 + 1.3$

Fig. 20.4

The approximate values of x where the curve cuts the x-axis are -1 and $+1$ (Fig. 20.4). For more accurate results we must plot the portion of the curve where it cuts the x-axis to a larger scale. We can see, however, both from the table of values and the graph, that the graph is symmetrical about the y-axis—so we need only plot one half. We will choose the portion to the right of the y-axis and draw up a table of values from $x = 0.7$ to $x = 1.3$

x	0.7	0.8	0.9	1.0	1.1	1.2	1.3
x^2 -1.3	0.49 -1.3	0.64 -1.3	0.81 -1.3	1.00 -1.3	1.21 -1.3	1.44 -1.3	1.69 -1.3
y	-0.81	-0.66	-0.49	-0.30	-0.09	0.14	0.39

These values are shown plotted in Fig. 20.5.

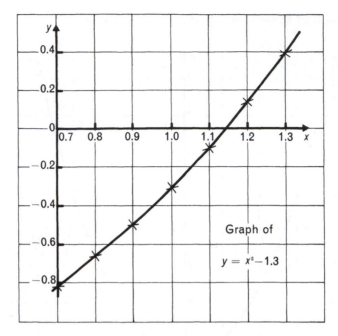

Fig. 20.5

The graph cuts the x-axis where $x = 1.14$ (Fig. 20.5) and we must not forget the other value of x where the curve cuts the x-axis to the left of the y-axis. This will be where $x = -1.14$ since the curve is symmetrical about the y-axis.

Hence the solutions of $x^2 - 1.3 = 0$ are

$$x = 1.14 \quad \text{and} \quad x = -1.14$$

EXAMPLE 20.5

Solve the equation $x^2 - 4x + 4 = 0$

We shall plot the graph of $y = x^2 - 4x + 4$ and find where it cuts the x-axis.

The graph is shown plotted in Fig. 20.6. In this case the curve does not actually cut the x-axis but touches it at the point where $x = 2$. Another way of looking at it is to say that the curve 'cuts' the x-axis at two points which lie on top of each other. The two points coincide and they are said to be coincident points. The roots are called repeated roots.

The only solution to $x^2 - 4x + 4 = 0$ is, therefore, $x = 2$

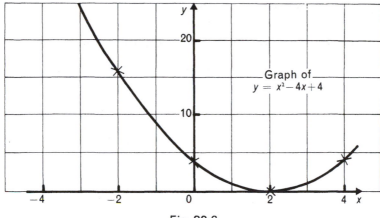

Fig. 20.6

EXAMPLE 20.6

Solve the equation $x^2 + x + 3 = 0$

We shall plot the graph of $y = x^2 + x + 3$ and find where it cuts the x-axis.

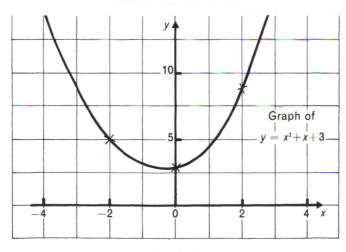

Fig. 20.7

The graph is shown plotted in Fig. 20.7. We can see that the curve does not cut the x-axis at all. This means there are no roots—in theory there are roots but they are complex or imaginary and have no arithmetical value.

Exercise 20.2

By plotting suitable graphs solve the following equations:

1) $x^2 - 7x + 12 = 0$ (plot between $x = 0$ and $x = 6$)

2) $x^2 + 16 = 8x$ (plot between $x = 1$ and $x = 7$)

3) $x^2 - 9 = 0$ (plot between $x = -4$ and $x = 4$)

4) $x^2 + 2x - 15 = 0$ 5) $3x^2 - 23x + 14 = 0$

6) $2x^2 + 13x + 15 = 0$ 7) $x^2 - 2x - 1 = 0$

8) $3x^2 - 7x + 1 = 0$ 9) $9x^2 - 5 = 0$

GRAPHICAL SOLUTION OF SIMULTANEOUS LINEAR AND QUADRATIC EQUATIONS

Since the solutions we require have to satisfy both the given equations they will be given by the values of x and y where the graphs of the equations intersect.

EXAMPLE 20.7

Solve simultaneously the equations:

$$y = x^2 + 3x - 4$$

and

$$y = 2x + 4$$

We must first draw up tables of values, and will use the range $x = -4$ to $x = +4$:

x	-4	-2	0	2	4
x^2 $+3x$ -4	16 -12 -4	4 -6 -4	0 0 -4	4 6 -4	16 12 -4
$y = x^2 + 3x - 4$	0	-6	-4	6	24
$2x$ $+4$	-8 4	-4 4	0 4	4 4	8 4
$y = 2x + 4$	-4	0	4	8	12

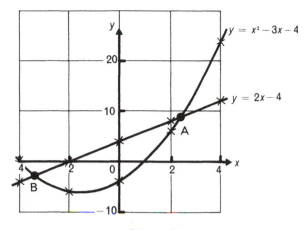

Fig. 20.8

The two graphs are shown plotted on the same axes in Fig. 20.8 and they intersect at the points A and B. Values of the x and y co-ordinates at these points will give the solutions of the given equations. We must be careful not to try to read the values too accurately, as the graphs have been plotted using only five values of x. In this case, even values to the first place of decimals cannot be guaranteed.

If more accurate answers are required then we must plot the portions of the graph containing points A and B using more values of x and also much larger scales.

Hence the required solutions are

x	2.4	-3.4
y	8.7	-2.7

Exercise 20.3

Solve simultaneously:

1) $y = x^2 - 2x - 2$
   $x - y + 2 = 0$

2) $y = x^2 - x + 5$
   $y = 2x + 5$

3) $y = 5x^2 + x - 3$
   $y = 5x - 2$

4) $y = 2x^2 - 2.3x + 1$
   $y = 3x - 0.25$

5) A rectangular plot of land has a perimeter of 280 m. The length of a diagonal drawn corner to corner is 100 m. If x and y are the length and width of the plot respectively show that:

$$x + y = 140 \qquad [1]$$

and $$x^2 + y^2 = 10\,000 \qquad [2]$$

Hence find the dimensions of the plot.

CUBIC EQUATIONS

An equation in which the highest power of the unknown is three, and containing no higher powers of the unknown is called a cubic equation. It is also known as an equation of the *third degree*. Thus

$$x^3 - 37 = 0$$
$$x^3 + 2x^2 + 1 = 0$$
$$3x^3 - 4x - 13 = 0$$
$$x^3 + x^2 + 7x - 10 = 0$$

are all examples of cubic equations.

The algebraic method of solving cubic equations is difficult and we shall solve cubic equations by a graphical method similar to that used for solving quadratic equations. This, in fact, is the usual way technicians solve cubic equations when they arise from practical problems.

EXAMPLE 20.8

Plot the graph of $y = x^3 - 1.5x^2 - 8.5x + 4.5$ from $x = -4$ to $x = +4$ at 1 unit intervals, and use the graph to solve the cubic equation $x^3 - 1.5x^2 - 8.5x + 4.5 = 0$

A table can be drawn up as follows giving values of y for the chosen values of x:

x	-4	-3	-2	-1	0
x^3	-64	-27	-8	-1	0
$-1.5x^2$	-24	-13.5	-6	-1.5	0
$-8.5x$	34	25.5	17	8.5	0
$+4.5$	4.5	4.5	4.5	4.5	4.5
y	-49.5	-10.5	7.5	10.5	4.5

x	1	2	3	4
x^3 $-1.5x^2$ $-8.5x$ $+4.5$	1 -1.5 -8.5 4.5	8 -6 -17 4.5	27 -13.5 -25.5 4.5	64 -24 -34 4.5
y	-4.5	-10.5	-7.5	10.5

The graph is shown plotted in Fig. 20.9. To solve the cubic equation we have to find the values of x when $y = 0$, that is, the value of x where the graph cuts the x-axis.

Hence the required solutions of $x^3 - 1.5x^2 - 8.5x + 4.5 = 0$ are

$$x = -2.5, \quad x = 0.5 \quad \text{and} \quad x = 3.5$$

We must remember that these values are only approximate, since the values of the first decimal place cannot be guaranteed. As in previous examples, if more accurate answers are required then we must plot the portions of the graph containing the points of intersection using more values of x and also much larger scales.

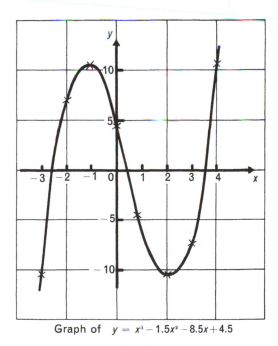

Graph of $y = x^3 - 1.5x^2 - 8.5x + 4.5$

Fig. 20.9

EXAMPLE 20.9

A domed roof is in the form of a cap of a sphere. Its base radius is 10 m. If the height of the dome is h metres and the volume of air space under the dome is 1525 m³. It can be shown that

$$h^3 + 300h - 2912 = 0$$

Plot the graph of $y = h^3 + 300h - 2912$ for values of h between 4 and 12 and hence find the value of h.

A table is drawn up giving values of y corresponding to the chosen values of h:

h	4	5	6	7	8
h^3 300h -2912	64 1200 -2912	125 1500 -2912	216 1800 -2912	343 2100 -2912	512 2400 -2912
y	-1648	-1287	-896	-469	0

h	9	10	11	12
h^3 300h -2912	729 2700 -2912	1000 3000 -2912	1331 3300 -2912	1728 3600 -2912
y	517	1088	1719	2416

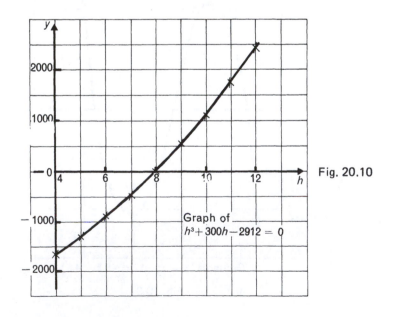

Fig. 20.10

The graph is drawn in Fig. 20.10 and the solution of the equation $h^3 + 300h - 2912 = 0$ is the value of h where the curve cuts the horizontal axis. Hence $h = 8$ is the solution. Note that this solution is exact since the table shows that when $h = 8$, $y = 0$

EXAMPLE 20.10

Plot the graph of $y = 5x^3 - 9x^2 + 3x + 1$ from $x = -0.4$ to $x = +1.4$ at intervals of 0.2 units, and hence find the values of the roots of the equation $5x^3 - 9x^2 + 3x + 1 = 0$

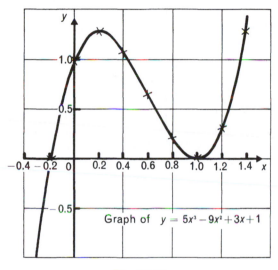

Fig. 20.11

The graph is shown plotted in Fig. 20.11. We can see that the curve cuts the x-axis where $x = -0.2$ and touches the x-axis where $x = 1$

As in Example 5.5 the point where $x = 1$ represents two coincident points and gives rise to two repeated roots.

Hence the solutions of the equation $5x^3 - 9x^2 + 3x + 1 = 0$ are

$$x = -0.2 \quad \text{and} \quad x = 1$$

EXAMPLE 20.11

Plot the graph of $y = x^3 - 1$ for x values from -1.0 to $+1.5$ at half unit intervals. Hence find the roots of $x^3 - 1 = 0$

The graph is shown plotted in Fig. 20.12 and it can be seen that the curve only cuts the x-axis at one point which gives the value of the only real root. The other two solutions are complex or imaginary and have no arithmetical meaning.

The only real solution of $x^3 - 1 = 0$ is, therefore, $x = 1$

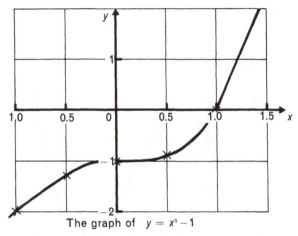

The graph of $y = x^3 - 1$

Fig. 20.12

Exercise 20.4

1) Plot the graph of $y = x^3 - 4x^2 - x + 4$ from $x = -2$ to $x = 5$ at one unit intervals. Hence solve the equation $x^3 - 4x^2 - x + 4 = 0$.

2) Plot the graph of $y = 3x^3 - 3.4x^2 - 6.4x + 2.4$ using values of x from -2 to $+3$ at half unit intervals. Hence find the roots of the cubic equation $3x^3 - 3.4x^2 - 6.4x + 2.4 = 0$.

3) Plot the graph of $y = x^3 + x^2 + x - 3$ from $x = -3$ to $x = +3$ at one unit intervals. Hence show that the cubic equation $x^3 + x^2 + x - 3 = 0$ has only one real root, and find its value.

4) Plot the graph of $y = x^3 - x^2 - x + 1$ taking values of x from -2 to $+2$ at half unit intervals. Hence find the roots of the equation $x^3 - x^2 - x + 1 = 1$.

5) Plot the graph of $y = 4x^3 - 15x^2 + 7x + 6$ from $x = -1$ to $x = 3.5$ at intervals of one half unit. Use this graph to solve the cubic equation $4x^3 - 15x^2 + 7x + 6 = 0$.

6) Plot the graph of $y = x^3 - x^2 - x - 2$ for values of x from -4 to $+4$ at one unit intervals, and hence find the roots of the equation $x^3 - x^2 - x - 2 = 0$.

7) Find the roots of the cubic equation $3x^3 + 4x^2 - 12x + 5 = 0$ by plotting a suitable graph taking values of x from -3.5 to $+1.5$ at half unit intervals.

8) By plotting the graph of $y = x^3 - x^2 - 8x + 12$ find the roots of the equation $x^3 - x^2 - 8x + 12 = 0$.

9) Find by graphical means the roots of $7x^3 - 6x^2 - 18x + 4 = 0$.

10) Show that the equation $2x^3 + 3x^2 + 2x + 1 = 0$ has only one real root, and find its value.

11) A spherical vessel has a radius of 8 m. It contains liquid to a height of h metres (Fig. 20.13). When the volume of the liquid in the vessel is 1.056 m³ the following equation applies:

$$\frac{3168}{\pi} = h^2(24 - h)$$

By plotting a suitable equation find the value of h (h lies between 3 and 8).

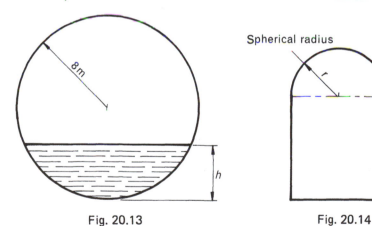

Fig. 20.13   Fig. 20.14

12) A pressure vessel (Fig. 20.14) has a capacity of 594 m³. The equation from which r may be found is

$$2r^3 + 45r^2 - 567 = 0$$

By drawing a suitable graph find the value of r (r lies between 0 m and 5 m).

13) A domed roof is in the form of a segment of a sphere. If the volume of air space under the dome is $497\,m^3$ and the radius of the sphere is $8\,m$, then

$$h^3 - 24h^2 + 475 = 0$$

Plot a suitable graph to find h, the height of the dome, given that its value is between $4\,m$ and $6\,m$.

14) A cement silo is in the form of a frustum of a cone with a height equal to the larger radius of the frustum. The following equation then applies:

$$R^3 + 2R^2 + 4R - 57 = 0$$

Find the value of R, the height of the silo, by plotting a suitable graph.

21. POLAR GRAPHS

Outcome:

1. *Define polar coordinates.*
2. *State the relationship between polar and Cartesian co-ordinates.*
3. *Convert Cartesian to polar co-ordinates and vice versa.*

4. *Plot graphs of functions defined in polar co-ordinates such as* r = a, θ = α, r = kθ, r = sin θ.

POLAR COORDINATES

You have met previously Cartesian (or rectangular) co-ordinates with which P may be given as the point (x, y) as shown in Fig. 21.1.

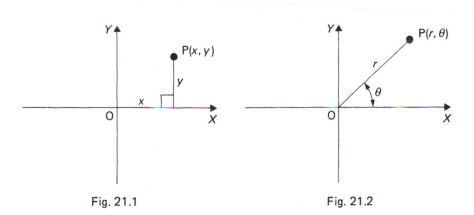

Fig. 21.1   Fig. 21.2

Another way of giving the location of P is by using its distance r from the origin O, together with angle $θ$ that OP makes with the horizontal axis OX (Fig. 21.2). Figure 21.3 shows five typical points, plotted on a polar grid, together with their respective polar co-ordinates.

319

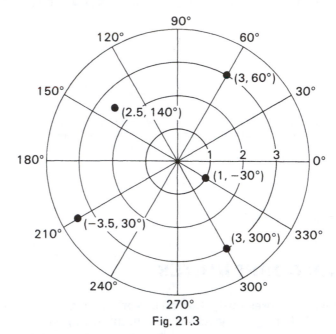

Fig. 21.3

Positive values of θ are always measured anticlockwise from OX whilst negative values are measured clockwise (see Fig. 21.3): you will not often meet the latter case. A negative value of r means that the 'radius length' is extended 'backwards' through O from the normal angle position: see point $(-3.5, 30°)$ in Fig. 21.3. Note that in polar co-ordinates it is possible to define a point in more than one way: for example the point $(3, 300°)$ in Fig. 21.3 could also be defined as $(-3, 120°)$.

Polar graph paper is available but is not so readily obtained as the common linear variety. However, it should be sufficient here for you to sketch the polar curves we meet, or perhaps draw your simple polar grid similar to that in Fig. 21.3.

A typical practical application of a polar plot would be to values of luminous intensity of an electric light bulb, to show how it varied around the bulb relative to the position of the filament etc.

RELATIONSHIP BETWEEN CARTESIAN AND POLAR COORDINATES

From the right-angled triangle POM in Fig. 21.4 we can see that:

$$x = r \cos \theta$$

and

$$y = r \sin \theta$$

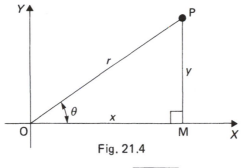

Fig. 21.4

Also $$r = \sqrt{x^2 + y^2}$$

and $$\tan \theta = \frac{y}{x}$$

Using the above relationships it is reasonably easy to convert from Cartesian to polar co-ordinates, and vice versa. Always make a sketch of the problem because this will enable you to see which quadrant you are dealing with.

EXAMPLE 21.1

Find the polar co-ordinates of the point $(-4, 3)$.

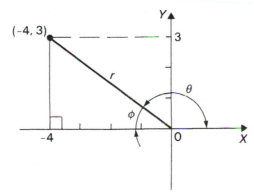

Fig. 21.5

From Fig. 21.5 $\tan \phi = \dfrac{3}{4} = 0.75$

$\therefore$ $\phi = 36.9°$

Thus $\theta = 180° - 36.9° = 143.1°$

Also $r = \sqrt{3^2 + 4^2} = 5$

Thus the point is $(5, 143.1°)$ in polar form.

EXAMPLE 21.2

Express in Cartesian form the point (6, 231°).

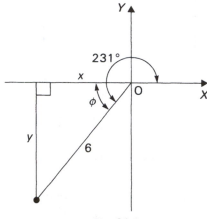

Fig. 21.6

From Fig. 21.6 $\phi = 231° - 180° = 51°$

Thus $x = 6\cos 51° = 3.78$

and $y = 6\sin 51° = 4.66$

Now, having drawn a diagram, we can see that both the x and y co-ordinates are negative.

Thus the Cartesian form of the point is $(-3.78, -4.66)$.

GRAPHS OF FUNCTIONS

When using Cartesian co-ordinates a graph may be drawn to illustrate y as a function of x. For example $y = mx + c$ represents a straight line graph. Similarly when using polar co-ordinates a graph may be drawn to illustrate r in terms of θ. In the examples which follow we will sketch some of the more common polar graphs and this should enable us to become familiar with their shapes.

EXAMPLE 21.3

Sketch the graph of $r = \sin \theta$ between $\theta = 0°$ and $\theta = 360°$.

From experience we know that $\sin \theta$ (and hence r) has a maximum value of $+1$, and a minimum value of -1. This will help initially in labelling our polar grid (Fig. 21.7). We may plot the graph from values obtained from our scientific calculator. We first measure off the angle θ and then measure off the length r along the angle boundary line.

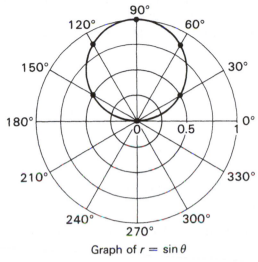

Graph of $r = \sin\theta$

Fig. 21.7

Plotting from $\theta = 1$ to $\theta = 180°$ is straightforward. However, between $180°$ and $360°$ the values of r are negative and we merely repeat the curve already plotted. This curve is, in fact, a circle.

For interest you may like to verify that this is the only curve for *any* value of θ however large — both positive and negative.

EXAMPLE 21.4

Sketch the graph of $r = 2$.

Here no mention is made of the angle θ. This means that $r = 2$ defines the graph whatever value is given for θ. Thus the graph is a circle of radius 2 as shown in Fig. 21.8.

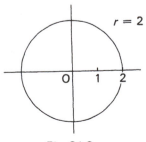

Fig. 21.8

EXAMPLE 21.5

Sketch the graph of $\theta = 40°$.

Here no mention is made of r. This means that $\theta = 40°$ defines the graph whatever value is given for r (whether positive or negative). Hence the graph is a straight line as shown in Fig. 21.9.

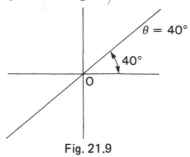

Fig. 21.9

EXAMPLE 21.6

Sketch the graph of $r = 3\theta$ for angle values equivalent to the range 0° to 540°

As is usual in mathematics when an angle is used directly in calculations its value must be in radians. But we find it more convenient to plot the angle values using degrees and so we must be careful to find the corresponding angle values in radians when calculating r values.

Your scientific calculator may convert directly from degrees to radians. If not, we know that $360° = 2\pi\,\text{rad}$ or $1° = \dfrac{2\pi}{360}\text{rad}$, and we may also use a constant multiplier facility to avoid entering this fraction for each r. A suitable sequence of operations would be:

For $\theta = 30°$:

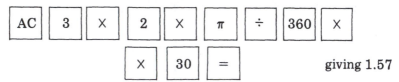

giving 1.57

and for other angles just enter its value,

e.g. for 60°:   [60] =   giving 3.14

and for the highest value of 540°:   [540] =   giving 28.27

The graph is known as an Archimedean spiral and is shown in Fig. 21.10.

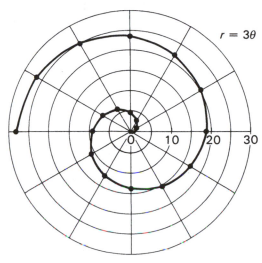

Fig. 21.10

Exercise 21.1

1) Find the polar co-ordinates of the following points:

(a) $(5, 7)$ (b) $(-2, 3)$ (c) $(2, -3)$ (d) $(-3, -4)$

2) Find the Cartesian co-ordinates of the following points:

(a) $(2, 35°)$ (b) $(3, 127°)$ (c) $(1.5, 240°)$
(d) $(0.6, 312°)$ (e) $(2.3, -21°)$ (f) $(-5, 130°)$

Sketch the graphs for the polar equations in the questions which follow:

3) $r = 2\cos\theta$ 4) $r = 3$ 5) $\theta = 120°$

6) $r = \theta$ 7) $r = \sin^2\theta$ 8) $r = \cos^2\theta$

9) $r = 3\sin 2\theta$ 10) $r = \cos 3\theta$ 11) $r = a(1 + \cos\theta)$

22. DATA COLLECTION AND ANALYSIS

Outcome:

1. Collect data and represent by suitable graphical displays.
2. Distinguish between discrete and continuous variables.
3. Record frequency distributions.
4. Determine class boundaries and class widths.
5. Draw a histogram for continuous and discrete variates.
6. Construct frequency curves.
7. Plot cumulative frequency curves.
8. Calculate the arithmetic mean, median and mode for set of numbers, histograms and frequency curves.
9. Understand when to use the mean, median or mode.
10. Recognise a normal distribution curve with varying mean and spread.
11. Calculate the mean and standard deviation.
12. Check the spread using the range.
13. Chance from the normal curve.
14. Appreciate the problem of curve fitting on scatter diagrams.
15. Use the method of least squares to find the best straight line.

INTRODUCTION

Statistics is the name given to the science of collecting and analysing data in the form of groups of numbers. These are often presented by means of tables and diagrams. We shall discuss ideas of analysing the figures and making practical use of the results obtained — for example taking samples of a factory production and hence being able to monitor quality control.

DISPLAYING INFORMATION

Suppose that in a certain factory the number of persons employed on various jobs is as given in the following table:

TABLE 22.1

Type of personnel	Number employed	Percentage
Machinists	140	35
Fitters	120	30
Clerical staff	80	20
Labourers	40	10
Draughtsmen	20	5
Total	400	100

326

The information in Table 22.1 can be represented pictorially in several ways:

(1) *The pie chart* (Fig. 22.1) displays the proportions as angles (or sector areas), the complete circle representing the total number employed. Thus for machinists the angle is $\frac{140}{400} \times 360 = 126°$ and for fitters $\frac{120}{400} \times 360 = 108°$ etc.

(2) *The bar chart* (Fig. 22.2) relies on heights (or areas) to convey the proportions; the total height of the diagram represents 100%.

(3) *The horizontal bar chart* (Fig. 22.3), or the *vertical bar chart* (Fig. 22.4), gives a better comparison of the various types of personnel employed but it does not readily display the total number employed in the factory.

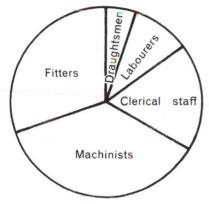

Fig. 22.1 The Pie Chart

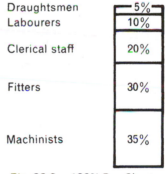

Fig. 22.2 100% Bar Chart

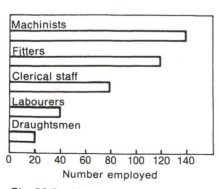

Fig. 22.3 Horizontal bar chart

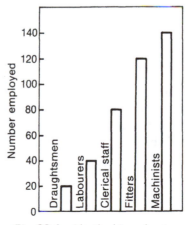

Fig. 22.4 Vertical bar chart

VARIABLES

Variables are measured quantities which can be expressed as numbers as, for instance, the number of cars in a garage or the height of a person. In statistics a variable is sometimes called a variate, particularly when dealing with histograms and frequency distributions.

Discrete Variables

Some variables can only take certain values, usually, but not always, whole numbers and are called discrete. Examples are the number of cars in a garage or the number of persons working in an office (you can hardly have parts of people implied by 21.7 say!). Shoe sizes are discrete although they are not all whole numbers since they include half sizes in their range: for example 5, $5\frac{1}{2}$, 6, $6\frac{1}{2}$, 7, $7\frac{1}{2}$, etc. But a size of 7.19 is not available.

Continuous Variables

These are all other non-exact numbers usually in a range between two given end values. A continuous variable is a resistance of 125 ± 0.2 ohm which may be anywhere between end values 124.8 ohm and 125.2 ohm. Most measurements in technology are continuous and thus occupy most of our attention.

THE POPULATION

'In 1978, 14 million people watched the Cup Final on BBC television.' Statements like this are made every day but how could the BBC be so confident that their figure is correct?

Clearly they cannot send researchers to every household in the country to see how many people are watching. What they do is to select a *sample* from the total population and then use the results of this sample to estimate the number watching.

In this case the parent population is the population of the country. But in the case of the data of Table 22.1, the parent population is the total workforce in the factory. Again, suppose a factory produces one million ball-bearings. This is the parent population of ball-bearings.

SAMPLING

It is rarely possible to examine every item making up a parent population and recourse has to be made to sampling. For the information obtained to be of value the sample must be representative of the population as a whole. We might take a sample of 100 ball-bearings and measure their diameters. The results obtained would then be regarded as being representative of the population as a whole.

FREQUENCY DISTRIBUTIONS

Suppose we measure the diameters of a sample of 100 ball-bearings. We might get the following reading in millimetres:

TABLE 22.2

6.2	6.3	5.8	5.8	6.0	6.1	6.0	6.1	5.9	6.2
5.9	6.3	6.2	6.1	6.1	6.2	6.4	5.8	6.1	6.2
5.8	6.1	6.1	6.1	5.9	6.0	6.0	6.0	6.1	6.1
6.1	6.3	5.8	5.9	5.9	5.8	6.0	5.7	6.0	6.2
6.0	6.1	6.0	5.9	6.0	6.0	6.2	5.6	6.1	5.8
6.1	6.0	6.1	6.0	6.1	5.9	6.1	6.0	5.9	6.2
5.9	6.1	6.0	6.1	6.0	5.9	5.8	5.7	5.9	6.0
5.8	5.7	6.0	5.9	5.8	6.3	5.9	6.3	6.0	5.9
5.7	6.2	6.3	6.3	5.9	6.0	5.9	5.9	5.6	6.4
5.9	6.1	6.0	6.0	6.0	6.3	5.8	5.9	6.1	5.9

These figures do not mean very much as they stand and so we rearrange them into what is called a frequency distribution. To do this we collect all the 5.6 mm readings together, all the 5.7 mm readings and so on. A tally chart (Table 22.3) is the best way of doing this. Each time a measurement arises a tally mark is placed opposite the appropriate measurement. The fifth tally mark is usually made in an oblique direction thus tying the tally marks into bundles of five to make counting easier. When the tally marks are complete the marks are counted and the numerical value recorded in the column headed 'frequency'. The frequency is the number of times each measurement occurs. From Table 22.3 it will be seen that the measurement 5.6 occurs twice (that is, it has a frequency of 2). The measurement 5.7 occurs four times (a frequency of 4) and so on.

TABLE 22.3

Measurement (mm)	Number of bars with this measurement	Frequency
5.6	//	2
5.7	////	4
5.8	LHT LHT /	11
5.9	LHT LHT LHT LHT	20
6.0	LHT LHT LHT LHT ///	23
6.1	LHT LHT LHT LHT /	21
6.2	LHT ////	9
6.3	LHT ///	8
6.4	//	2

CLASS BOUNDARIES AND WIDTH

In Fig. 22.5 the measurement 5.8 mm represents a group of figures from the halfway point between 5.7 and 5.8, to the halfway point between 5.8 and 5.9, or between 5.75 and 5.85 mm.

This represents a class having a *lower* class boundary of 5.75 mm and an *upper* class boundary of 5.85 mm.

The class width = upper class boundary − lower class boundary

$$= 5.85 - 5.75 = 0.1 \text{ mm}$$

The next class would have lower and upper boundaries of 5.85 mm and 5.95 mm and a class width also of 0.1 mm.

THE HISTOGRAM

The frequency distribution of Table 22.3 becomes even more understandable if we draw a diagram to represent it. The best type of diagram is the histogram (Fig. 22.5) which consists of a set of rectangles whose areas represent the frequencies.

If all the class widths are the same, which is usually the case, then the heights of the rectangles represent the frequencies. Note also that the left hand edge of each rectangle represents the lower boundary and the right hand edge represents the upper class boundary.

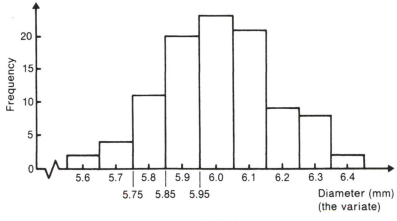

Fig. 22.5

On studying the histogram the pattern of the variation becomes clear, most of the measurements being grouped near the centre of the diagram with a few values more widely dispersed.

GROUPED DATA

When dealing with a large number of observations it is often useful to group them into classes or categories. We can then determine the number of items which belong to each class thus obtaining the class frequency.

EXAMPLE 22.1

The following gives the heights, in centimetres, of 100 small coniferous trees:

53	62	68	54	51	68	62	53	61	67
57	54	65	56	63	65	60	71	70	73
64	58	63	55	67	62	68	53	61	66
63	62	63	63	63	64	65	63	62	62
63	61	62	62	58	62	67	58	69	68
68	56	60	58	69	56	58	57	68	63
64	62	64	63	62	61	69	64	61	56
67	68	67	59	64	74	68	65	59	65
68	59	62	65	68	54	63	62	57	50
65	69	64	63	63	63	69	60	64	67

Draw up a tally chart for the classes 50–54, 55–59, etc.

The tally chart is shown in Table 22.4.

TABLE 22.4

Class (cm)	Tally	Frequency
50–54	LHT ///	8
55–59	LHT LHT LHT /	16
60–64	LHT LHT LHT LHT LHT LHT LHT LHT ///	43
65–69	LHT LHT LHT LHT LHT ////	29
70–74	////	4
	Total	100

The main advantage of grouping is that it produces a clear overall picture of the distribution.

The number of classes chosen depends upon the amount of original data. However, it should be borne in mind that too many groups will destroy the pattern of the data, whilst too few will destroy much of the detail contained in the original data.

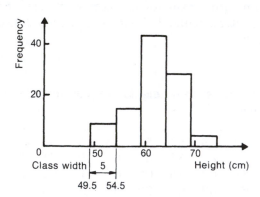

Fig. 22.6

Note that for the first class (50–54) the lower boundary is 49.5 cm and the upper boundary is 54.5, the class width being 5 cm. For the fourth class (65–69) the lower and upper boundaries are 64.5 and 69.5 cm respectively. The width of each class is the same and hence the frequencies of the various classes are represented by the heights of the rectangles in Fig. 22.6.

DISCRETE DISTRIBUTIONS

The histogram shown in Fig. 22.6 represents a distribution in which the variable is continuous. The data in Example 22.2 is discrete and we shall see how a discrete distribution is represented.

EXAMPLE 22.2

Five coins were tossed 100 times and after each toss the number of heads was recorded. The table below gives the number of tosses during which 0, 1, 2, 3, 4 and 5 heads were obtained. Represent this data in a suitable diagram.

Number of heads	0	1	2	3	4	5
Number of tosses (frequency)	4	15	34	29	16	2

Since the data is discrete (there cannot be 2.3 or 3.6 heads), Fig. 22.7 seems the most natural diagram to use. This diagram is in the form of a vertical bar chart in which the bars have zero width. Fig. 22.8 shows the same data represented as a histogram.

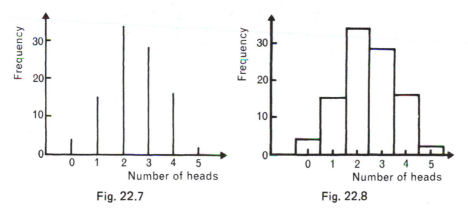

Fig. 22.7   Fig. 22.8

Note that the area under the diagram gives the total frequency of 100 which is as it should be. Discrete data is often represented as a histogram as was done in Fig. 22.8, despite the fact that in doing this we are treating the data as though it was continuous.

Exercise 22.1

1) Which of the following area discrete variables and which are continuous variables:

(a) The size of men's shirts.

(b) The length of plastic rod being produced in quantity.
(c) The masses of castings being produced in a foundry.
(d) The temperature of a furnace.
(e) The number of electric motors produced per day in a factory.

2) The following figures are the hottest daily temperatures (°C) during June at a particular coastal resort:

> 20, 21, 19, 22, 22, 23, 23, 23, 24, 25, 25, 26, 27, 28, 25
> 24, 24, 23, 22, 21, 22, 23, 23, 24, 25, 24, 25, 26, 27, 26

With the aid of a frequency table draw a histogram for these temperatures.

3) During trials of one variety of broad bean plant a seed merchant noted the number of beans in each pod as listed below:

> 7, 10, 6, 7, 10, 7, 8, 9, 7, 7, 6, 5, 7, 7, 3,
> 6, 9, 5, 11, 4, 7, 5, 4, 10, 7, 7, 6, 8, 5, 6,
> 9, 9, 8, 8, 7, 4, 8, 6, 5, 8, 7, 7, 4, 6, 2,
> 8, 5, 7, 9, 5, 5, 8, 5, 6, 6, 6, 8, 8, 9, 5,
> 8, 6, 6, 9, 9, 6, 9, 8, 9, 6, 7, 6, 8, 6, 8,
> 6, 8, 6, 8, 7, 7, 7, 7, 7, 9, 7, 6, 7, 6, 7,
> 7, 10, 10, 7, 7, 9, 6, 8, 11, 8

Draw up a frequency table and hence construct a histogram for the number of beans/pod.

4) Group the distribution in Question 3 into 5 classes. Hence construct an amended frequency table and histogram.

5) For the grouped frequency distribution given below, draw a histogram and state the class width for each of the classes.

Resistance (ohms)	110–112	113–115	116–118	199–121	122–124
Frequency	2	8	15	9	3

6) The data below was obtained by measuring the frequencies (in kilohertz) of 60 tuned circuits. Construct a frequency distribution and hence draw a histogram to represent the distribution.

12.37	12.29	12.40	12.41	12.31	12.35	12.37	12.35
12.33	12.36	12.32	12.36	12.40	12.38	12.33	12.35
12.30	12.30	12.34	12.39	12.44	12.32	12.27	12.32
12.41	12.40	12.37	12.46	12.35	12.34	12.38	12.43
12.36	12.35	12.26	12.28	12.36	12.24	12.42	12.39
12.45	12.42	12.28	12.25	12.34	12.33	12.32	12.39
12.38	12.27	12.35	12.35	12.34	12.36	12.36	12.32
12.31	12.35	12.29	12.30				

7) The table below gives a grouped frequency distribution for the compressive strength of a certain type of load-carrying brick.

Strength (N/mm²)	59.4–59.6	59.7–59.9	60.0–60.2
Frequency	8	37	90

Strength (N/mm²)	60.3–60.5	60.6–60.8
Frequency	52	13

Determine the class width and draw a histogram for this distribution.

FREQUENCY CURVES

In a grouped frequency distribution represented by a histogram we could make the class intervals smaller and smaller. The widths of the rectangles would also become smaller and smaller and the jumps in frequency between one class and another would become very tiny. Eventually the outline of the histogram would have the appearance of a smooth curve. This frequency curve is obtained by plotting frequency against the corresponding class mid-point, as shown in Fig. 22.9 for the data in Table 22.5.

TABLE 22.5

Resistance (ohm)	Frequency
7.8–8.0	4
8.1–8.3	19
8.4–8.6	45
8.7–8.9	26
9.0–9.2	6

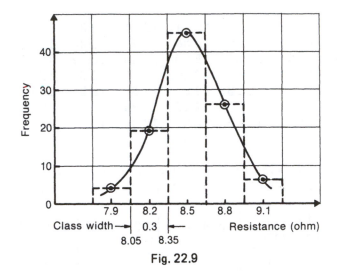

Fig. 22.9

CUMULATIVE FREQUENCY DISTRIBUTION

An alternative display may be obtained by using cumulative (or running total) frequencies, as shown in the next example.

EXAMPLE 22.3

The results of 80 compression tests on cube specimens were:

Crushing strength (N/mm^2)	8.7	8.8	8.9	9.0	9.1	9.2
Frequency	5	9	19	25	18	4

Plot a cumulative frequency distribution diagram and use it to estimate:

a) the number of cubes having crushing strength below $8.9\,N/mm^2$,

b) the median crushing strength,

c) the upper and lower quartiles.

Table 22.6 shows the cumulative frequencies corresponding to the appropriate class boundaries.

A smooth curve, called an ogive, is drawn (Fig. 22.10) through the plotted points.

a) Reading off from $8.9\,N/mm^2$ we see that the cumulative frequency is 22, and this indicates the number of cubes of strength below $8.9\,N/mm^2$.

TABLE 22.6

Crushing strength (N/mm²)	Cumulative frequency
Not more than 8.75	5
Not more than 8.85	5 + 9 = 14
Not more than 8.95	14 + 19 = 33
Not more than 9.05	33 + 25 = 58
Not more than 9.15	58 + 18 = 76
Not more than 9.25	76 + 4 = 80

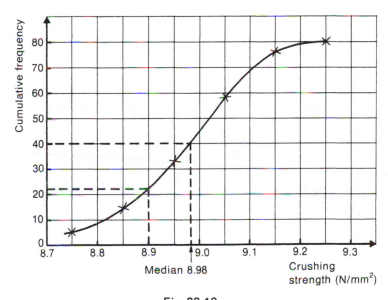

Fig. 22.10

b) The median (see also p. 339) is the strength which corresponds to half the total frequency — here 40. This gives the median as 8.98 N/mm².

c) Quartiles divide the frequency into four — they are values of the variate, here crushing strength, which occur at $\frac{1}{4}$, $\frac{1}{2}$ and $\frac{3}{4}$ of the total frequency.

The second quartile coincides with the median, and reading off from the curve the lower (or first) quartile is 8.89 N/mm² at a frequency of $\frac{1}{4} \times 80$ or 20. Also the higher (or third) quartile is 9.06 N/mm² at a frequency of $\frac{3}{4} \times 80$ or 60.

THE AVERAGE

Mean, median and mode are the more common types of average.

THE MEAN (OR ARITHMETIC MEAN)

The arithmetic mean is found by adding up all the observations in a set and dividing the result by the number of observations. That is,

$$\text{Arithmetic mean} = \frac{\text{Sum of the observations}}{\text{Number of observations}}$$

EXAMPLE 22.4

Five turned bars are measured and their diameters were found to be: 15.03, 15.02, 15.02, 15.00 and 15.03 mm. What is their mean diameter?

$$\text{Mean diameter} = \frac{15.03 + 15.02 + 15.02 + 15.00 + 15.03}{5}$$

$$= \frac{75.10}{5} = 15.02 \, \text{mm}$$

THE MEAN OF A FREQUENCY DISTRIBUTION

The mean of a frequency distribution must take into account the frequencies as well as the measured observations.

If $x_1, x_2, x_3 \ldots x_n$ are measured observations which have frequencies $f_1, f_2, f_3 \ldots f_n$ then the mean of the distribution is

$$\bar{x} = \frac{x_1 f_1 + x_2 f_3 + x_3 f_3 + \ldots + x_n f_n}{f_1 + f_2 + f_3 \ldots + f_n} = \frac{\Sigma \, xf}{\Sigma \, f}$$

The symbol Σ simply means the 'sum of'. Thus Σxf tells us to multiply together corresponding values of x and f and add the results together.

Finding the mean of grouped frequency distributions follows later in the chapter.

EXAMPLE 22.5

Five castings have a mass of 20.01 kg each, three have mass of 19.98 kg each and two have mass of 20.03 kg each. What is the mean mass of the ten castings?

The total mass is

$$(5 \times 20.01) + (3 \times 19.98) + (2 \times 20.03) = 200.05 \, \text{kg}$$

$$\text{Mean mass} = \frac{\text{Total mass of the castings}}{\text{Number of castings}} = \frac{200.05}{10} = 20.005 \, \text{kg}$$

THE MEDIAN

If a set of numbers is arranged in ascending (or descending) order of size, the median is the value which lies half-way along the set. Thus for the set

$$3, 4, 4, 5, 6, 7, 7, 9, 10$$

the median is 6

If there is an even number of values the median is found by taking the average of the two middle values. Thus for the set

$$3, 3, 5, 7, 9, 10, 13, 15$$

the median is $\frac{1}{2}(7 + 9) = 8$

EXAMPLE 22.6

The hourly wages of five employees in an office are £3.04, £5.92, £4,56, £16.40 and £5.50. Find the median.

Arranging the amounts in ascending order we have

$$£3.04, \quad £4.56, \quad £5.50, \quad £5.92, \quad £16.40$$

The median is therefore £5.50.

The Median of a Frequency Distribution

This is shown in Example 22.3 (p. 337).

THE MODE

The mode of a set of numbers is the number which occurs most frequently. Thus the mode of

$$2 \ 3 \ 3 \ 4 \ 4 \ 4 \ 5 \ 5 \ 6 \ 6 \ 7 \ 8$$

is 4, since this number occurs three times which is more than any of the other numbers in the set.

For a set of numbers the mode may not exist. Thus the set of numbers

$$4\ 5\ 6\ 8\ 9\ 10\ 12$$

has no mode.

It is possible for there to be more than one mode. The set of numbers

$$2\ 3\ 3\ 5\ 5\ 5\ 6\ 6\ 7\ 8\ 8\ 8\ 9\ 10$$

has two modes, 5 and 8. The set of numbers is said to be *bimodal*. If there is only one mode, then the set of numbers is said to be *unimodal*.

The Mode from a Histogram

Using the tallest rectangle the construction is as shown in Fig. 22.11 for a typical histogram representation of a frequency distribution.

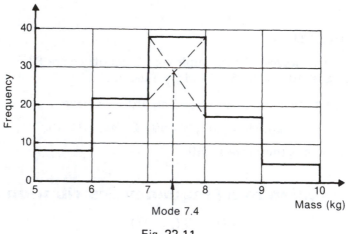

Fig. 22.11

DISCUSSION ON THE MEAN, MEDIAN AND MODE

Which of the above averages to use will depend upon the particular application.

The arithmetic mean is the most familiar kind of average and it is used extensively in statistical work. However it can be misleading especially if it is affected by extreme values. Again the mean size of screws is not of much use to a purchasing officer, because it might be at some point between stock sizes — better here to use the mode.

The median is the value halfway along a set of numbers and simple to find. Not often favoured in scientific work.

The mode will indicate the most popular item(s) and so will help the stock purchasing officer, and is useful in applications of this nature.

Exercise 22.2

1) Find the mode of the following set of numbers:

$$3, 5, 2, 7, 5, 8, 5, 2, 7, 6$$

2) Find the mode of: 38.7, 29.6, 32.1, 35.8, 43.2

3) Find the modes of: 8, 4, 9, 3, 5, 3, 8, 5, 3, 8, 9, 5, 6, 7

4) The data below relates to the resistance in ohms of an electrical part. Find the mode of this distribution, by drawing a histogram.

Resistance (ohms)	119	120	121	122	123	124
Frequency	5	9	19	25	18	4

5) The information below shows the distribution of the diameters of rivet heads for rivets manufactured by a certain company.

Diameter (mm)	18.407–18.412	18.413–18.418	18.419–18.424
Frequency	2	6	8

Diameter (mm)	18.425–18.430	18.431–18.436	18.437–18.442
Frequency	12	7	3

Find the mode of this distribution.

6) Find the median of the following set of numbers:

$$9, 2, 7, 3, 8, 5, 4$$

7) A student receives the following marks in an examination in five subjects: 84, 77, 95, 80 and 97. What is the median mark?

8) The following are the weekly wages earned by six people working in a small factory: £114, £213, £177, £189, £174 and £204. What is the median wage?

9) Find the mean and the median for the following set of observations: 15.63, 14.95, 16.00, 12.04, 15.88 and 16.04 ohms. Which of the two, the median or the mean, is, in your opinion, the better to use for these observations?

10) Draw up a cumulative frequency distribution for the distribution given in Table 22.7. Hence find the median and the 1st and 3rd quartiles for the distribution.

TABLE 22.7

Resistance (ohms)	Frequency
115	3
116	7
117	12
118	20
119	15
120	8
121	2

11) By drawing an ogive for the distribution of Table 22.8 determine the values of the quartiles.

TABLE 22.8

Diameter (mm)	Frequency
20.00–20.03	4
20.04–20.07	12
20.08–20.11	23
20.12–20.15	11
20.16–20.19	2

12) The diameters of eight pipes were measured with the following results: 109.23, 109.21, 108.98, 109.03, 108.98. 109.22, 109.20, 108.91 mm. What is the mean diameter of the pipes?

13) 22 bricks have a mean mass of 24 kg and 18 similar bricks have a mean mass of 23.7 kg. What is the mean mass of the 40 bricks?

14) A sample of 100 lengths of timber was measured with the following results:

Length (m)	9.61	9.62	9.63	9.64	9.65	9.66	9.67	9.68	9.69
Frequency	2	4	12	18	31	22	8	2	1

Calculate the mean length of the timber.

15) The table below shows the distribution of the maximum loads supported by certain cables.

Max. load (kN)	19.2–19.5	19.6–19.9	20.0–20.3	20.4–20.7
Frequency	4	12	18	3

Calculate the mean load which the cables will support.

THE NORMAL DISTRIBUTION

It will be recalled that a frequency distribution may be represented by a frequency curve (Fig. 22.9). Where the data has been obtained by actual measurement this approximates to a symmetrical bell-shaped curve. This will only be so if sufficient measurements are made — usually 100 will suffice. This curve is called a normal distribution curve and it may be defined in terms of total frequency, the arithmetic mean and the standard deviation.

Since the normal curve is symmetrical about its vertical centre-line, then this centre-line represents the mean of the distribution. This mean locates the position of the curve from the reference axis as shown in Fig. 22.12 which displays similar distributions with different means.

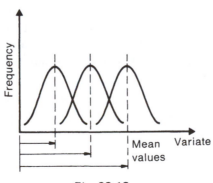

Fig. 22.12

Spread or Dispersion

The most valuable and widely used measure for this is the standard deviation, always represented by the Greek letter σ (sigma), which gives an idea of dispersion about the mean (Fig. 22.13).

Although the normal curve extends to infinity on either side of the mean, for most practical purposes it may be assumed to terminate at three standard deviations on either side of the mean (Fig. 22.14).

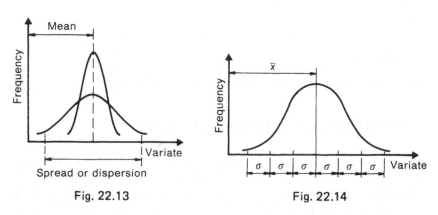

Fig. 22.13   Fig. 22.14

Calculation of the Mean and the Standard Deviation

The formulae used are:

$$\text{Mean } \bar{x} = \frac{\Sigma xf}{\Sigma f} \qquad \text{Standard deviation } \sigma = \sqrt{\frac{\Sigma x^2 f}{f} - \bar{x}^2}$$

A scientific calculator will generally have the facilities for you to enter the values of the variate x and the corresponding values of frequency f. The results may then be found from the keys labelled $\bar{x}$ and σ_n. Some machines have another standard deviation labelled σ_{n-1} which is based on a slightly different formula — this is *not* the one we require here.

Each make of calculator will require a different procedure and will be given in the instruction manual which accompanies the machine.

Alternatively the two examples which follow show a tabulation method. This is called the coded method and this uses the idea of choosing a mean value and then working out by how much the true mean value varies.

EXAMPLE 22.7

Calculate the mean and standard deviation for the following frequency distribution.

Resistance (ohm)	5.37	5.38	5.39	5.40	5.41	5.42	5.43	5.44
Frequency	4	10	14	24	34	18	10	6

Chosen value of $x = 5.40$ Unit size $= 0.01$ ohm

x	x_c	f	$x_c f$	$x_c^2 f$
5.37	-3	4	-12	36
5.38	-2	10	-20	40
5.39	-1	14	-14	14
5.40	0	24	0	0
5.41	1	34	34	34
5.42	2	18	36	72
5.43	3	10	30	90
5.44	4	6	24	96
		120	78	382

Now $\bar{x}_c = \dfrac{\Sigma x_c f}{\Sigma f} = \dfrac{78}{120} = 0.65$

$\therefore \quad \bar{x} = 5.40 + 0.65 \times 0.01 = 5.4065$ ohm

and $\sigma_c = \sqrt{\dfrac{\Sigma x_c^2 f}{\Sigma f} - (\bar{x}_c)^2} = \sqrt{\dfrac{382}{120} - (0.65)^2} = \sqrt{2.7068}$

$\qquad = 1.662$

$\therefore \quad \sigma = \sigma_c \times \text{unit size} = 1.662 \times 0.01 = 0.016\,62$ ohm

Rough Check on Standard Deviation Using the Range

The range is the difference between the largest value and the smallest value of a set. It gives some idea of the spread but it depends solely on end values — it gives no idea of the distribution of data and is never used as a measure for calculation purposes.

In the last example: Range $= 5.44 - 5.37 = 0.07$

Hence: Approximate $\sigma = \dfrac{0.07}{6} = 0.012$ ohm

This does not verify the accuracy of the calculated value 0.016 62 but it does show it is of the right order (i.e. not wildly incorrect).

EXAMPLE 22.8

The table indicates experimental results from a sample of resistors.

Resistance (ohm)	24.92–24.94	24.95–24.97	24.98–25.00
Frequency	2	3	9

Resistance (ohm)	25.01–25.03	25.04–25.06	25.07–25.09
Frequency	23	18	5

Calculate the standard deviation.

Chosen value of $x = 25.02$ ohm   Unit size $= 0.03$ ohm

Class	x	x_c	f	$x_c f$	$x_c^2 f$
24.92–24.94	24.93	-3	2	-6	18
24.95–24.97	24.96	-2	3	-6	12
24.98–25.00	24.99	-1	9	-9	9
25.01–25.03	25.02	0	23	0	0
25.04–25.06	25.05	1	18	18	18
25.07–25.09	25.08	2	5	10	20
			60	7	77

A unit size of 0.03 ohm has been chosen because each value of x differs from its preceding value by 0.03. Making the unit size as large as possible simplifies the calculation of the standard deviation.

Now $\bar{x}_c = \dfrac{+7}{60} = +0.116$

∴ $\bar{x} = 25.02 + (0.03)(0.116) = 25.0235$ ohm

Also $\sigma_c = \sqrt{\dfrac{77}{60} - (0.116)^2} = \sqrt{1.2697} = 1.1268$

∴ $\sigma = 1.1268 \times 0.03 = 0.0338$ ohm

(A rough check for σ gives $\dfrac{25.09 - 24.92}{6} = 0.0283$, which is of the same order as the value calculated above.)

CHANCE (OR PROBABILITY) FROM THE NORMAL CURVE

If a sample — the bigger the better but at least 100 measurements — has been taken from a production line, the results will generally follow the profile of a normal curve. Fig. 22.15 shows the area of each vertical strip as a percentage of the total area under the normal curve. As you will see, these areas may be used to estimate the chances of something happening, which is extremely useful, for example, in quality control.

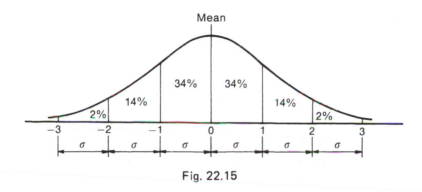

Fig. 22.15

For instance, there is a 34% chance of something happening between the mean and one standard deviation from the mean. Tables are available for the areas of any width of strip, but those shown in Fig. 22.15 will suffice to give you the general idea.

On the base scale of -3 to $+3$ the value corresponding to a particular value of variate x is given by $(x - \bar{x})/\sigma$.

EXAMPLE 22.9

By measuring a large number of components produced on an automatic lathe it was found that the mean length was 20.10 mm with a standard deviation of 0.03 mm. Find

a) Within what limits would you expect the lengths for the whole of the components to lie.

b) The chance that one component taken at random would have:
    (i)   A length greater than 20.16 mm,
    (ii) A length less than 20.07 mm,
    (iii) A length between 20.04 mm and 20.13 mm.

a) For most practical purposes the normal curve may be regarded as terminating at three standard deviations either side of the mean. Thus we would expect the lengths for the whole of the components to lie between

$$\text{Mean} \pm 3\sigma = 20.10 \pm 3 \times 0.03$$

$$= 20.10 \pm 0.09$$

$$= \text{between } 20.01 \text{ and } 20.19 \text{ mm}$$

b) (i) For $x = 20.16$:

$$\text{Scale value} = \frac{20.16 - 20.10}{0.03} = 2$$

Now the area is 2% between 2 and the upper limit, so we would expect 2% of all components to have lengths greater than 20.16 mm.

(ii) For $x = 20.07$:

$$\text{Scale value} = \frac{20.07 - 20.10}{0.03} = -1$$

Now the area between -1 and the lower limit is $2\% + 14\% = 16\%$, so we would expect 16% of all components to have lengths less than 20.07 mm.

(iii) For $x = 20.04$:

$$\text{Scale value} = \frac{20.04 - 20.10}{0.03} = -2$$

for $x = 20.13$:

$$\text{Scale value} = \frac{20.13 - 20.10}{0.03} = 1$$

Now the area between -2 and 1 is $14\% + 34\% + 34\% = 82\%$, so we would expect 82% of all components to have lengths between 20.04 and 20.13 mm.

CURVE FITTING

Readings which are obtained as a result of an experiment usually contain errors in measurement and observation. When the points are plotted on a graph it is usually possible to visualise a straight line or a curve which approximates to the data. Thus in Fig. 22.16 the data appears to be approximated by a straight line whilst in Fig. 22.17 the data is approximated by a smooth curve.

Figs. 22.16 and 22.17 are called *scatter diagrams*. The problem is to find equations of curves or straight lines which approximately fit the plotted data. Finding equations for the approximating curves or straight lines is called *curve fitting*.

Individual judgement may be used to draw the approximating straight line or curve but this has the disadvantage that different individuals will obtain different straight lines or curves and hence different equations. To avoid this disadvantage the method of least squares is usually used to obtain the equation of the approximating curve or straight line. Here we shall only consider a straight line.

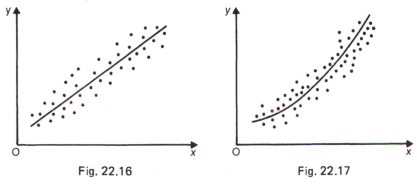

Fig. 22.16   Fig. 22.17

Method of Least Squares

Consider Fig. 22.18 where an approximating straight line has been drawn to fit the given data. There is a deviation between the point (x_i, y_i) of the given data and the point (x_i, Y_i) which lies on the approximating straight line. This deviation is:

$$D_i = Y_i - y_i$$

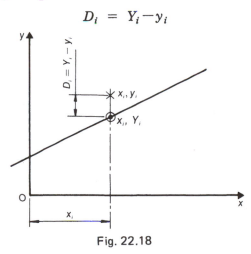

Fig. 22.18

The straight line having the property that:

$$\Sigma D_i^2 \;=\; D_1^2 + D_2^2 + D_3^2 + \ldots$$

is a minimum, is called the best fitting straight line

The Least Square Line

The best straight line approximating to a set of points (x_1, y_1), $(x_2, y_2) \ldots (x_n, y_n)$ has the equation

$$y \;=\; a + bx$$

It can be shown that

$$\Sigma y \;=\; an + b\Sigma x \qquad\qquad [1]$$

$$\Sigma xy \;=\; a\Sigma x + b\Sigma x^2 \qquad\qquad [2]$$

These are a pair of simultaneous equations from which the constants a and b may be found, n being the number of points. The equation obtained is called the *least square line*.

Some scientific calculators have the facilities for you to enter corresponding values of x and y, and then obtain the values of constants a and b to find the best fitting straight line.

Alternatively the following example shows a suitable method.

EXAMPLE 22.10

The table shows the results of an experiment to establish the relationship between the resistance of a conductor (R ohm) and its temperature ($t°C$).

t	25	50	75	100	125	150
R	20.7	21.6	22.2	23.0	23.9	24.6

a) Find the equation connecting R and t if it is of the type $R = a + bt$.
b) Estimate the value of R when $t = 121.1°C$.

a) Here R is the dependent variable (corresponding to y) and t is the independent variable (corresponding to x).

t	R	tR	t^2
25	20.7	517.5	625
50	21.6	1080.0	2500
75	22.2	1665.0	5625
100	23.0	2300.0	10 000
125	23.9	2987.5	15 625
150	24.6	3690.0	22 500
525	136.0	12 240.0	56 875

$\Sigma R = 136$, $\Sigma t = 525$ and $n = 6$

$$136 = 6a + 525b \qquad [1]$$

$\Sigma tR = 12\,240$, $\Sigma t = 525$ and $\Sigma t^2 = 56\,875$

$$12\,240 = 525a + 56\,875b \qquad [2]$$

Now from equation [1] and rearranging for a we have:

$$a = \frac{136}{6} - \left(\frac{525}{6}\right)b$$

and substituting this value for a into equation [2] we have:

$$12\,240 = 525\left[\frac{136}{6} - \left(\frac{525}{6}\right)b\right] + 56\,875b$$

To simplify divide through by 525 giving:

$$23.31 = 22.67 - 87.50b + 108.33b$$

from which $\qquad b = 0.0307$

Substituting this value into equation [1] we have:

$$a = \frac{136}{6} - \left(\frac{525}{6}\right)(0.0307)$$

giving $\qquad a = 19.98$

Hence $\qquad R = 19.98 + 0.0307t$

b) When $\qquad t = 121.1°C$

$$R = 19.98 + (0.0307 \times 121.1)$$
$$= 23.7 \text{ ohm}$$

Exercise 22.3

1) In a water absorption test on 100 bricks the following results were obtained:

% absorption	7	8	9	10	11	12	13	14
Frequency	1	4	9	24	30	26	5	1

Calculate the mean and standard deviation.

2) Find the mean and standard deviation for the following distribution which relates to the strength of load carrying bricks:

Strength (N/mm^2)	11.46	11.47	11.48	11.49	11.50	11.51	11.52	11.53
Frequency	1	4	12	15	11	6	3	1

3) 100 watt is the nominal value of the sample of electric light bulbs given below. Calculate the mean and standard deviation from the sample.

Power (watt)	99.6	99.7	99.8	99.9	100.0	100.1	100.2	100.3
Frequency	3	8	13	18	15	9	6	3

4) A brand of washing powder tested prior to marketing revealed its capacity to launder woollen garments of equivalent sizes.

Laundered garments	5–7	8–10	11–13	14–16	17–19
Frequency	1	5	11	7	3

From the data above calculate the mean and standard deviation.

5) It is considered that a person uses, on average, 40 gallons of water daily. Using the following data calculate the mean water consumption and the standard deviation.

Daily consumption (gallons)	30–34	35–39	40–44	45–49	50–54	55–59
Number of users	3	19	43	26	7	2

6) Measurement from a large batch of mass produced components showed a mean diameter of 18.60 mm, with a standard deviation of 0.02 mm. Find:

(a) within what limits the diameters of the whole of the components would be expected to lie,

(b) The chance that one component taken at random would have
    (i)   a diameter greater than 18.62 mm,
    (ii) a diameter less than 18.56 mm.

7) In mass production of bushes it was found that the average bore was 12.5 mm with a standard deviation of 0.015 mm. If 2000 bushes are produced find the number of bushes that are expected to have dimensions between 12.47 and 12.53 mm.

8) 70 000 components for a motor vehicle are being produced. A batch of 300 was picked at random and lengths were checked to the nearest 0.01 mm with the following results:

Length (mm)	9.96	9.97	9.98	9.99	10.00
Frequency	1	6	25	72	93

Length (mm)	10.01	10.02	10.03	10.04
Frequency	69	27	6	1

Using results correct to four decimal places, find how many components of the 70 000 produced will be expected to have

(a) lengths less than 9.9744 mm,

(b) lengths between 10.0128 and 10.0256 mm.

9) A test on a metal filament lamp gave the following values of resistance (R ohm) at various voltages (V volt):

V	62	75	89	100	120
R	100	118	136	149	176

R and V are connected by an equation of the type $R = aV + b$. Determine the equation by finding the least square line.

10) During a test with a thermocouple pyrometer the e.m.f. (E millivolt) was measured against the temperature of the hot junction ($t°C$) and the following results were obtained:

t	200	300	400	500	600	700	800	900	1000
E	6	9.1	12.0	14.8	18.2	21.0	24.1	26.8	30.2

By finding the least square line determine an equation of the standard straight line form which gives E in terms of t. Hence find the value of E when $t = 840°C$.

23. DIFFERENTIATION

THE GRADIENT OF A CURVE

Graphical Method

In mathematics and technology we often need to know the rate of change of one variable with respect to another. For instance, velocity is the rate of change of distance with respect to time, and acceleration is the rate of change of velocity with respect to time.

Consider the graph of $y = x^2$, part of which is shown in Fig. 23.1. As the values of x increase so do the values of y, but they do not increase at the same rate. A glance at the portion of the curve shown shows that the values of y increase faster when x is large, because the gradient of the curve is increasing.

To find the rate of change of y with respect to x at a particular point we need to find the gradient of the curve at that point.

If we draw a tangent to the curve at the point, the gradient of the tangent will be the same as the gradient of the curve.

EXAMPLE 23.1

Find the gradient of the curve $y = x^2$ at the point where $x = 2$

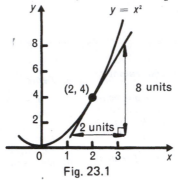

Fig. 23.1

The point where $x = 2$ is the point $(2, 4)$. We draw a tangent at this point, as shown in Fig. 23.1. Then by constructing a right-angled triangle the gradient is found to be $\frac{8}{2} = 4$. This gradient is positive, in accordance with our previous work, since the tangent slopes upwards from left to right.

EXAMPLE 23.2

Draw the graph of $y = x^2 - 3x + 7$ between $x = -4$ and $x = 3$ and hence find the gradient at: a) the point $x = -3$, b) the point $x = 2$.

a) At the point where $x = -3$,
$$y = (-3)^2 - 3(-3) + 7 = 25$$

At the point $(-3, 25)$ draw a tangent as shown in Fig. 23.2. The gradient is found by drawing a right-angled triangle (which should be as large as conveniently possible for accuracy) as shown, and measuring its height and base.

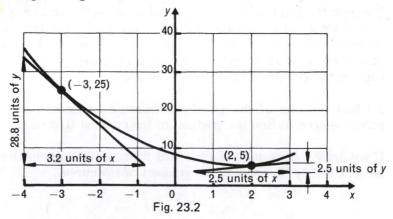

Fig. 23.2

Hence: Gradient at point $(-3, 25) = -\dfrac{28.8}{3.2} = -9$

the negative sign indicating a downward slope from left to right.

b) At the point where $x = 2$,
$$y = 2^2 - 3(2) + 7 = 5$$

Hence by drawing a tangent and a right-angled triangle at the point $(2, 5)$ in a similar manner to above,

$$\text{Gradient at point } (2, 5) = \dfrac{2.5}{2.5} = 1$$

being positive as the tangent slopes upwards from left to right.

Exercise 23.1

1) Draw the graph of $2x^2 - 5$ for values of x between -2 and $+3$. Draw, as accurately as possible, the tangents to the curve at the points where $x = -1$ and $x = +2$ and hence find the gradient of the curve at these points.

2) Draw the curve $y = x^2 - 3x + 2$ from $x = 2.5$ to $x = 3.5$ and find its gradient at the point where $x = 3$

3) Draw the curve $y = x - \dfrac{1}{x}$ from $x = 0.8$ to 1.2. Find its gradient at $x = 1$

Numerical Method

The gradient of a curve may always be found by graphical means but this method is often inconvenient. A numerical method will now be developed.

Consider the curve $y = x^2$, part of this is shown in Fig. 23.3. Let P be the point on the curve at which $x = 1$ and $y = 1$. Q is a variable point on the curve, which will be considered to start at the point $(2, 4)$ and move down the curve towards P, rather like a bead slides down a wire.

The symbol δx will be used to represent an increment of x, and δy will be used to represent the corresponding increment of y. The gradient of the chord PQ is then $\dfrac{\delta y}{\delta x}$

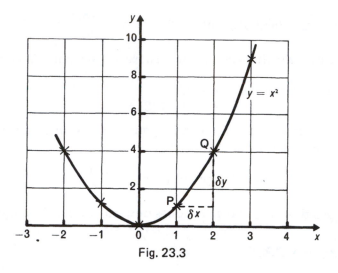

Fig. 23.3

When Q is at the point $(2, 4)$ then $\delta x = 1$, and $\delta y = 3$

$$\therefore \qquad \frac{\delta y}{\delta x} = \frac{3}{1} = 3$$

The following table shows how $\dfrac{\delta y}{\delta x}$ alters as Q moves nearer and nearer to P.

Co-ordinates of Q				Gradient of
x	y	δy	δx	$PQ = \dfrac{\delta y}{\delta x}$
2	4	3	1	3
1.5	2.25	1.25	0.5	2.5
1.4	1.96	0.96	0.4	2.4
1.3	1.69	0.69	0.3	2.3
1.2	1.44	0.44	0.2	2.2
1.1	1.21	0.21	0.1	2.1
1.01	1.0201	0.0201	0.01	2.01
1.001	1.002001	0.002001	0.001	2.001

It will be seen that as Q approaches nearer and nearer to P, the value of $\dfrac{\delta y}{\delta x}$ approaches 2. It is reasonable to suppose that eventually when Q coincides with P (that is, when the chord PQ becomes a tangent to the curve at P) the gradient of the tangent will be exactly equal to 2. The gradient of the tangent will give us the gradient of the curve at P.

Now as Q approaches P, δx tends to zero and the gradient of the chord, $\dfrac{\delta y}{\delta x}$, tends, in the limit (as we say), to the gradient of the tangent. We denote the gradient of the tangent as $\dfrac{dy}{dx}$. We can write all this as

$$\underset{\delta x \to 0}{\text{Limit}} \frac{\delta y}{\delta x} = \frac{dy}{dx}$$

DIFFERENTIATION FROM FIRST PRINCIPLES

Instead of selecting special values for δy and δx let us now consider the general case, so that P has the co-ordinates (x, y) and Q has the co-ordinates $(x + \delta x, y + \delta y)$, (Fig. 23.4). Q is taken very close to P, so that δx is a very small quantity.

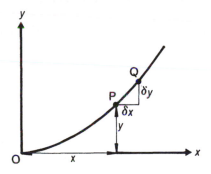

Fig. 23.4

Now
$$y = x^2$$

and as $Q(x + \delta x, y + \delta y)$ lies on the curve, then
$$y + \delta y = (x + \delta x)^2$$
$\therefore$
$$y + \delta y = x^2 + 2x\,\delta x + (\delta x)^2$$

But $y = x^2$, so
$$\delta y = 2x\,\delta x + (\delta x)^2$$

and, by dividing both sides by δx, the gradient of chord PQ is
$$\frac{\delta y}{\delta x} = 2x + \delta x$$

As Q approaches P, δx tends to zero and $\dfrac{\delta y}{\delta x}$ tends, in the limit, to the gradient of the tangent of the curve at P.

Thus $\qquad \underset{\delta x \to 0}{\text{Limit}} \dfrac{\delta y}{\delta x} = \dfrac{dy}{dx} = 2\acute{x}$

The process of finding $\dfrac{dy}{dx}$ is called *differentiation*.

The symbol $\dfrac{dy}{dx}$ means the differential coefficient of y with respect to x.

We can now check our assumption regarding the gradient of the curve at P. Since at P the value of $x = 1$, then substituting in the expression $\dfrac{dy}{dx} = 2x$ we get $\dfrac{dy}{dx} = 2 \times 1 = 2$

and we see that our assumption was correct.

DIFFERENTIAL COEFFICIENT OF x^n

It can be shown, by a method similar to that used for finding the differential coefficient of x^2, that

If	$y = x^n$
then	$\dfrac{dy}{dx} = nx^{n-1}$

This is true for all values of n including negative and fractional indices. When we use it as a formula it enables us to avoid having to differentiate each time from first principles.

EXAMPLE 23.3

$y = x^3$

$\therefore \dfrac{dy}{dx} = 3x^2$

$y = \dfrac{1}{x} = x^{-1}$

$\therefore \dfrac{dy}{dx} = -x^{-2} = -\dfrac{1}{x^2}$

$y = \sqrt{x} = x^{1/2}$

$\therefore \dfrac{dy}{dx} = \dfrac{1}{2}x^{1/2} = \dfrac{1}{2}\dfrac{1}{x^{1/2}} = \dfrac{1}{2\sqrt{x}}$

$y = \sqrt[5]{x^2} = x^{2/5}$

$\therefore \dfrac{dy}{dx} = \dfrac{2}{5}x^{2/5-1} = \dfrac{2}{5}x^{-3/5} = \dfrac{2}{5(\sqrt[5]{x^3})}$

When a power of x is multiplied by a constant, that constant remains unchanged by the process of differentiation.

Hence if $\qquad\qquad y = ax^n$

then $\qquad\qquad \dfrac{dy}{dx} = anx^{n-1}$

EXAMPLE 23.4

$$y = 2x^{1.3}$$

$$\therefore \; \frac{dy}{dx} = 2(1.3)x^{0.3} = 2.6x^{0.3}$$

$$y = \tfrac{1}{5}x^7$$

$$\therefore \; \frac{dy}{dx} = \frac{1}{5} \times 7x^6 = \frac{7}{5}x^6$$

$$y = \tfrac{3}{4}\sqrt[3]{x} = \tfrac{3}{4}x^{1/3}$$

$$\therefore \; \frac{dy}{dx} = \frac{3}{4} \times \frac{1}{3}x^{-2/3} = \frac{1}{4}x^{-2/3}$$

$$y = \frac{4}{x^2} = 4x^{-2}$$

$$\therefore \; \frac{dy}{dx} = 4(-2)x^{-3} = -8x^{-3}$$

When a numerical constant is differentiated the result is zero. This can be seen since $x^0 = 1$ and we can write, for example, constant 4 as $4x^0$, then differentiating with respect to x we get

$$4(0)x^{-1} = 0$$

If, as an alternative method, we plot the graph of $y = 4$ we get a straight line parallel with the x-axis as shown in Fig. 23.5.

The gradient of the line is zero: that is, $\dfrac{dy}{dx} = 0$

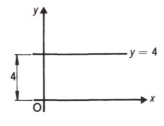

Fig. 23.5

To differentiate an expression containing the sum of several terms, differentiate each individual term separately.

EXAMPLE 23.5

a) $y = 3x^2 + 2x + 3$

∴ $\dfrac{dy}{dx} = 3(2)x + 2(1)x^0 + 0 = 6x + 2$

b) $y = ax^3 + bx^2 + cx + d$ where a, b, c and d are constants,

∴ $\dfrac{dy}{dx} = 3ax^2 + 2bx + c$

So far our differentiation has been in terms of x and y only. But they are only letters representing variables and we may choose other letters or symbols.

c) $s = \sqrt{t} + \dfrac{1}{\sqrt{t}} = t^{1/2} + t^{-1/2}$

∴ $\dfrac{ds}{dt} = \dfrac{1}{2}t^{-1/2} + \left(-\dfrac{1}{2}\right)t^{-3/2} = \dfrac{1}{2\sqrt{t}} - \dfrac{1}{2\sqrt{t^3}}$

d) $v = 3.1u^{1.4} - \dfrac{3}{u} + 5 = 3.1u^{1.4} - 3u^{-1} + 5$

∴ $\dfrac{dv}{du} = (3.1)(1.4)u^{0.4} - 3(-1)u^{-2} = 4.34u^{0.4} + \dfrac{3}{u^2}$

Finding the Gradient of a Curve by Differentiation

EXAMPLE 23.6

Find the gradient of the graph $y = 3x^2 - 3x + 4$:

a) when $x = 3$, and,   b) when $x = -2$

The gradient at a point is expressed by $\dfrac{dy}{dx}$.

If $y = 3x^2 - 3x + 4$

then $\dfrac{dy}{dx} = 6x - 3$

a) When $x = 3$ b) When $x = -2$

$\dfrac{dy}{dx} = 6(3) - 3 = 15$ $\dfrac{dy}{dx} = 6(-2) - 3 = -15$

Exercise 23.2

Differentiate the following:

1) $y = x^2$
2) $y = x^7$
3) $y = 4x^3$

4) $y = 6x^5$
5) $s = 0.5t^3$
6) $A = \pi R^2$

7) $y = x^{1/2}$
8) $y = 4x^{3/2}$
9) $y = 2\sqrt{x}$

10) $y = 3 \times \sqrt[3]{x^2}$
11) $y = \dfrac{1}{x^2}$
12) $y = \dfrac{1}{x}$

13) $y = \dfrac{3}{5x}$
14) $y = \dfrac{2}{x^3}$
15) $y = \dfrac{1}{\sqrt{x}}$

16) $y = \dfrac{2}{3\sqrt{x}}$
17) $y = \dfrac{5}{x\sqrt{x}}$
18) $s = \dfrac{3\sqrt{t}}{5}$

19) $K = \dfrac{0.01}{h}$
20) $y = \dfrac{5}{x^7}$

21) $y = 4x^2 - 3x + 2$
22) $s = 3t^3 - 2t^2 + 5t - 3$

23) $q = 2u^2 - u + 7$
24) $y = 5x^4 - 7x^3 + 3x^2 - 2x + 5$

25) $s = 7t^5 - 3t^2 + 7$
26) $y = \dfrac{x + x^3}{\sqrt{x}}$

27) $y = \dfrac{3 + x^2}{x}$
28) $y = \sqrt{x} + \dfrac{1}{\sqrt{x}}$
29) $y = x^3 + \dfrac{3}{\sqrt{x}}$

30) $s = t^{1.3} - \dfrac{1}{4t^{2.3}}$
31) $y = \dfrac{3x^3}{5} - \dfrac{2x^2}{7} + \sqrt{x}$

32) $y = 0.08 + \dfrac{0.01}{x}$
33) $y = 3.1x^{1.5} - 2.4x^{0.6}$

34) $y = \dfrac{x^3}{2} - \dfrac{5}{x} + 3$
35) $s = 10 - 6t + 7t^2 - 2t^3$

36) Find the gradient of the curve $y = 3x^2 + 7x + 3$ at the points where $x = -2$ and $x = 2$

37) Find the gradient of the curve $y = 2x^3 - 7x^2 + 5x - 3$ at the points where $x = -1.5$, $x = 0$ and $x = 3$

38) Find the values of x for which the gradient of the curve $y = 3 + 4x - x^2$ is equal to:
(a) -1 (b) 0 (c) 2?

To Find $\dfrac{d}{d\theta}$ (sin θ) or the Rate of Change of sin θ

The rate of change of a curve at any point is the gradient of the tangent at that point. We shall, therefore, find the gradient at various points on the graph of sin θ and then plot the values of these gradients to obtain a new graph.

It is suggested that the reader follows the method given, plotting his own curves on graph paper.

First, we plot the graph of $y = \sin\theta$ from $\theta = 0°$ to $\theta = 90°$ using values of sin θ obtained from tables which are:

θ	0°	15°	30°	45°	60°	75°	90°
$y = \sin\theta$	0	0.259	0.500	0.707	0.866	0.966	1.000

These values are shown plotted in Fig. 23.6.

Consider point P on the curve, where $\theta = 45°$, and draw the tangent APM.

We can find the gradient of the tangent by constructing a suitable right-angled triangle AMN (which should be as large as conveniently possible for accuracy) and finding the value of $\dfrac{MN}{AN}$.

Using the scale on the y-axis gives, by measurement, MN = 1.29.
Using the scale on the θ-axis gives, by measurement, AN = 104°.

In calculations of this type it is necessary to obtain AN in radians.

Since $\qquad\qquad 360° = 2\pi$ radians

then $\qquad\qquad 1° = \dfrac{2\pi}{360}$ radians

and $\qquad\qquad 104° = \dfrac{2\pi}{360} \times 104 = 1.81$ radians

Hence $\qquad$ Gradient at P $= \dfrac{MN}{AN} = \dfrac{1.29}{1.81} = 0.71$

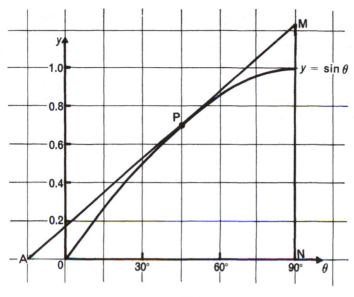

Fig. 23.6

The value 0.71 is used as the y-value at $\theta = 45°$ to plot a point on a new graph using the same scales as before. This new graph could be plotted on the same axes as $y = \sin \theta$ but for clarity it has been shown on new axes in Fig. 23.7.

Fig. 23.7

This procedure is repeated for points on the $\sin\theta$ curve at θ values $0°$, $15°$, $30°$, $60°$, $75°$ and $90°$ and the new curve obtained will be as shown in Fig. 23.7. This is the graph of the gradients of the sine curve at various points.

If we now plot a graph of $\cos\theta$, taking values from tables, on the axes in Fig. 23.7 we shall find that the two curves coincide—any difference will be due to errors from drawing the tangents.

Hence the gradient of the $\sin\theta$ curve at any value of θ is the same as the value of $\cos\theta$.

In other words the rate of change of $\sin\theta$ is $\cos\theta$, provided that the angle θ is in radians.

In the above work we have only considered the graphs between $0°$ and $90°$ but the results are true for all values of the angle.

Hence if $\qquad y = \sin\theta \quad$ then $\quad \dfrac{dy}{dx} = \cos\theta$

or $\qquad \boxed{\dfrac{d}{d\theta}\left(\sin\theta\right) = \cos\theta} \qquad$ provided that θ is in radians

The same procedure may be used to show that

$$\boxed{\dfrac{d}{d\theta}\left(\cos\theta\right) = -\sin\theta} \qquad \text{provided that } \theta \text{ is in radians.}$$

EXAMPLE 23.7

Find $\dfrac{dy}{dx}$ if $y = 3\sin x + 2\cos x$

We have $\qquad y = \sin x + \sin x + \sin x + \cos x + \cos x$

Since this is a sum of terms we may differentiate each in turn.

Thus $\qquad \dfrac{dy}{dx} = \cos x + \cos x + \cos x + (-\sin x) + (-\sin x)$

$\therefore \qquad \dfrac{dy}{dx} = 3\cos x - 2\sin x$

We can see from the above result that when either $\sin x$ or $\cos x$ are preceded by a constant multiplier it does not affect the differentiation but merely remains there.

EXAMPLE 23.8

Find the value of $\dfrac{d}{d\theta}(5 \sin \theta + 3 \cos \theta)$ if θ has a value equivalent to $25°$.

Let $\qquad y = 5 \sin \theta + 3 \cos \theta$

$\therefore \qquad \dfrac{dy}{d\theta} = 5 \cos \theta + 3(-\sin \theta)$

$\qquad\qquad = 5 \cos \theta - 3 \sin \theta$

Thus when $\theta = 25°$

$\qquad \dfrac{dy}{d\theta} = 5 \cos 25° - 3 \sin 25°$

A suitable sequence of operation on the calculating machine is:

Set the machine for calculations in degrees:

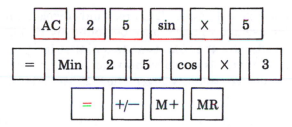

giving an answer 3.264 correct to four significant figures.

Exercise 23.3

1) Draw the curve of $\cos \theta$ between $\theta = 0°$ and $90°$ and show by graphical differentiation that the differential coefficient of $\cos \theta$ is $-\sin \theta$.

2) If $y = 3 \cos \theta - 2 \sin \theta$, find the value of $\dfrac{dy}{d\theta}$ if θ has a value equivalent to $20°$.

3) Find $\dfrac{dy}{dx}$ if $y = 5 \sin x + 3 \cos x$

4) Find $\dfrac{d}{d\theta}(4 \sin \theta - 2 \cos \theta)$

24. APPLICATION OF CALCULUS TO PROBLEMS

Outcome:

1. Understand the relationships between distance, time, velocity and acceleration.
2. Understand similar relationships for angular motion.
3. Calculate velocity and acceleration from distance–time equations, and similar for angular motion.
4. Appreciate small change approximations.
5. Calculate these small changes and percentage errors.
6. Define the turning point of a curve, and discover whether maximum or minimum using the gradient on either side of the point.
7. Solve problems involving maxima and minima relevant to technology.

VELOCITY AND ACCELERATION

Suppose that a vehicle starts from rest and travels 60 metres in 12 seconds. The average velocity may be found by dividing the total distance travelled by the total time taken, that is $\frac{60}{12} = 5$ m/s. This is *not* the *instantaneous* velocity, however, *at* a time of 12 seconds, but is the *average velocity* over the 12 seconds as calculated previously.

Fig. 24.1 shows a graph of distance s against time t. The average velocity over a period is given by the gradient of the chord which meets the curve at the extremes of the period. Thus in the diagram the gradient of the dotted chord QR gives the average velocity between $t = 2$ s and $t = 6$ s. It is found to be $\frac{13}{4} = 3.25$ m/s.

The velocity at any point is the rate of change of s with respect to t and may be found by finding the gradient of the curve at that point. In mathematical notation this is given by $\dfrac{ds}{dt}$.

Suppose we know that the relationship between s and t is

$$s = 0.417t^2$$

Then velocity $\qquad v = \dfrac{ds}{dt} = 0.834t$

and hence when $t = 12$ seconds, $v = 0.834 \times 12 = 10$ m/s.

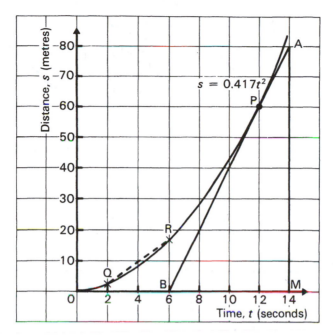

Fig. 24.1

This result may be found graphically by drawing the tangent to the curve of s against t at the point P and constructing a suitable right-angled triangle ABM.

Hence the velocity at $P = \dfrac{AM}{BM} = \dfrac{80}{8} = 10 \text{ m/s}$ which verifies the theoretical result.

Similarly, the rate of change of velocity with respect to time is called acceleration and is given by the gradient of the velocity–time graph at any point. In mathematical notation this is given by $\dfrac{dv}{dt}$.

Thus   Acceleration $a = \dfrac{dv}{dt}$

The above reasoning was applied to linear motion, but it could also have been used for angular motion. The essential difference is that distance s is replaced by angle turned through, θ rad.

Both sets of results are summarised in Fig. 24.2.

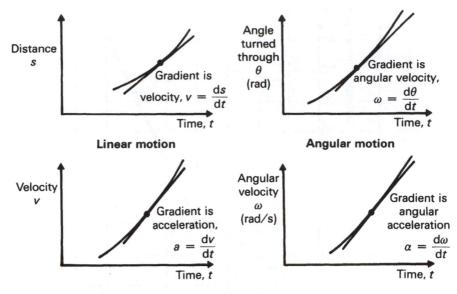

Fig. 24.2

EXAMPLE 24.1

A body moves a distance s metres in a time of t seconds so that $s = 2t^3 - 9t^2 + 12t + 6$. Find:

a) its velocity after 3 seconds,
b) its acceleration after 3 seconds,
c) when the velocity is zero.

We have $\qquad s = 2t^3 - 9t^2 + 12t + 6$

$\therefore \qquad v = \dfrac{ds}{dt} = 6t^2 - 18t + 12$

and $\qquad a = \dfrac{dv}{dt} = 12t - 18$

a) When $t = 3$, then the velocity

$$v = \frac{ds}{dt} = 6(3)^2 - 18(3) + 12 = 12\,\text{m/s}$$

b) When $t = 3$, then the acceleration

$$a = \frac{dy}{dt} = 12(3) - 18 = 18\,\text{m/s}^2.$$

c) When the velocity is zero then $\dfrac{ds}{dt} = 0$.

That is

$$6t^2 - 18t + 12 = 0$$

$\therefore$

$$t^2 - 3t + 2 = 0$$

$\therefore$

$$(t-1)(t-2) = 0$$

$\therefore$ either

$$t - 1 = 0 \quad \text{or} \quad t - 2 = 0$$

$\therefore$ either

$$t = 1 \text{ second} \quad \text{or} \quad t = 2 \text{ seconds}$$

EXAMPLE 24.2

The angle θ radians is connected with the time t seconds by the relationship $\theta = 20 + 5t^2 - t^3$. Find:

a) the angular velocity when $t = 2$ seconds,
b) the value of t when the angular deceleration is 4 rad/s^2.

We have

$$\theta = 20 + 5t^2 - t^3$$

$\therefore$

$$\omega = \frac{d\theta}{dt} = 10t - 3t^2$$

and

$$\alpha = \frac{d\omega}{dt} = 10 - 6t$$

a) When $t = 2$, then the angular velocity

$$\omega = \frac{d\theta}{dt} = 10(2) - 3(2)^2 = 8 \text{ rad/s}$$

b) An angular deceleration of 4 rad/s^2 may be called an angular acceleration of -4 rad/s^2.

$\therefore$ when $\alpha = \dfrac{d\omega}{dt} = -4$ then $-4 = 10 - 6t$

$$\text{or} \quad t = 2.33 \text{ seconds}$$

Exercise 24.1

1) If $s = 10 + 50t - 2t^2$, where s metres is the distance travelled in t seconds by a body, what is the velocity of the body after 2 seconds?

2) If $v = 5 + 24t - 3t^2$ where v m/s is the velocity of a body at a time t seconds, what is the acceleration when $t = 3$?

3) A body moves s metres in t seconds where $s = t^3 - 3t^2 - 3t + 8$. Find.

(a) its velocity at the end of 3 seconds,
(b) when its velocity is zero,
(c) its acceleration at the end of 2 seconds,
(d) when its acceleration is zero.

4) A body moves s metres in t seconds, where $s = \dfrac{1}{t^2}$. Find the velocity and acceleration after 3 seconds.

5) The distance s metres travelled by a falling body starting from rest after a time t seconds is given by $s = 5t^2$. Find its velocity after 1 second and after 3 seconds.

6) The distance s metres moved by the end of a lever after a time t seconds is given by the formula $s = 6t^2$. Find the velocity of the end of the lever when it has moved a distance $\frac{1}{2}$ metre.

7) The angular displacement θ radians of the spoke of a wheel is given by the expression $\theta = \frac{1}{2}t^4 - t^3$ where t seconds is the time. Find:

(a) the angular velocity after 2 seconds,
(b) the angular acceleration after 3 seconds,
(c) when the angular acceleration is zero.

8) An angular displacement θ radians in time t seconds is given by the equation $\theta = \sin 3t$. Find:

(a) the angular velocity when $t = 1$ second,
(b) the smallest positive value of t for which the angular velocity is 2 rad/s,
(c) the angular acceleration when $t = 0.5$ seconds,
(d) the smallest positive value of t for which the angular acceleration is 9 rad/s^2.

9) A mass of 5000 kg moves along a straight line so that the distance s metres travelled in a time t seconds is given by $s = 3t^2 + 2t + 3$. If v m/s is its velocity and m kg is its mass, then its kinetic energy is given by the formula $\frac{1}{2}mv^2$. Find its kinetic energy at a time $t = 0.5$ seconds, remembering that the joule (J) is the unit of energy.

SMALL CHANGE APPROXIMATIONS

If y is a function of x then the change δy in y corresponding to a small change δx in x is given by:

$$\frac{\delta y}{\delta x} \approx \frac{dy}{dx} \text{ (see p. 359)}$$

from which
$$\boxed{\delta y \approx \left(\frac{dy}{dx}\right)\delta x}$$

EXAMPLE 24.3

Find the approximate increase in the volume of a spherical container if the radius increases by 0.5 mm from 1.1 m.

The volume of a sphere $V = \frac{4}{3}\pi r^3$

from which $\dfrac{dV}{dr} = 4\pi r^2$

Now using $\delta V \approx \left(\dfrac{dV}{dr}\right)\delta r$

Substituting for $\dfrac{dV}{dr}$: $\delta V \approx 4\pi r^2 \delta r$

Hence approximate $\Big\}$ increase in volume $\delta V \approx 4\pi(1.1)^2\left(\dfrac{0.5}{1000}\right)$

$$\approx 0.007\,60 \text{ m}^3$$

EXAMPLE 24.4

Find the approximate error in calculating the area of a triangle if two sides 60 mm and 80 mm long and an included angle measured incorrectly as $26.2°$ instead of the true $26.5°$

If we let side $a = 60$ mm and side $b = 80$ mm then the included

angle $C = 26.5 \left(\dfrac{\pi}{180}\right)$ rad,

and $\delta C = (26.5 - 26.2)° = 0.3\left(\dfrac{\pi}{180}\right)$ rad

Note how the angles must be changed from degrees to radians.

Now triangular area $A = \frac{1}{2}ab \sin C$

from which $\dfrac{dA}{dC} = \frac{1}{2}ab \cos C$

Now using $\delta A \approx \left(\dfrac{dA}{dV}\right)\delta C$

Substituting for $\dfrac{dA}{dC}$:

$$\delta A \approx \tfrac{1}{2}ab\,(\cos C)\,\delta C \approx \tfrac{1}{2}(60)(80)\left\{\cos\left(26.5\times\dfrac{\pi}{180}\right)\right\}\left(0.3\times\dfrac{\pi}{180}\right)$$

Thus error in area $\delta A \approx 11.2\,\text{mm}^2$ too small.

Exercise 24.2

1) Find the approximate increase in the volume of a cone, whose height is 24 mm and radius 30 mm, if the radius increases by 0.3 mm.

2) A voltage, v, is given by $v = 3 \sin t$ volts. Find the approximate error in v if t is measured incorrectly as 0.034 instead of the correct 0.032 seconds.

3) The distance, s metres, travelled by an object falling freely, from rest, under the influence of gravity, is given by $s = \frac{1}{2}gt^2$ where $g\,\text{m/s}^2$ is the gravitational constant and t s is the time taken. Find the approximate change in the distance travelled when

(a) g is correct at 9.81 m/s^2 but t is taken wrongly as 1.96 instead of the correct 2.00 seconds, and
(b) t is taken correctly as 2.00 seconds but g is taken to be approximately 10 m/s^3 when its correct value is 9.81 m/s^2.

4) Find the approximate error in the total surface area of a fibre-glass hemispherical dome of nominal diameter 4 metres if the measured diameter is taken as 8 mm too small.

5) Find the approximate percentage error in the outer surface area of a cylindrical copper container 1.5 m high and base radius 0.5 m and having an open top. The height measurement is correct but the percentage error in the base radius is 2%.

Hint: Percentage error $= \dfrac{\text{Actual error}}{\text{True measurement}} \times 100.$

TURNING POINTS

Consider the graph of $y = x^3 + 3x^2 - 9x + 6$.

To plot the graph we draw up a table in the usual way:

x	-5	-4	-3	-2	-1	0	1	2	3
$y = x^3 + 3x^2 - 9x + 6$	1	26	33	28	11	6	1	8	33

The graph is shown plotted in Fig. 24.3.

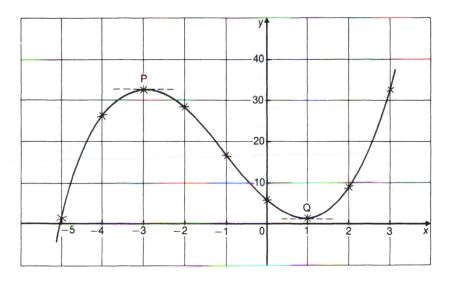

Fig. 24.3

The points P and Q are called turning points since the gradient of the tangent is zero at each of them. Also P is called a maximum and Q is called a minimum. It will be seen from the graph that the value of y at P is not the greatest value that y can have — nor the value of y at Q the least. The terms maximum and minimum values apply only in the vicinity of the turning points and not to the values of y in general.

So at P, when $x = -3$, we have a maximum value of y of 33 and at Q, when $x = 1$, we have a minimum value of y of 1.

MAXIMUM AND MINIMUM USING CALCULUS

It is not always convenient to draw the full graph to find the turning points. At the turning points the gradient (slope) is zero, i.e. $\dfrac{dy}{dx} = 0$ in calculus notation (Fig. 24.4). As you will see this enables us to find the turning points.

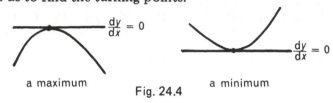

a maximum

a minimum

Fig. 24.4

Maximum or Minimum?

It is usually necessary to determine if a turning point is a maximum or minimum. Fig. 24.5 shows how the gradient of a curve changes in the vicinity of a turning point.

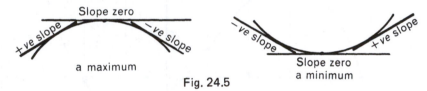

a maximum

a minimum

Fig. 24.5

EXAMPLE 24.5

Check the results for maximum and minimum values of y, found previously by a graphical means, given that $y = x^3 + 3x^2 - 9x + 6$.

We have

$$y = x^3 + 3x^2 - 9x + 6$$

$\therefore$

$$\frac{dy}{dx} = 3x^2 + 6x - 9$$

At a turning point $\dfrac{dy}{dx} = 0$

$\therefore \qquad 3x^2 + 6x - 9 = 0$

$\therefore \qquad x^2 + 2x - 3 = 0$ by dividing through by 3

$\therefore \qquad (x - 1)(x + 3) = 0$

$\therefore$ either $\qquad x - 1 = 0$ or $x + 3 = 0$

$\therefore$ either $\qquad x = 1$ or $\qquad x = -3$

Test for Maximum or Minimum

At the turning point where $x = 1$, we know that $\frac{dy}{dx} = 0$, i.e. zero slope, and using a value of x slightly less than 1, say $x = 0.5$, gives $\frac{dy}{dx} = 3(0.5)^2 + 6(0.5) - 9 = -5.25$, i.e. negative slope, and using a value of x slightly greater than 1, say $x = 1.5$, gives $\frac{dy}{dx} = 3(1.5)^2 + 6(1.5) - 9 = +6.75$, i.e. positive slope.

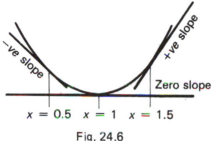

Fig. 24.6

These results are best shown by means of a diagram (Fig. 24.6) which indicates clearly that when $x = 1$ we have a minimum.

The minimum value of y may be found by substituting $x = 1$ into the given equation. Hence

$$y_{min} = (1)^3 + 3(1)^2 - 9(1) + 6 = 1$$

Similarly for the turning point at $x = -3$, at values of $x = -3.5$ (slightly below -3) and $x = -2.5$ (slightly above -3) we obtain the result shown in Fig. 24.7.

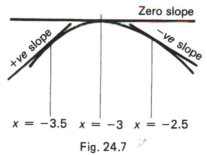

Fig. 24.7

Fig. 24.7 indicates that when $x = -3$ we have a maximum.

The maximum value of y may be found by substituting $x = -3$ into the given equation. Hence

$$y_{max} = (-3)^3 + 3(-3)^2 - 9(-3) + 6 = 33$$

APPLICATIONS IN TECHNOLOGY

In order to find maxima and minima using calculus we need an equation connecting the quantity for which a maximum or minimum is required in terms of another variable. It may well be necessary to form this equation and it may help to draw a diagram representing the problem.

EXAMPLE 24.6

An electric current is represented by $i = 5 \sin \theta$ where θ is in radians. Find the maximum value of i between $\theta = 0$ and π radians.

We have
$$i = 5 \sin \theta$$

then
$$\frac{di}{d\theta} = 5 \cos \theta$$

At a turning point
$$\frac{di}{d\theta} = 0$$

i.e.
$$5 \cos \theta = 0$$

or
$$\cos \theta = 0$$

giving
$$\theta = \frac{\pi}{2} \text{rad}$$

This is the only solution between 0 and π radians.

Test for Maximum or Minimum

At the turning point where $\theta = \frac{\pi}{2} = 1.57 \, \text{rad}$ we know that $\frac{di}{d\theta} = 0$, i.e. zero slope

and using a value of θ slightly less than 1.57, say $\theta = 1.50$, gives $\frac{di}{d\theta} = 5 \cos 1.50 = 0.35$, i.e. positive slope

and using a value of θ slightly greater than 1.57, say $\theta = 1.60$, gives $\frac{di}{d\theta} = 5 \cos 1.60 = -0.15$, i.e. a negative slope.

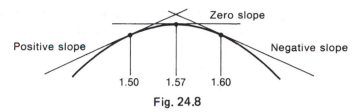

Fig. 24.8

Fig. 24.8 indicates a maximum at $\theta = \dfrac{\pi}{2}$ rad, and since $\sin\dfrac{\pi}{2} = 1$

then the maximum value of current $i_{max} = 5\sin\dfrac{\pi}{2} = 5$

EXAMPLE 24.7

A rectangular sheet of metal 360 mm by 240 mm has four equal squares cut out at the corners. The sides are then turned up to form a rectangular box. Find the length of the sides of the squares cut out so that the volume of the box may be as great as possible, and find this maximum volume.

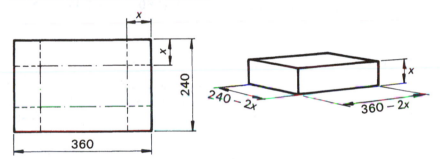

Fig. 24.9

Let the length of the side of each cut away square be x mm as shown in Fig. 24.9.

Hence the volume is
$$V = x(240 - 2x)(360 - 2x)$$
$$= 4x^3 - 1200x^2 + 86\,400x$$

$\therefore$
$$\frac{dV}{dx} = 12x^2 - 2400x + 86\,400$$

At a turning point
$$\frac{dV}{dx} = 0$$

$\therefore$ $12x^2 - 2400x + 86\,400 = 0$

or $x^2 - 200x + 7200 = 0$ by dividing through by 12

Now this is a quadratic equation which does not factorise so we will have to solve using the formula for the standard quadratic $ax^2 + bx + c = 0$ which gives $x = \dfrac{-b \pm \sqrt{b^2 - 4ac}}{2a}$.

Hence the solution of our equation is

$$x = \frac{-(-200) \pm \sqrt{(-200)^2 - 4 \times 1 \times 7200}}{2 \times 1}$$

∴ either $\qquad\qquad x = 152.9 \quad$ or $\quad x = 47.1$

However, from the physical sizes of the sheet, it is not possible for x to be 152.9 mm (since one side is only 240 mm long) so we reject this solution. Hence $x = 47.1$ mm.

We will leave you to check that the turning point at $x = 47.1$ is a maximum.

It only remains to find the maximum volume by substituting $x = 47.1$ into the equation for V. Therefore

$$V_{max} = 47.1(240 - 2 \times 47.1)(360 - 2 \times 47.1)$$
$$= 1.825 \times 10^6 \, \text{mm}^3$$

EXAMPLE 24.8

A cylinder with an open top has a capacity of $2 \, \text{m}^3$ and is made from sheet metal. Neglecting any overlaps at the joints find the dimensions of the cylinder so that the amount of sheet steel used is a minimum.

Let the height of the cylinder be h metres and the radius of the base be r metres as shown in Fig. 24.10.

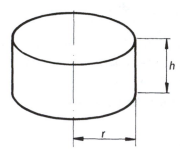

Fig. 24.10

Now the total area of metal = area of base + area of curved side

$$A = \pi r^2 + 2\pi rh$$

We cannot proceed to differentiate as there are two variables on the right hand side of the equation. It is possible, however, to find a connection between r and h using the fact that the volume is $2 \, \text{m}^3$.

Now   Volume of a cylinder $= \pi r^2 h$

$\therefore$ $2 = \pi r^2 h$

from which $h = \dfrac{2}{\pi r^2}$

We may now substitute for h in the equation for A.

$\therefore$ $A = \pi r^2 + 2\pi r \left(\dfrac{2}{\pi r^2} \right)$

$= \pi r^2 + \dfrac{4}{r}$

$= \pi r^2 + 4r^{-1}$

$\therefore$ $\dfrac{\mathrm{d}A}{\mathrm{d}r} = 2\pi r - 4r^{-2}$

Now for a turning point $\dfrac{\mathrm{d}A}{\mathrm{d}r} = 0$

or $2\pi r - 4r^{-2} = 0$

$\therefore$ $2\pi r - \dfrac{4}{r^2} = 0$

$\therefore$ $2\pi r = \dfrac{4}{r^2}$

$\therefore$ $r^3 = \dfrac{2}{\pi} = 0.637$

$r = \sqrt[3]{0.637} = 0.860$

Again we leave you to check that the turning point at $r = 0.86$ is a minimum.

Hence $r = 0.86$ m makes A a minimum as required.

We may find the corresponding value of h by substituting $r = 0.86$ into the equation found previously for h in terms of r.

$$h = \frac{2}{\pi (0.86)^2} = 0.86$$

Hence for the minimum amount of metal to be used the radius is 0.86 m and the height is 0.86 m.

Exercise 24.3

1) Find the maximum and minimum values of:
(a) $y = 2x^3 - 3x^2 - 12x + 4$
(b) $y = x^3 - 3x^2 + 4$
(c) $y = 6x^2 + x^3$.

2) Given that $y = 60x + 3x^2 - 4x^3$, calculate:
(a) the gradient of the tangent to the curve of y at the point where $x = 1$,
(b) the value of x for which y has its maximum value,
(c) the value of x for which y has its minimum value.

3) Calculate the coordinates of the points on the curve

$$y = x^3 - 3x^2 - 9x + 12$$

at each of which the tangent to the curve is parallel to the x-axis.

4) A curve has the equation $y = 8 + 2x - x^2$. Find:
(a) the value of x for which the gradient of the curve is 6,
(b) the value of x which gives the maximum value of y,
(c) the maximum value of y.

5) The curve $y = 2x^2 + \dfrac{k}{x}$ has a gradient of 5 when $x = 2$.

Calculate:

(a) the value of k,
(b) the minimum value of y.

6) Find the maximum and minimum values of the voltage v given by the expression $v = 3(\cos\theta) + 7$ between, and including, values of $\theta \doteq 0$ and $\theta = 2\pi$ radians.

7) In an electric circuit the number of heat units, H, produced when a current, i, is flowing is given by $H = Ei - Ri^2$ where E is the e.m.f. and R is the resistance. Find the maximum heat that can be produced when $E = 8$ and $R = 2$.

8) The power, p watts, available from the rotor of an undershot waterwheel is given by $p = \dot{m}(v_j - u)u$. The water mass flow rate $\dot{m} = 7.85\,\text{kg/s}$ and the velocity of the vanes $v_j = 25\,\text{m/s}$. Find the velocity, u, of the vanes for maximum power. Find also the value of the maximum power.

9) From a rectangular sheet of metal measuring 120 mm by 75 mm equal squares of side x are cut from each of the corners. The remaining flaps are then folded upwards to form an open box.

Show that the volume of the box is given by

$$V = 9000x - 390x^2 + 4x^3$$

Find the value of x such that the volume is a maximum.

10) An open rectangular tank of height h metres with a square base of side x metres is to be constructed so that it has a capacity of 500 cubic metres. Show that the surface area of the four walls and the base will be $\dfrac{2000}{x} + x^2$ square metres. Find the value of x for this expression to be a minimum.

11) The volume of a cone is given by the formula $V = \frac{1}{3}\pi r^2 h$, where h is the height of the cone and r its radius. If $h = 6 - r$, calculate the value of r for which the volume is a maximum.

12) A box without a lid has a square base of side x mm and rectangular sides of height h mm. It is made from $10\,800$ mm^2 of sheet metal of negligible thickness. Show that $h = \dfrac{10\,800 - x^2}{4x}$ and that the volume of the box is $(2700x - \frac{1}{4}x^3)$. Hence calculate the maximum volume of the box.

13) A cylindrical lemonade can made from thin metal has to hold 0.5 litres. Find its dimensions if the area of metal used is a minimum.

14) A cooling tank is to be made with the trapezoidal section as shown:

Its cross-sectional area is to be $300\,000$ mm^2. Show that the width of material needed to form, from one sheet, the bottom and folded-up sides is $w = \dfrac{300\,000}{h} + 1.828h$. Hence find the height h of the tank so that the width of material needed is a minimum.

15) A cylindrical cup is to be drawn from a disc of metal of 50 mm diameter. Assuming that the surface area of the cup is the same as that of the disc find the dimensions of the cup so that its volume is a maximum.

25. INTEGRATION

INTEGRATION

Integration as the Inverse of Differentiation

We have previously discovered how to obtain the differential coefficients of various functions. Our objective in this section is to find out how to reverse the process. That is, being given the differential coefficient of a function we try to discover the original function.

If
$$y = \frac{x^4}{4}$$

then
$$\frac{dy}{dx} = x^3$$

or we may write
$$dy = x^3\,dx$$

The expression $x^3\,dx$ is called the differential of $\dfrac{x^4}{4}$.

Reversing the process of differentiation is called *integration*.

It is indicated by using the integration sign $\int$ in front of the differential.

Thus, if:
$$dy = x^3\,dx$$

then reversing the process
$$y = \int x^3\,dx = \frac{x^4}{4}$$

384

Similarly if $$y = \frac{x^5}{5}$$

then $$\frac{dy}{dx} = x^4$$

or $$dy = x^4\,dx$$

and reversing the process $$y = \int x^4\,dx = \frac{x^5}{5}$$

Also, if $$y = \frac{x^{n+1}}{n+1}$$

then $$\frac{dy}{dx} = x^n$$

or $$dy = x^n\,dx$$

from which $$y = \int x^n\,dx = \frac{x^{n+1}}{n+1}$$

Now $\dfrac{x^{n+1}}{n+1}$ is called the integral of $x^n\,dx$.

This rule applies to all indices, positive, negative and fractional except for $\int x^{-1}\,dx$, which is a special case beyond the scope of this book.

Since the differential coefficient of $\sin x$ is $\cos x$ it follows that the integral of $\cos x$, with respect to x, is $\sin x$

Similarly the integral of $\sin x$, with respect to x, is $-\cos x$

Summarising all these results we have:

$\int x^n\,dx = \dfrac{x^{n+1}}{n+1}$	$\int \sin x\,dx = -\cos x$	$\int \cos x\,dx = \sin x$

THE CONSTANT OF INTEGRATION

We know that the differential of $\frac{1}{2}x^2$ is $x\,dx$. Therefore if we are asked to integrate $x\,dx$, $\frac{1}{2}x^2$ is one answer; but it is not the only possible answer because $\frac{1}{2}x^2 + 2$, $\frac{1}{2}x^2 + 5$, $\frac{1}{2}x^2 + 19$, etc. are all expressions whose differential is $x\,dx$. The general expression for $\int x\,dx$ is therefore $\frac{1}{2}x^2 + c$, where c is a constant known as the constant of integration. Each time we integrate the constant of integration must be added.

EXAMPLE 25.1

a) $\displaystyle\int x^5\,dx = \frac{x^{5+1}}{5+1} + c = \frac{x^6}{6} + c$ b) $\displaystyle\int x\,dx = \frac{x^{1+1}}{1+1} + c = \frac{x^2}{2} + c$

c) $\displaystyle\int \sqrt{x}\,dx = \int x^{1/2}\,dx = \frac{x^{3/2}}{3/2} + c = \frac{2x^{3/2}}{3} + c$

d) $\displaystyle\int \frac{dx}{x^3} = \int x^{-3}\,dx = \frac{x^{-2}}{-2} + c = -\frac{1}{2x^2} + c$

e) $\displaystyle\int \cos x\,dx = \sin x + c$

A Constant Coefficient May be Taken Outside the Integral Sign

EXAMPLE 25.2

a) $\displaystyle\int 3x^2\,dx = 3\int x^2\,dx = 3\frac{x^3}{3} + c = x^3 + c$

b) $\displaystyle\int 4\sin\theta\,d\theta = 4\int \sin\theta\,d\theta = 4(-\cos\theta) + c = -4\cos\theta + c$

The Integral of a Sum is the Sum of Their Separate Integrals

EXAMPLE 25.3

a) $\displaystyle\int (x^2 + x)\,dx$

Integrate each term separately:

$$\int x^2\,dx = \frac{x^3}{3} \quad \text{and} \quad \int x\,dx = \frac{x^2}{2}$$

Thus $\int (x^2 + x)\,dx = \frac{x^3}{3} + \frac{x^2}{2} + c$

b) $\int (3x^4 - 6)\,dx = \frac{3}{5}x^5 - 6x + c$

c) $\int (3\sin t - 5\cos t)\,dt = -3\cos t - 5\sin t + c$

d) $\int (2x + 5)^2\,dx = \int (4x^2 + 20x + 25)\,dx = \frac{4x^3}{3} + 10x^2 + 25x + c$

Exercise 25.1

Integrate with respect to x:

1) x^2 2) x^8 3) $\sqrt{x}$

4) $\dfrac{1}{x^2}$ 5) $\dfrac{1}{x^4}$ 6) $\dfrac{1}{\sqrt{x}}$

7) $3x^4$ 8) $5x^8 + e^x$ 9) $x^2 + x + 3$

10) $2x^3 - 7x - 4$ 11) $x^2 - 5x + \dfrac{1}{\sqrt{x}} + \dfrac{2}{x^2}$

12) $\dfrac{8}{x^3} - \dfrac{2}{x^2} + \sqrt{x}$ 13) $(x - 2)(x - 1)$ 14) $(x + 3)^2$

15) $(2x - 7)^2$ 16) $2\cos x + 3\sin x$

Evaluating the Constant of Integration

The value of the constant of integration may be found provided a corresponding pair of values of x and y are known.

EXAMPLE 25.4

The gradient of the curve which passes through the point $(2, 3)$ is given by x^2. Find the equation of the curve.

We are given $\qquad\qquad\qquad \dfrac{dy}{dx} = x^2$

$$\therefore \qquad\qquad y = \int x^2 \, dx = \frac{x^3}{3} + c$$

We are also given that when $x = 2, \quad y = 3$

Substituting these values in $y = \dfrac{x^3}{3} + c$

we have $3 = \dfrac{2^3}{3} + c$

$$\therefore \qquad\qquad c = \frac{1}{3}$$

Hence the equation of the curve is

$$y = \frac{x^3}{3} + \frac{1}{3}$$

or $y = \dfrac{1}{3}(x^3 + 1)$

Exercise 25.2

1) The gradient of the curve which passes through the point $(2, 3)$ is given by x. Find the equation of the curve.

2) The gradient of the curve which passes through the point $(3, 8)$ is given by $(x^2 + 3)$. Find the value of y when $x = 5$.

3) It is known that for a certain curve $\dfrac{dy}{dx} = 3 - 2x$ and the curve cuts the x-axis where $x = 5$. Express y in terms of x. State the length of the intercept on the y-axis.

4) Find the equation of the curve which passes through the point $(1, 4)$ and is such that $\dfrac{dy}{dx} = 2x^2 + 3x + 2$.

5) If $\dfrac{dp}{dt} = (3 - t)^2$ find p in terms of t given that $p = 3$ when $t = 2$.

6) The gradient of a curve is $ax + b$ at all points, where a and b are constants. Find the equation of the curve given that it passes through the points $(0, 4)$ and $(1, 3)$ and that the tangent at $(1, 3)$ is parallel to the x-axis.

7) Find the equation of the curve which passes through the point $(1, 2)$ and has the property of $\dfrac{dy}{dx}$ $\cos x$.

8) A curve is such that $\dfrac{dy}{d\theta} = \cos\theta$, and also $y = 1$ when $\theta = \dfrac{\pi}{2}$ radians. Find the equation of the curve.

9) At any point on a curve $\dfrac{dy}{dt} = 3\sin t$. Find the equation of the curve given that $y = 2$ when t has a value equivalent to 25 degrees.

THE DEFINITE INTEGRAL

It has been shown that $\displaystyle\int x^n \, dx = \dfrac{x^{n+1}}{n+1} + c.$

Since the expression contains an arbitrary constant c, the value of which is not known, it is called an indefinite integral.

A definite integral has a specific numerical answer without an unknown constant. An example of a definite integral is $\displaystyle\int_a^b x^n \, dx.$

a and b are called limits, a being the lower limit and b the upper limit. The method of evaluating a definite integral is shown in the following examples.

EXAMPLE 25.5

Find the value of $\displaystyle\int_2^3 x^2 \, dx$

$$\int_2^3 x^2 \, dx = \left[\frac{x^3}{3} + c \right]_2^3$$

$$= \left(\text{Value of } \frac{x^3}{3} + c \text{ when } x \text{ is put equal to } 3 \right)$$

$$- \left(\text{Value of } \frac{x^3}{3} + c \text{ when } x \text{ is put equal to } 2 \right)$$

$$= \left(\frac{3^3}{3} + c \right) - \left(\frac{2^3}{3} + c \right)$$

$$= \frac{27}{3} + c - \frac{8}{3} - c$$

$$= \frac{19}{3} = 6.33$$

In integration the use of the square brackets, as in the above solution, has a specific meaning, that is *the integration of each term has been completed and the next step is to substitute the values of the limits for x.*

We should also note that the constant c cancelled out. This will always happen and in solving definite integrals it is usual to omit c as shown in the next example.

EXAMPLE 25.6

Find the value of $\int_{1}^{2} (3x^2 - 2x + 5)\, dx$

$$\int_{1}^{2} (3x^2 - 2x + 5) = \left[x^3 - x^2 + 5x \right]_{1}^{2}$$

$$= (2^3 - 2^2 + 5 \times 2) - (1^3 - 1^2 + 5 \times 1)$$

$$= 14 - 5 = 9$$

EXAMPLE 25.7

Find the value of $\int_{0}^{\pi/2} \sin \theta \, d\theta$

$$\int_{0}^{\pi/2} \sin \theta \, d\theta = \left[-\cos \theta \right]_{0}^{\pi/2} = \left(-\cos \frac{\pi}{2} \right) - \left(-\cos 0 \right)$$

$$= 0 - (-1) = 1$$

Exercise 25.3

Evaluate the following definite integrals:

1) $\int_{1}^{2} x^2 \, dx$

2) $\int_{2}^{3} (2x + 3) \, dx$

3) $\int_{0}^{2} (x^2 + 3) \, dx$

4) $\int_{1}^{2} (3x^2 - 4x + 3) \, dx$

5) $\int_{1}^{2} x(2x - 1) \, dx$

6) $\int_{0}^{2} \sqrt{x} \, dx$

7) $\displaystyle\int_1^3 \frac{1}{x^2}\, \mathrm{d}x$

8) $\displaystyle\int_2^4 (x-1)(x-3)\, \mathrm{d}x$

9) $\displaystyle\int_0^{\pi/2} \cos\phi \, \mathrm{d}\phi$

AREA UNDER A CURVE

Suppose that we wish to find the shaded area shown in Fig. 25.1. P, whose co-ordinates are (x,y) is a point on the curve.

Let us now draw, below P, a vertical strip whose width δx is very small. Since the width of the strip is very small we may consider

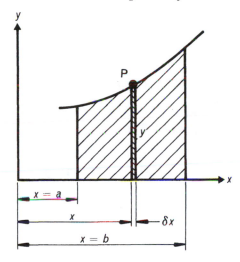

Fig. 25.1

the strip to be a rectangle with height y. Hence the area of the strip is approximately $y \times \delta x$. Such a strip is called an elementary strip and we will consider that the shaded area is made up from many elementary strips. Hence the required area is the sum of all the elementary strip areas between the values $x = a$ and $x = b$. In mathematical notation this may be stated as:

$$\text{Area} = \sum_{x=a}^{x=b} y \times \delta x \quad \text{approximately}$$

The process of integration may be considered to sum up an infinite number of elementary strips and hence gives an exact result.

$\therefore$ $\displaystyle\text{Area} = \int_a^b y \, \mathrm{d}x \quad \text{exactly}$

EXAMPLE 25.8

Find the area bounded by the curve $y = x^3 + 3$, the x-axis and the lines $x = 1$ and $x = 3$

It is always wise to sketch the graph of given curve and show the area required together with an elementary strip, as shown in Fig. 25.2.

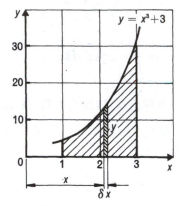

Fig. 25.2

The required area $= \displaystyle\sum_{x=1}^{x=3} y \times \delta x$ approximately

$$= \int_1^3 y \, dx \quad \text{exactly}$$

$$= \int_1^3 (x^3 + 3) \, dx$$

$$= \left[\frac{x^4}{4} + 3x \right]_1^3 = \left(\frac{3^4}{4} + 3 \times 3 \right) - \left(\frac{1^4}{4} + 3 \times 1 \right)$$

$$= 26 \text{ square units}$$

EXAMPLE 25.9

Find the area under the curve of $2 \cos \theta$ between $\theta = 20°$ and $\theta = 60°$.

The curve of $2 \cos \theta$ is shown in Fig. 25.3 from $0°$ to $90°$ and the required area together with an elementary strip.

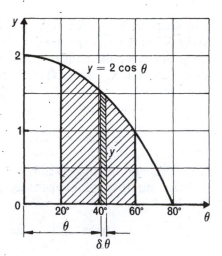

Fig. 25.3

The required area $= \sum\limits_{\theta=20°}^{\theta=60°} y \times \delta\theta$ approximately

$$= 2 \int_{20°}^{60°} \cos\theta \, d\theta \quad \text{exactly}$$

$$= 2\left[\sin\theta\right]_{20°}^{60°} = 2\left(\sin 60° - \sin 20°\right)$$

$$= 1.05 \text{ square units}$$

Exercise 25.4

1) Find the area between the curve $y = x^3$, the x-axis and the lines $x = 5$ and $x = 3$.

2) Find the area between the curve $y = 3 + 2x + 3x^2$, the x-axis and the lines $x = 1$ and $x = 4$.

3) Find the area between the curve $y = x^2(2x - 1)$, the x-axis and the lines $x = 1$ and $x = 2$.

4) Find the area between the curve $y = \dfrac{1}{x^2}$, the x-axis and the lines $x = 1$ and $x = 3$.

5) Find the area between the curve $y = 5x - x^3$, the x-axis and the lines $x = 1$ and $x = 2$.

6) Evaluate the integral $\int_{0}^{2\pi} \sin\theta \, d\theta$ and explain the result with reference to a sketched graph.

7) Find the area under the curve $2\sin\theta + 3\cos\theta$ between $\theta = 0$ and $\theta = \pi$ radians.

8) Find the area under the curve of $y = \sin\phi$ between $\phi = 0$ and $\phi = \pi$ radians.

26. VECTORS

SCALARS

A scalar quantity is one that is fully defined by magnitude alone. Some examples of scalar quantities are time (e.g. 30 seconds), temperature (e.g. 8 degrees Celsius) and mass (e.g. 7 kilograms).

VECTORS

A vector quantity needs magnitude, direction and sense to describe it fully. A vector may be represented by a straight line, its length representing the magnitude of the vector, its direction being that of the vector and suitable notation giving the sense of the vector (Fig. 26.1).

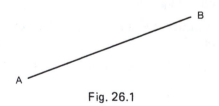

Fig. 26.1

The usual way of naming a vector is to name its end points. The vector in Fig. 26.1 starts at A and ends at B and we write $\overrightarrow{AB}$ which means 'the vector from A to B' — this gives the sense.

394

Some examples of vector quantities are:

(1) A displacement in a given direction, e.g. 15 metres due west.

(2) A velocity in a given direction, e.g. 40 km/h due north.

(3) A force of 20 kN acting vertically downwards.

GRAPHICAL REPRESENTATION

A vector may be represented by a straight line drawn to scale.

EXAMPLE 26.1

A man walked a distance of 8 km due east. Draw the vector.

We first choose a suitable scale to represent the magnitude of the vector. In Fig. 26.2 a scale of 10 mm = 2 km has been chosen as convenient. We then draw a horizontal line 40 mm long and label the ends as shown. So $\overrightarrow{AB}$ represents the vector 8 km due east. Sometimes we add an arrow to confirm the sense, but it is not necessary.

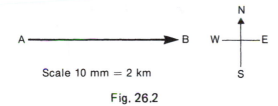

Scale 10 mm = 2 km

Fig. 26.2

CARTESIAN COMPONENTS

Fig. 26.3 shows vector $\overrightarrow{AB}$ of magnitude 6 m/s and making an angle of 30° with the horizontal. An alternative is to define the vector AB by resolving it into its horizontal component $\overrightarrow{AC}$, and its vertical component $\overrightarrow{CB}$. These are said to be the Cartesian components of the vector.

The magnitude of
Component $\overrightarrow{AC}$ } = 6 × cos 30° = 5.20 m/s

The magnitude of
Component $\overrightarrow{CB}$ $\Big\}$ $= 6 \times \sin 30° = 3 \, \text{m/s}$

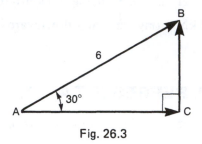

Fig. 26.3

EXAMPLE 26.2

The vector $\overrightarrow{AB}$ in Fig. 26.4 has a vertical component of 3 N and a horizontal component of 4 N. Calculate its magnitude and direction.

Using Pythagoras' theorem the magnitude of $\overrightarrow{AB} = \sqrt{4^2 + 3^3} = 5 \, \text{N}$.

To state its direction we find the size of angle θ.

$$\tan \theta = \tfrac{3}{4} \quad \text{giving} \quad \theta = 36.9°$$

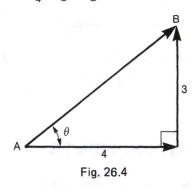

Fig. 26.4

ADDITION OF VECTORS

Three points P, Q and R are marked out in a field, and are shown to scale in Fig. 26.5. A man walks from P to Q (i.e. he describes $\overrightarrow{PQ}$), and then walks on from Q to R (i.e. he describes $\overrightarrow{QR}$). Instead the man could have walked directly from P to R, thus describing the vector PR.

Now going from P to R directly has the same result as going from P to Q and then from Q to R. We therefore call $\overrightarrow{PR}$ the *resultant* of the sum of the vectors $\overrightarrow{PQ}$ and $\overrightarrow{QR}$, and we write this as

$$\overrightarrow{PR} \; = \; \overrightarrow{PQ} + \overrightarrow{QR}$$

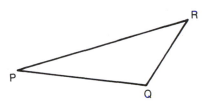

Fig. 26.5

EXAMPLE 26.3

Two forces act at a point O as shown in Fig. 26.6. Find the resultant force at O

a) by making a scale drawing,
b) by calculation.

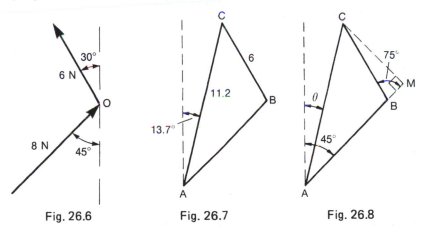

Fig. 26.6   Fig. 26.7   Fig. 26.8

a) The vector diagram is shown in Fig. 26.7. Choose your own scale and measure AC to find the magnitude of the resultant — its direction is given by the angle it makes with the vertical.

Hence the resultant of the two forces is 11.2 N acting at 13.8° to the vertical.

b) We extend AB so that the line CM may be drawn perpendicular to AB as shown in Fig. 26.8. We are only concerned with lengths and angles in these calculations.

In right-angled $\triangle$BMC

$$CM = CB\sin\widehat{CBM} = 6\sin 75° = 5.80$$

$$BM = CB\cos\widehat{CBM} = 6\cos 75° = 1.55$$

Thus $\quad AM = AB + BM = 8 + 1.55 = 9.55$

Using Pythagoras' theorem in $\triangle$ACM

$$AC = \sqrt{(5.80)^2 + (9.55)^2} = 11.2\,N$$

Also in $\triangle$ACM

$$\tan\widehat{CAM} = \frac{CM}{AM} = \frac{5.80}{9.55} = 0.607$$

giving $\quad \widehat{CAM} = 31.3°$

So the angle AC makes with the vertical $= 45° - 31.3° = 13.7°$.

These results confirm those obtained by drawing and measurement.

VECTOR ADDITION USING A PARALLELOGRAM

In the previous example vector addition made use of a triangle. A parallelogram may also be considered as in Fig. 26.9.

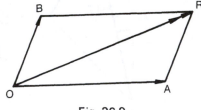

Fig. 26.9

We have $\qquad \overrightarrow{OA} + \overrightarrow{OB} = \overrightarrow{OR}$

Using the triangle OAR we would have said that $\overrightarrow{OA} + \overrightarrow{AR} = \overrightarrow{OR}$; this is a similar statement since $\overrightarrow{OB} = \overrightarrow{AR}$ because they are equal vectors having the same magnitude, direction and sense. Any numerical calculations would be carried out using $\triangle$OAR as in Example 26.3.

SUBTRACTION OF VECTORS

The inverse of a vector is one having the same magnitude and direction, but of opposite sense. Thus $\overrightarrow{BA}$ is the inverse of $\overrightarrow{AB}$, or $\overrightarrow{BA} = -\overrightarrow{AB}$.

To subtract a vector we add its inverse. Hence if we wish to subtract vector $\overrightarrow{OB}$ from $\overrightarrow{OA}$ we have:

$$\overrightarrow{OA} - \overrightarrow{OB} = \overrightarrow{OA} + \overrightarrow{BO} = \overrightarrow{BO} + \overrightarrow{OA} = \overrightarrow{BA}$$

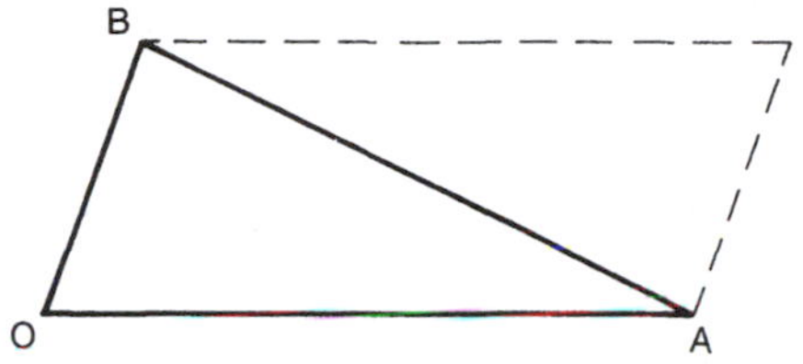

Fig. 26.10

In Fig. 26.10 we can see that from the $\triangle OAB$ a vector sum gives $\overrightarrow{BO} + \overrightarrow{OA} = \overrightarrow{BA}$. Thus:

$$\text{Vector difference} \qquad \overrightarrow{OA} - \overrightarrow{OB} = \overrightarrow{BA}$$

This vector difference is represented by the other diagonal on the parallelogram than that used for the vector sum.

PHASORS

In electrical engineering currents and voltages are represented by *phasors* in a similar manner to that in which vectors may be used to represent forces and velocities. The methods of adding and subtracting phasors and vectors are similar.

EXAMPLE 26.4

Phasor OA represents 7 V and phasor OB represents 5 V as shown in Fig. 26.11. Find the phasor difference $\overrightarrow{OA} - \overrightarrow{OB}$.

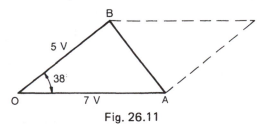

Fig. 26.11

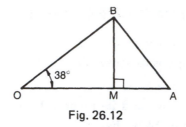

Fig. 26.12

We may draw a diagram to scale and measure the results — you may wish to do this and check the answers obtained below by calculation. We have redrawn △OAB in Fig. 26.12 and added a perpendicular BM to OA.

Now $\overrightarrow{OA} - \overrightarrow{OB} = \overrightarrow{BA}$, but in the calculations we shall only be concerned with angles and lengths.

In right-angled △OBM

$$BM = OB\sin 38° = 5\sin 38° = 3.08$$

and $\qquad OM = OB\cos 38° = 5\cos 38° = 3.94$

Thus $\qquad MA = OA - OM = 7 - 3.94 = 3.06$

From △BMA using the theorem of Pythagoras:

$$BA = \sqrt{(3.08)^2 + (3.06)^2} = 4.34$$

Thus the vector difference $\overrightarrow{OA} - \overrightarrow{OB}$ has a magnitude of 4.34 V and, if required its direction may be found by using right-angled △BMA to find $B\widehat{A}M$, the angle it makes with OA.

Exercise 26.1

In Questions 1–4 find the values of the horizontal and vertical components illustrating your results in each case on a suitable vector diagram.

1)

2)

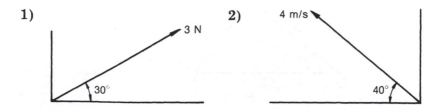

3)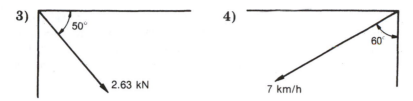
50°

2.63 kN

4)
60°

7 km/h

5) A block is being pulled up an incline, as shown in Fig. 26.13, using a rope making an angle of 25° with the incline. If the pull in the rope is 50 kN draw a suitable vector diagram indicating the magnitudes of the components of this force parallel, and at right angles, to the incline.

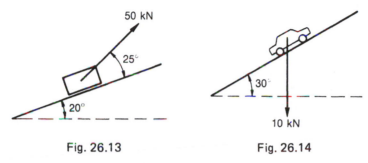

50 kN

25°

20°

Fig. 26.13

30°

10 kN

Fig. 26.14

6) A motor car weighing 10 kN is shown on a 30° slope in Fig. 26.14. Resolve this weight into components along the slope and at right angles to it.

7) A garden roller is being pushed forwards with a force of 95 N as shown in Fig. 26.15. Find the horizontal and vertical components of this force.

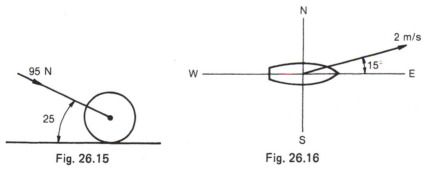

95 N

25

Fig. 26.15

N

2 m/s

W

15°

E

S

Fig. 26.16

8) A ship is being steered in an easterly direction as shown in plan view in Fig. 26.16. However, owing to a S to N cross-wind its motion and velocity are as shown. Find the forward speed of the vessel.

9) A sledge is being pulled on the level by two horizontal ropes which are at right angles to each other. If the respective tensions in the ropes are 30 N and 50 N find the resultant force on the sledge and the angle it makes with the rope having the greater tension.

10) A plane is flying due north with a speed of 700 km/h. If there is a west to east cross-wind of 70 km/h, determine the resultant velocity of the plane and give its direction relative to north.

11) A dinghy is drifting under the influence of a 0.5 m/s tide flowing from north-east to south-west, and a wind of 0.8 m/s blowing from south to north. Find the resultant velocity of the dinghy and its direction relative to north.

12) The plan of a truck on rails is shown in Fig. 26.17. Find the resultant pull of the two ropes and the direction it acts in relative to the line of the rails.

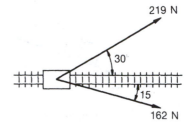

Fig. 26.17

13) Using the results obtained to Question 12 find the force component pulling the truck along the track, and the force between the wheel flanges and the rails.

In Questions 14–17 find the resultants of the given phasors: in each case giving the direction relative to the larger given value.

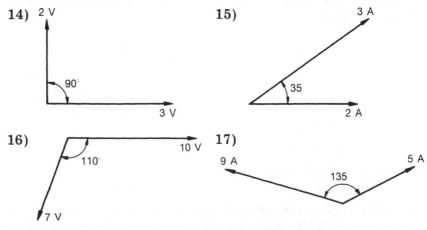

27. MATRICES AND DETERMINANTS

INTRODUCTION

The block within which a printer sets his type, and a car radiator could each be called a matrix.

A matrix in mathematics is any rectangular array of numbers, usually enclosed in brackets.

Before using matrices in practical problems we first look at how they are ordered, added, subtracted and manipulated in other ways.

ELEMENT

Each number or symbol in a matrix is called an *element* of the matrix.

ORDER

The *dimension* or *order* of a matrix is stated by the number of rows followed by the number of columns in the rectangular array.

e.g.

Matrix	$\begin{pmatrix} 1 & 2 \\ 3 & 4 \end{pmatrix}$	$\begin{pmatrix} a & 2 & -3 \\ 4 & b & x \end{pmatrix}$	$\begin{pmatrix} \sin\theta & 1 \\ \cos\theta & 2 \\ \tan\theta & 3 \end{pmatrix}$	(6)
Order	2×2	2×3	3×2	1×1

EQUALITY

If two matrices are equal, then they must be of the same order and their corresponding elements must be equal.

Thus if $\begin{pmatrix} 2 & 3 & x \\ a & 5 & -2 \end{pmatrix} = \begin{pmatrix} 2 & 3 & 4 \\ -1 & 5 & -2 \end{pmatrix}$ then $x = 4$ and $a = -1$

ADDITION AND SUBTRACTION

Two matrices may be added or subtracted only if they are of the *same order*. We say the matrices are *conformable* for addition (or subtraction) and we add (or subtract) by combining corresponding elements.

EXAMPLE 27.1

If $A = \begin{pmatrix} 3 & 4 \\ 5 & 6 \end{pmatrix}$ and $B = \begin{pmatrix} 0 & 6 \\ 5 & 2 \end{pmatrix}$ determine: a) $C = A + B$
and   b) $D = A - B$

a) $C = \begin{pmatrix} 3 & 4 \\ 5 & 6 \end{pmatrix} + \begin{pmatrix} 0 & 6 \\ 5 & 2 \end{pmatrix} = \begin{pmatrix} 3+0 & 4+6 \\ 5+5 & 6+2 \end{pmatrix} = \begin{pmatrix} 3 & 10 \\ 10 & 8 \end{pmatrix}$

b) $D = \begin{pmatrix} 3 & 4 \\ 5 & 6 \end{pmatrix} - \begin{pmatrix} 0 & 6 \\ 5 & 2 \end{pmatrix} = \begin{pmatrix} 3-0 & 4-6 \\ 5-5 & 6-2 \end{pmatrix} = \begin{pmatrix} 3 & -2 \\ 0 & 4 \end{pmatrix}$

ZERO OR NULL MATRIX

A *zero* or *null* matrix, denoted by O, is one in which all the elements are zero. It may be of any order.

Thus $\begin{pmatrix} 0 & 0 \\ 0 & 0 \end{pmatrix}$ is a zero matrix of order 2. It behaves like zero in the real number system.

IDENTITY OR UNIT MATRIX

The *identity* matrix can be of any suitable order with all the main diagonal elements 1 and the remaining elements 0. It is denoted by I and behaves like unity in the real number system.

Thus $\begin{pmatrix} 1 & 0 \\ 0 & 1 \end{pmatrix}$ is a unit matrix of order 2

TRANSPOSE

The *transpose* of a matrix A is written as A' or A^T. When the row of a matrix is interchanged with its corresponding column, that is row 1 becomes column 1 and row 2 becomes column 2 and so on then the matrix is transposed.

Thus if $A = \begin{pmatrix} 1 & 2 & -3 \\ 4 & 7 & 0 \end{pmatrix}$ then $A' = \begin{pmatrix} 1 & 4 \\ 2 & 7 \\ -3 & 0 \end{pmatrix}$

Exercise 27.1

1) State the order of each of the following matrices:

(a) $\begin{pmatrix} 1 & 2 \\ 3 & 4 \end{pmatrix}$ (b) $\begin{pmatrix} 5 \\ -6 \end{pmatrix}$ (c) $\begin{pmatrix} a & b & 4 \\ 2 & 3 & 5 \\ x & -6 & 0 \end{pmatrix}$

(d) $\begin{pmatrix} 1 & -2 & -3 & -4 \\ 6 & 2 & 0 & -1 \end{pmatrix}$

2) How many elements are there in:

(a) a 3×3 matrix   (b) a 2×2 matrix
(c) a square matrix of order n?

3) Write down the transpose of each matrix in Question 1).

4) Combine the following matrices:

(a) $\begin{pmatrix} 2 & 1 \\ 3 & 2 \end{pmatrix} + \begin{pmatrix} -2 & -1 \\ 6 & 0 \end{pmatrix}$ (b) $\begin{pmatrix} 2 & 1 \\ 3 & 2 \end{pmatrix} - \begin{pmatrix} -2 & -1 \\ 6 & 0 \end{pmatrix}$

(c) $\begin{pmatrix} \frac{1}{2} & 1 \\ \frac{1}{3} & \frac{1}{5} \end{pmatrix} + \begin{pmatrix} \frac{1}{3} & -\frac{1}{2} \\ \frac{1}{2} & \frac{4}{5} \end{pmatrix}$

5) Determine a, b and c if $(a \quad b \quad c) - (-3 \quad 4 \quad 1) = (-5 \quad 1 \quad 0)$.

6) Complete $\begin{pmatrix} \frac{1}{2} & \frac{1}{4} \\ \frac{1}{5} & \frac{1}{6} \end{pmatrix} - \begin{pmatrix} \frac{1}{6} & \frac{1}{5} \\ \frac{1}{6} & \frac{1}{9} \end{pmatrix}$

7) Solve the equation $X - \begin{pmatrix} 1 & 3 \\ 5 & -2 \end{pmatrix} = \begin{pmatrix} 4 & 5 \\ 7 & 0 \end{pmatrix}$ where X is a 2×2 matrix.

8) If $\begin{pmatrix} 4 \\ 5 \end{pmatrix} + \begin{pmatrix} x \\ y \end{pmatrix} = \begin{pmatrix} 4 \\ 10 \end{pmatrix}$, determine $\begin{pmatrix} x \\ y \end{pmatrix}$

MULTIPLICATION OF A MATRIX BY A REAL NUMBER

A matrix may be multiplied by a number in the following way:

$$4\begin{pmatrix} 2 & 3 \\ 7 & -1 \end{pmatrix} = \begin{pmatrix} 4 \times 2 & 4 \times 3 \\ 4 \times 7 & 4 \times (-1) \end{pmatrix} = \begin{pmatrix} 8 & 12 \\ 28 & -4 \end{pmatrix}$$

Conversely the common factor of each element in a matrix may be written outside the matrix. Thus $\begin{pmatrix} 9 & 3 \\ 42 & 15 \end{pmatrix} = 3\begin{pmatrix} 3 & 1 \\ 14 & 5 \end{pmatrix}$

MATRIX MULTIPLICATION

Two matrices can only be multiplied together if the number of columns in the first matrix is equal to the number of rows in the second matrix. We say that the matrices are *conformable* for multiplication. The method for multiplying together a pair of 2×2 matrices is as follows

$$\begin{pmatrix} a & b \\ c & d \end{pmatrix} \times \begin{pmatrix} e & f \\ g & h \end{pmatrix} = \begin{pmatrix} ae + bg & af + bh \\ ce + dg & cf + dh \end{pmatrix}$$

EXAMPLE 27.2

a) $\begin{pmatrix} 2 & 3 \\ 4 & 5 \end{pmatrix} \times \begin{pmatrix} 7 & 1 \\ 0 & 6 \end{pmatrix} = \begin{pmatrix} (2 \times 7) + (3 \times 0) & (2 \times 1) + (3 \times 6) \\ (4 \times 7) + (5 \times 0) & (4 \times 1) + (5 \times 6) \end{pmatrix}$

$$= \begin{pmatrix} 14 & 20 \\ 28 & 34 \end{pmatrix}$$

b) $\begin{pmatrix} 3 & 4 \\ 2 & 5 \end{pmatrix} \times \begin{pmatrix} 6 \\ 7 \end{pmatrix} = \begin{pmatrix} (3 \times 6) + (4 \times 7) \\ (2 \times 6) + (5 \times 7) \end{pmatrix} = \begin{pmatrix} 46 \\ 47 \end{pmatrix}$

c) $\begin{pmatrix} 3 \\ 2 \end{pmatrix} \times \begin{pmatrix} 4 & 6 \\ 5 & 7 \end{pmatrix}$ This is not possible since the matrices are not comformable.

EXAMPLE 27.3

Form the products AB and BA given that $A = \begin{pmatrix} 1 & 2 \\ 3 & 4 \end{pmatrix}$ and

$B = \begin{pmatrix} 5 & 6 \\ 7 & 8 \end{pmatrix}$ and hence show that $AB \neq BA$.

$$AB = \begin{pmatrix} 1 & 2 \\ 3 & 4 \end{pmatrix}\begin{pmatrix} 5 & 6 \\ 7 & 8 \end{pmatrix} = \begin{pmatrix} (1 \times 5)+(2 \times 7) & (1 \times 6)+(2 \times 8) \\ (3 \times 5)+(4 \times 7) & (3 \times 6)+(4 \times 8) \end{pmatrix}$$

$$= \begin{pmatrix} 19 & 22 \\ 43 & 50 \end{pmatrix}$$

$$BA = \begin{pmatrix} 5 & 6 \\ 7 & 8 \end{pmatrix}\begin{pmatrix} 1 & 2 \\ 3 & 4 \end{pmatrix} = \begin{pmatrix} (5 \times 1)+(6 \times 3) & (5 \times 2)+(6 \times 4) \\ (7 \times 1)+(8 \times 3) & (7 \times 2)+(8 \times 4) \end{pmatrix}$$

$$= \begin{pmatrix} 23 & 34 \\ 31 & 46 \end{pmatrix}$$

As we see the results are different and, in general, matrix multiplication is non-commutative.

Exercise 27.2

1) If $A = \begin{pmatrix} 3 & 0 \\ -2 & 1 \end{pmatrix}$ and $B = \begin{pmatrix} -4 & 1 \\ 3 & -2 \end{pmatrix}$ determine:

(a) $2A$ (b) $3B$ (c) $2A + 3B$ (d) $2A - 3B$

2) Calculate the following products:

(a) $\begin{pmatrix} 3 & 1 \\ 2 & 0 \end{pmatrix}\begin{pmatrix} 4 & -1 \\ 2 & 3 \end{pmatrix}$ (b) $\begin{pmatrix} 2 & 1 \\ 3 & 1 \end{pmatrix}\begin{pmatrix} 1 & 0 \\ 0 & 1 \end{pmatrix}$ (c) $\begin{pmatrix} 2 & 1 \\ 4 & 2 \end{pmatrix}\begin{pmatrix} 2 & 3 \\ 1 & 5 \end{pmatrix}$

(d) $\begin{pmatrix} 1 & 0 \\ 0 & 1 \end{pmatrix}\begin{pmatrix} a & b \\ c & d \end{pmatrix}$ (e) $\begin{pmatrix} k & 0 \\ 0 & k \end{pmatrix}\begin{pmatrix} a & b \\ c & d \end{pmatrix}$

3) If $A = \begin{pmatrix} 1 & 2 \\ 3 & 4 \end{pmatrix}$ and $B = \begin{pmatrix} 2 & -1 \\ 1 & 3 \end{pmatrix}$ calculate:

(a) A^2 (that is $A \times A$)   (b) B^2 (c) $2AB$

(d) $A^2 + B^2 + 2AB$ (e) $(A + B)^2$

DETERMINANT OF A SQUARE MATRIX OF ORDER 2

If matrix $A = \begin{pmatrix} a & b \\ c & d \end{pmatrix}$ then its *determinant* is denoted by $|A|$ or $\det A$ and the result is a *number* given by

$$|A| = \begin{vmatrix} a & b \\ c & d \end{vmatrix} = ad - bc$$

EXAMPLE 27.4

Evaluate $|A|$ if $A = \begin{pmatrix} 1 & -2 \\ 3 & 4 \end{pmatrix}$

$$|A| = \begin{vmatrix} 1 & -2 \\ 3 & 4 \end{vmatrix} = 1 \times 4 - (-2) \times 3 = 10$$

SOLUTION OF SIMULTANEOUS LINEAR EQUATIONS USING DETERMINANTS

To solve simultaneous linear equations with two unknowns using determinants, the following procedure is used.

(1) Write out the two equations in order:
$$a_1 x + b_1 y = c_1$$
$$a_2 x + b_2 y = c_2$$

(2) Calculate $\Delta = \begin{vmatrix} a_1 & b_1 \\ a_2 & b_2 \end{vmatrix}$

(3) Then $x = \dfrac{\begin{vmatrix} c_1 & b_1 \\ c_2 & b_2 \end{vmatrix}}{\Delta}$ and $y = \dfrac{\begin{vmatrix} a_1 & c_1 \\ a_2 & c_2 \end{vmatrix}}{\Delta}$

EXAMPLE 27.5

By using determinants, solve the simultaneous equations
$$3x + 4y = 22$$
$$2x + 5y = 24$$

Now $\Delta = \begin{vmatrix} 3 & 4 \\ 2 & 5 \end{vmatrix} = (3 \times 5) - (4 \times 2) = 7$

Thus $x = \dfrac{\begin{vmatrix} 22 & 4 \\ 24 & 5 \end{vmatrix}}{7} = \dfrac{(22 \times 5) - (4 \times 24)}{7} = \dfrac{14}{7} = 2$

And $y = \dfrac{\begin{vmatrix} 3 & 22 \\ 2 & 24 \end{vmatrix}}{7} = \dfrac{(3 \times 24) - (22 \times 2)}{7} = \dfrac{28}{7} = 4$

Exercise 27.3

1) Evaluate the following determinants:

(a) $\begin{vmatrix} 5 & 2 \\ 3 & 6 \end{vmatrix}$ (b) $\begin{vmatrix} 7 & 4 \\ 5 & 2 \end{vmatrix}$ (c) $\begin{vmatrix} 6 & 8 \\ 2 & 5 \end{vmatrix}$

2) Solve the following simultaneous equations by using determinants:

(a) $3x + 4y = 11$ (b) $5x + 3y = 29$ (c) $4x - 6y = -2.5$
    $x + 7y = 15$ $4x + 7y = 37$ $7x - 5y = -0.25$

THE INVERSE OF A SQUARE MATRIX OF ORDER 2

Instead of dividing a number by 5 we can multiply by $\frac{1}{5}$ and obtain the same result.

Thus $\frac{1}{5}$ is the multiplicative inverse of 5. That is $5 \times \frac{1}{5} = 1$

In matrix algebra we never divide by a matrix but multiply instead by the inverse. The inverse of matrix A is denoted by A^{-1} and is such that

$$AA^{-1} = \begin{pmatrix} 1 & 0 \\ 0 & 1 \end{pmatrix} = I, \quad \text{the identity matrix.}$$

To find the inverse, A^{-1}, of the square matrix $A = \begin{pmatrix} a & b \\ c & d \end{pmatrix}$

we use the expression: $A^{-1} = \dfrac{1}{|A|} \begin{pmatrix} d & -b \\ -c & a \end{pmatrix} = \dfrac{1}{ad - bc} \begin{pmatrix} d & -b \\ -c & a \end{pmatrix}$

EXAMPLE 27.6

Determine the inverse of $A = \begin{pmatrix} 1 & -2 \\ 3 & 4 \end{pmatrix}$ and verify the result.

Now $|A| = \begin{vmatrix} 1 & -2 \\ 3 & 4 \end{vmatrix} = (1 \times 4) - (3 \times -2) = 10$

Hence $A^{-1} = \frac{1}{10} \begin{pmatrix} 4 & 2 \\ -3 & 1 \end{pmatrix} = \begin{pmatrix} 0.4 & 0.2 \\ -0.3 & 0.1 \end{pmatrix}$

To verify the result we have

$$AA^{-1} = \begin{pmatrix} 1 & -2 \\ 3 & 4 \end{pmatrix} \begin{pmatrix} 0.4 & 0.2 \\ -0.3 & 0.1 \end{pmatrix} = \begin{pmatrix} 1 & 0 \\ 0 & 1 \end{pmatrix} = I$$

The Property $AA^{-1} = A^{-1}A = I$

In Example 27.6 we have

$$A^{-1}A = \begin{pmatrix} 0.4 & 0.2 \\ -0.3 & 0.1 \end{pmatrix} \begin{pmatrix} 1 & -2 \\ 3 & 4 \end{pmatrix} = \begin{pmatrix} 1 & 0 \\ 0 & 1 \end{pmatrix} = I$$

and in general $$\boxed{AA^{-1} = A^{-1}A = I}$$

SINGULAR MATRIX

A matrix which does not have an inverse is called a *singular matrix*. This happens when $|A| = 0$

For example, since $$\begin{vmatrix} 3 & 6 \\ 1 & 2 \end{vmatrix} = (3 \times 2) - (6 \times 1) = 0$$

then $\begin{pmatrix} 3 & 6 \\ 1 & 2 \end{pmatrix}$ is a singular matrix.

Exercise 27.4

Decide whether each of the matrices in Question 1-9 has an inverse. If the inverse exists, find it.

1) $\begin{pmatrix} 2 & 5 \\ 1 & 4 \end{pmatrix}$ 2) $\begin{pmatrix} 2 & 5 \\ 1 & 3 \end{pmatrix}$ 3) $\begin{pmatrix} 3 & 2 \\ 1 & 2 \end{pmatrix}$ 4) $\begin{pmatrix} 4 & 10 \\ 2 & 5 \end{pmatrix}$

5) $\begin{pmatrix} 224 & 24 \\ 24 & 4 \end{pmatrix}$ 6) $\begin{pmatrix} a & -b \\ -a & b \end{pmatrix}$ 7) $\begin{pmatrix} 2 & 3 \\ -1 & 1 \end{pmatrix}$ 8) $\begin{pmatrix} 2 & -3 \\ 1 & 5 \end{pmatrix}$

9) $\begin{pmatrix} 1 & 1 \\ 0 & 1 \end{pmatrix}$

10) Given that $A = \begin{pmatrix} 1 & 0 \\ 3 & 2 \end{pmatrix}$ and $B = \begin{pmatrix} 3 & 5 \\ 1 & 2 \end{pmatrix}$, calculate:

(a) A^{-1} (b) B^{-1} (c) $B^{-1}A^{-1}$ (d) AB

(e) $(AB)^{-1}$ (f) Compare the answers to (c) and (e)

SYSTEMS OF LINEAR EQUATIONS

Given the system of equations $\left. \begin{matrix} 5x + y = 7 \\ 3x - 4y = 18 \end{matrix} \right\}$

we can rewrite it in the form $\begin{pmatrix} 5x + y \\ 3x - 4y \end{pmatrix} = \begin{pmatrix} 7 \\ 18 \end{pmatrix}$

or
$$\begin{pmatrix} 5 & 1 \\ 3 & -4 \end{pmatrix} \begin{pmatrix} x \\ y \end{pmatrix} = \begin{pmatrix} 7 \\ 18 \end{pmatrix}$$

That is
$$\begin{pmatrix} \text{Matrix of} \\ \text{coefficients} \end{pmatrix} \begin{pmatrix} \text{Matrix of} \\ \text{variables} \end{pmatrix} = \begin{pmatrix} \text{Matrix of} \\ \text{constants} \end{pmatrix}$$

Denote the matrix of coefficients by C and its inverse by C^{-1}.

Then
$$|C| = \begin{vmatrix} 5 & 1 \\ 3 & -4 \end{vmatrix} = 5 \times (-4) - 1 \times 3 = -23$$

and
$$C^{-1} = \frac{1}{-23} \begin{pmatrix} -4 & -1 \\ -3 & 5 \end{pmatrix} = \frac{1}{23} \begin{pmatrix} 4 & 1 \\ 3 & -5 \end{pmatrix}$$

Now
$$C \begin{pmatrix} x \\ y \end{pmatrix} = \begin{pmatrix} 7 \\ 18 \end{pmatrix}$$

and multiplying both sides by C^{-1} gives

$$C^{-1} C \begin{pmatrix} x \\ y \end{pmatrix} = C^{-1} \begin{pmatrix} 7 \\ 18 \end{pmatrix}$$

$$\therefore \qquad I \begin{pmatrix} x \\ y \end{pmatrix} = C^{-1} \begin{pmatrix} 7 \\ 18 \end{pmatrix}$$

or
$$\begin{pmatrix} 1 & 0 \\ 0 & 1 \end{pmatrix} \times \begin{pmatrix} x \\ y \end{pmatrix} = \frac{1}{23} \begin{pmatrix} 4 & 1 \\ 3 & -5 \end{pmatrix} \times \begin{pmatrix} 7 \\ 18 \end{pmatrix}$$

$$\therefore \qquad \begin{pmatrix} 1 \times x & 0 \times y \\ 0 \times x & 1 \times y \end{pmatrix} = \frac{1}{23} \begin{pmatrix} 4 \times 7 + & 1 \times 18 \\ 3 \times 7 + & (-5) \times 18 \end{pmatrix}$$

$$\therefore \qquad \begin{pmatrix} x \\ y \end{pmatrix} = \frac{1}{23} \begin{pmatrix} 46 \\ -69 \end{pmatrix}$$

$$\therefore \qquad \begin{pmatrix} x \\ y \end{pmatrix} = \begin{pmatrix} 2 \\ -3 \end{pmatrix}$$

Thus comparing the matrices shows that $x = 2$ and $y = -3$

We would not normally perform multiplication by the unit matrix. We did so here to illustrate that when a matrix, here $\begin{pmatrix} x \\ y \end{pmatrix}$, is multiplied by the unit matrix then it is unaltered.

This confirms that the unit matrix performs as unity (the number one) in normal arithmetic.

Exercise 27.5

Use matrix methods to solve each of the following systems of equations.

1) $\left.\begin{array}{r} x + y = 1 \\ 3x + 2y = 8 \end{array}\right\}$

2) $\left.\begin{array}{r} x + y = 6 \\ 3x - 2y = -7 \end{array}\right\}$

3) $\left.\begin{array}{r} 5x - 2y = 17 \\ 2x + 3y = 3 \end{array}\right\}$

4) $\left.\begin{array}{r} 3x - 2y = 12 \\ 4x + y = 5 \end{array}\right\}$

5) $\left.\begin{array}{r} 3x + 2y = 6 \\ 4x - y = 5 \end{array}\right\}$

6) $\left.\begin{array}{r} 3x - 4y = 26 \\ 5x + 6y = -20 \end{array}\right\}$

28. COMPLEX NUMBERS

Outcome:

1. *Understand the necessity of extending the number system to include the square roots of negative numbers.*
2. *Define j as $\sqrt{-1}$.*
3. *Define a complex number as consisting of a real part and an imaginary part.*
4. *Define a complex number z in the algebraic form $x + jy$.*
5. *Determine the complex roots of $ax^2 + bx + c = 0$ when $b^2 < 4ac$ using the quadratic formula.*
6. *Perform the addition and subtraction of complex numbers in algebraic form.*
7. *Define the conjugate of a complex number in algebraic form.*
8. *Perform the multiplication and division of complex numbers in algebraic form.*
9. *Represent the algebraic form of a complex number on an Argand diagram, and show how it may be represented as a phasor.*
10. *Deduce that j may be considered to be an operator, such that when the phasor representing $x + jy$ is multiplied by j it rotates the phasor through $90°$ anticlockwise.*
11. *Understand how phasors on an Argand diagram may be added and subtracted in a manner similar to the addition and subtraction of vectors.*
12. *Show that the full polar form of a complex number is $(\cos\theta + j\sin\theta)$ which may be abbreviated to r/θ.*
13. *Perform the operations involved in the conversion of complex numbers in algebraic form to polar form and vice versa.*

INTRODUCTION

The solution of the quadratic equation $ax^2 + bx + c = 0$ is given by the formula

$$x = \frac{-b \pm \sqrt{b^2 - 4ac}}{2a}$$

When we use this formula most of the quadratic equations we meet, when solving technology problems, are found to have roots which are ordinary positive or negative numbers.

Consider now the equation $x^2 - 4x + 13 = 0$.

Then
$$x = \frac{-(-4) \pm \sqrt{(-4)^2 - 4 \times 1 \times 13}}{2 \times 1}$$

$$= \frac{4 \pm \sqrt{16 - 52}}{2}$$

$$= \frac{4 \pm \sqrt{-36}}{2}$$

413

$$= \frac{4 \pm \sqrt{(-1)(36)}}{2}$$

$$= \frac{4 \pm \sqrt{(-1)} \times \sqrt{(36)}}{2}$$

$$= \frac{4}{2} \pm \sqrt{-1} \times \frac{6}{2}$$

$$= 2 \pm \sqrt{-1} \times 3$$

It is not possible to find the value of the square root of a negative number.

In order to try to find a meaning for roots of this type we represent $\sqrt{-1}$ by the symbol j.

(Books on pure mathematics often use the symbol i, but in technology j is preferred as i is used for the instantaneous value of a current.)

Thus the roots of the above equation become $2 + j3$ and $2 - j3$.

DEFINITIONS

Expressions such as $2 + j3$ are called **complex numbers.** The number 2 is called the **real part** and j3 is called the **imaginary part.**

The general expression for a complex number is $x + jy$, which has a real part equal to x and an imaginary part equal to jy. The form $x + jy$ is said to be the **algebraic form** of a complex number. It may also be called the **Cartesian form** or **rectangular notation.**

POWERS OF j

We have defined j such that

$$j = \sqrt{-1}$$

∴ squaring both sides of the equation gives

$$j^2 = (\sqrt{-1})^2 = -1$$

Hence $j^3 = j^2 \times j = (-1) \times j = -j$

and $j^4 = (j^2)^2 = (-1)^2 = 1$

and $j^5 = j^4 \times j = 1 \times j = j$

and $$j^6 = (j^2)^3 = (-1)^3 = -1$$

and so on.

The most used of the above relationships is $j^2 = -1$.

ADDITION AND SUBTRACTION OF COMPLEX NUMBERS IN ALGEBRAIC FORM

The real and imaginary parts must be treated separately. The real parts may be added and subtracted and also the imaginary parts may be added and subtracted, both obeying the ordinary laws of algebra.

Thus
$$(3+j2)+(5+j6) = 3+j2+5+j6$$
$$= (3+5)+j(2+6)$$
$$= 8+j8$$

and
$$(1-j2)-(-4+j) = 1-j2+4-j$$
$$= (1+4)-j(2+1)$$
$$= 5-j3$$

EXAMPLE 28.1

If z_1, z_2 and z_3 represent three complex numbers such that $z_1 = 1.6+j2.3$, $z_2 = 4.3-j0.6$ and $z_3 = -1.1-j0.9$ find the complex numbers which represent:

(a) $z_1+z_2+z_3$,

(b) $z_1-z_2-z_3$.

(a) $z_1+z_2+z_3 = (1.6+j2.3)+(4.3-j0.6)+(-1.1-j0.9)$
$$= 1.6+j2.3+4.3-j0.6-1.1-j0.9$$
$$= (1.6+4.3-1.1)+j(2.3-0.6-0.9)$$
$$= 4.8+j0.8$$

(b) $z_1-z_2-z_3 = (1.6+j2.3)-(4.3-j0.6)-(-1.1-j0.9)$
$$= 1.6+j2.3-4.3+j0.6+1.1+j0.9$$
$$= (1.6-4.3+1.1)+j(2.3+0.6+0.9)$$
$$= -1.6+j3.8$$

MULTIPLICATION OF COMPLEX NUMBERS IN ALGEBRAIC FORM

Consider the product of two complex numbers, $(3 + j2)(4 + j)$.

The brackets are treated in exactly the same way as they are in ordinary algebra, such that

$$(a + b)(c + d) = ac + bc + ad + bd$$

Hence
$$
\begin{aligned}
(3 + j2)(4 + j) &= 3 \times 4 + j2 \times 4 + 3 \times j + j2 \times j \\
&= 12 + j8 + j3 + j^2 2 \\
&= 12 + j8 + j3 - 2 \qquad \text{since} \qquad j^2 = -1 \\
&= (12 - 2) + j(8 + 3) \\
&= 10 + j11
\end{aligned}
$$

EXAMPLE 28.2

Express the product of $2 + j$, $-3 + j2$, and $1 - j$ as a single complex number.

Then

$$
\begin{aligned}
(2 + j)(-3 + j2)(1 - j) &= (2 + j)(-3 + j2 + j3 - j^2 2) \\
&= (2 + j)(-1 + j5) \qquad \text{since} \qquad j^2 = -1 \\
&= -2 - j + j10 + j^2 5 \\
&= -7 + j9 \qquad \text{since} \qquad j^2 = -1
\end{aligned}
$$

CONJUGATE COMPLEX NUMBERS

Consider
$$
\begin{aligned}
(x + jy)(x - jy) &= x^2 + jxy - jxy - j^2 y \\
&= x^2 - (-1)y^2 \\
&= x^2 + y^2
\end{aligned}
$$

Hence we have the product of two complex numbers which produces a real number since it does not have a j term. If $x + jy$ represents a complex number then $x - jy$ is known as its **conjugate** (and vice versa). For example, the conjugate of $(3 + j4)$ is $(3 - j4)$ and their product is

$$(3 + j4)(3 - j4) = 9 + j12 - j12 - j^2 16 = 9 - (-1)16$$

$$= 25 \quad \text{which is a real number}$$

DIVISION OF COMPLEX NUMBERS IN ALGEBRAIC FORM

Consider $\dfrac{(4+j5)}{(1-j)}$.

We use the method of rationalising the denominator.

This means removing the j terms from the bottom line of the fraction. If we multiply $(1-j)$ by its conjugate $(1+j)$ the result will be a real number. Hence, in order not to alter the value of the given expression, we multiply both the numerator and the denominator by $(1+j)$.

Thus

$$\frac{(4+j5)}{(1-j)} = \frac{(4+j5)(1+j)}{(1-j)(1+j)}$$

$$= \frac{4+j5+j4+j^2 5}{1-j+j-j^2}$$

$$= \frac{4+j9+(-1)5}{1-(-1)}$$

$$= \frac{-1+j9}{2}$$

$$= -\frac{1}{2}+j\frac{9}{2}$$

$$= -0.5+j4.5$$

EXAMPLE 28.3

The impedance Z of a circuit having a resistance and inductive reactance in series is given by the complex number $Z = 5 + j6$.

Find the admittance Y of the circuit if $Y = \dfrac{1}{Z}$.

Now

$$Y = \frac{1}{Z} = \frac{1}{5+j6}$$

The conjugate of the denominator is $5 - j6$ and we therefore multiply both the numerator and denominator by $5 - j6$.

Then

$$Y = \frac{(5-j6)}{(5+j6)(5-j6)}$$

$$= \frac{5-j6}{25+j30-j30-j^2 36}$$

$$= \frac{5-j6}{25-(-1)36} = \frac{5-j6}{61} = \frac{5}{61} - j\frac{6}{61} = 0.082 - j0.098$$

EXAMPLE 28.4

Two impedances Z_1 and Z_2 are given by the complex numbers $Z_1 = 1 + j5$ and $Z_2 = j7$. Find the equivalent impedance Z if

a) $Z = Z_1 + Z_2$ where Z_1 and Z_2 are in series,

b) $\dfrac{1}{Z} = \dfrac{1}{Z_1} + \dfrac{1}{Z_2}$ when Z_1 and Z_2 are in parallel.

a)
$$Z = Z_1 + Z_2 = (1+j5)+j7$$
$$= 1+j5+j7$$
$$= 1+j12$$

b)
$$\frac{1}{Z} = \frac{1}{Z_1} + \frac{1}{Z_2} = \frac{1}{(1+j5)} + \frac{1}{j7}$$

$$= \frac{j7+(1+j5)}{(1+j5)j7}$$

$$= \frac{1+j12}{j7+j^2 35}$$

$$= \frac{1+j12}{j7+(-1)35}$$

Thus
$$Z = \frac{j7-35}{1+j12}$$

$$= \frac{(j7-35)(1-j12)}{(1+j12)(1-j12)}$$

$$= \frac{j7-35-j^2 84+j420}{1+j12-j12-j^2 144}$$

$$= \frac{j427-35-(-1)84}{1-(-1)144}$$

$$= \frac{49+j427}{145}$$

$$= 0.338+j2.945$$

Exercise 28.1

1) Add the following complex numbers:

(a) $3+j5$, $7+j3$ and $8+j2$
(b) $2-j7$, $3+j8$ and $-5-j2$
(c) $4-j2$, $7+j3$, $-5-j6$ and $2-j5$.

2) Subtract the following complex numbers:

(a) $3+j5$ from $2+j8$
(b) $7-j6$ from $3-j9$
(c) $-3-j5$ from $7-j8$.

3) Simplify the following expressions giving the answers in the form $x+jy$:

(a) $(3+j3)(2+j5)$
(b) $(2-j6)(3-j7)$
(c) $(4+j5)^2$
(d) $(5+j3)(5-j3)$
(e) $(-5-j2)(5+j2)$
(f) $(3-j5)(3-j3)(1+j)$
(g) $\dfrac{1}{2+j5}$
(h) $\dfrac{2+j5}{2-j5}$
(i) $\dfrac{-2-j3}{5-j2}$
(j) $\dfrac{7+j3}{8-j3}$
(k) $\dfrac{(1+j2)(2-j)}{(1+j)}$
(l) $\dfrac{4+j2}{(2+j)(1-j3)}$

4) Find the real and imaginary parts of:

(a) $1+\dfrac{j}{2}$
(b) $j3+\dfrac{2}{j^3}$
(c) $(j2)^2+3(j)^5-j(j)$

5) Solve the following equations giving the answers in the form $x+jy$:

(a) $x^2+2x+2=0$
(b) $x^2+9=0$

6) Find the admittance Y of a circuit if $Y=\dfrac{1}{Z}$ where $Z=1.3+j0.6$.

7) Three impedances Z_1, Z_2 and Z_3 are represented by the complex numbers $Z_1=2+j$, $Z_2=1+j$ and $Z_3=j2$. Find the equivalent impedance Z if:

(a) $Z=Z_1+Z_2+Z_3$
(b) $\dfrac{1}{Z}=\dfrac{1}{Z_1}+\dfrac{1}{Z_2}+\dfrac{1}{Z_3}$
(c) $Z=\dfrac{1}{\dfrac{1}{Z_1}+\dfrac{1}{Z_2}}+Z_3$

THE ARGAND DIAGRAM

When plotting a graph, Cartesian co-ordinates are generally used to plot the points. Thus the position of the point P (Fig. 28.1) is defined by the co-ordinates $(3, 2)$ meaning that $x = 3$ and $y = 2$.

Complex numbers may be represented in a similar way on the Argand diagram. The real part of the complex number is plotted along the horizontal real-axis whilst the imaginary part is plotted along the vertical imaginary, or j-axis.

However, a complex number is denoted not by a point but as a **phasor.** A phasor is a line in which regard is paid both to its magnitude and to its direction. Hence in Fig. 28.2 the complex number $4 + \text{j}3$ is represented by the phasor $\overrightarrow{\text{OQ}}$, the end Q of the line being found by plotting 4 units along the real-axis and 3 units along the j-axis.

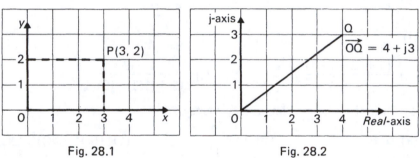

Fig. 28.1   Fig. 28.2

A single letter, the favourite being z, is often used to denote a phasor which represents a complex number. Thus if $z = x + \text{j}y$ it is understood that z represents a phasor and not a simple numerical value.

Four typical complex numbers z_1, z_2, z_3 and z_4 are shown on the Argand diagram in Fig. 28.3.

A real number such as 2.7 may be regarded as a complex number with a zero imaginary part, i.e. $2.7 + \text{j}0$, and may be represented on the Argand diagram (Fig. 28.4) as the phasor $z = 2.7$ denoted by $\overrightarrow{\text{OA}}$ in the diagram.

A number such as j3 is said to be wholly imaginary and may be regarded as a complex number having a zero real part, i.e. $0 + \text{j}3$, and may be represented on the Argand diagram (Fig. 28.4) as the phasor $z = \text{j}3$ denoted by $\overrightarrow{\text{OB}}$ in the diagram.

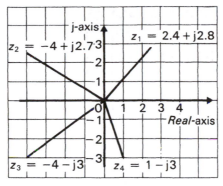

Fig. 28.3

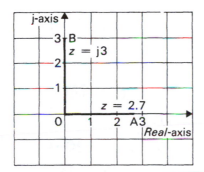

Fig. 28.4

THE j-OPERATOR

Consider the real number 3 shown on the Argand diagram, in Fig. 28.5.

It may be denoted by $\overrightarrow{OA}$. (This is a phasor because it has magnitude and direction.)

If we now multiply the real number 3 by j we obtain the complex number j3 which may be represented by the phasor $\overrightarrow{OB}$.

It follows that the effect of j on phasor $\overrightarrow{OA}$ is to make it become phasor $\overrightarrow{OB}$,

that is $$\overrightarrow{OB} = j\overrightarrow{OA}$$

Hence j is known as an **operator** (called the **j-operator**) which, when applied to a phasor, alters its direction by $90°$ in an anti-clockwise direction without changing its magnitude.

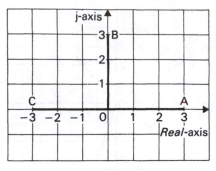

Fig. 28.5

If we now operate on the phasor $\overrightarrow{OB}$ we shall obtain, therefore, phasor $\overrightarrow{OC}$.

In equation form this is

$$OC = j\overrightarrow{OB}$$

but since $\overrightarrow{OB} = j\overrightarrow{OA}$, then

$$\overrightarrow{OC} = j(j\overrightarrow{OA})$$
$$= j^2\overrightarrow{OA}$$
$$= -\overrightarrow{OA} \quad \text{since} \quad j^2 = -1$$

This is true since it may be seen from the phasor diagram that phasor $\overrightarrow{OC}$ is equal in magnitude, but opposite in direction, to phasor $\overrightarrow{OA}$.

Consider now the effect of the j-operator on the complex number $5 + j3$.

In the equation from this is

$$j(5 + j3) = j5 + j(j3)$$
$$= j5 + j^2 3$$
$$= j5 + (-1)3$$
$$= -3 + j5$$

If phasor $z_1 = 5 + j3$ and phasor $z_2 = -3 + j5$, it may be seen from the Argand diagram in Fig. 28.6 that their magnitudes are the same but the effect of the operator j on z_1 has been to alter its direction by $90°$ anticlockwise to give phasor z_2.

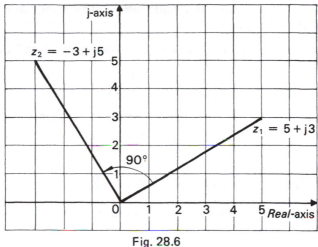

Fig. 28.6

ADDITION OF PHASORS

Consider the addition of the two complex numbers $2 + j3$ and $4 + j2$.

We have

$$(2 + j3) + (4 + j2) = 2 + j3 + 4 + j2$$
$$= (2 + 4) + j(3 + 2)$$
$$= 6 + j5$$

On the Argand diagram shown in Fig. 28.7, the complex number $2 + j3$ is represented by the phasor $\overrightarrow{OA}$, whilst $4 + j2$ is represented by phasor $\overrightarrow{OB}$. The addition of the real parts is performed

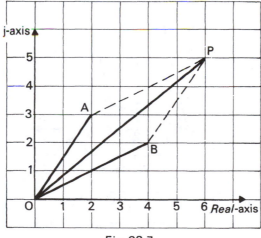

Fig. 28.7

along the real-axis and the addition of the imaginary parts is carried out on the j-axis.

Hence the complex number $6+j5$ is represented by the phasor $\overrightarrow{OP}$.

It follows that $\qquad \overrightarrow{OP} = \overrightarrow{OB} + \overrightarrow{OA}$

Hence the addition of phasors is similar to vector addition used when dealing with forces or velocities.

SUBTRACTION OF PHASORS

Consider the difference of the two complex number, $4+j5$ and $1+j4$.

We have $\qquad (4+j5)-(1+j4) = 4+j5-1-j4$

$$= (4-1)+j(5-4)$$

$$= 3+j$$

On the Argand diagram shown in Fig. 28.8, the complex number $4+j5$ is represented by the phasor $\overrightarrow{OC}$, whilst $1+j4$ is represented by the phasor $\overrightarrow{OD}$. The subtraction of the real parts is performed along the real-axis, and the subtraction of the imaginary parts is carried out along the j-axis. Now let $(4+j5)-(1+j4) = 3+j$ be represented by the phasor $\overrightarrow{OQ}$.

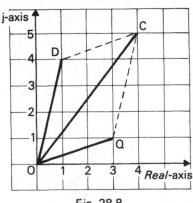

Fig. 28.8

It follows that $\qquad \overrightarrow{OQ} = \overrightarrow{OC} - \overrightarrow{OD}$

As for phasor addition, the subtraction of phasors is similar to the subtraction of vectors.

THE POLAR FORM OF A COMPLEX NUMBER

Let z denote the complex number represented by the phasor $\overrightarrow{OP}$ shown in Fig. 28.9. Then from the right-angled triangle PMO we have

$$z = x + jy$$
$$= r\cos\theta + j(r\sin\theta)$$
$$= r(\cos\theta + j\sin\theta)$$

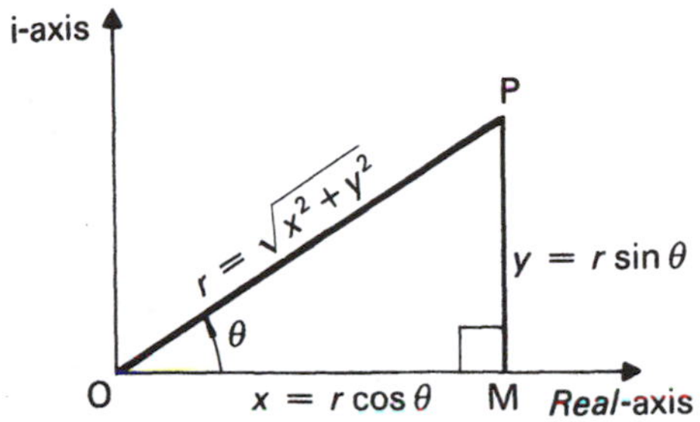

Fig. 28.9

The expression $r(\cos\theta + j\sin\theta)$ is known as the **polar form** of the complex number z. Using conventional notation it may be shown abbreviated as $r\underline{/\theta}$.

r is called the **modulus** of the complex number z and is denoted by $\mathrm{mod}\,z$ or $|z|$.

Hence, from the diagram, $|z| = r = \sqrt{x^2 + y^2}$, using the theorem of Pythagoras for right-angled triangle PMO.

It should be noted that the plural of 'modulus' is 'moduli'.

The angle θ is called the **argument** (or amplitude) of the complex number z, and is denoted by $\arg z$ (or $\mathrm{amp}\,z$).

Hence $\arg z = \theta$

and, from the diagram $\tan\theta = \dfrac{y}{x}$

There are an infinite number of angles whose tangents are the same, and so it is necessary to define which value of θ to state when solving the equation $\tan\theta = \dfrac{y}{x}$. It is called the **principal value** of the angle and lies between $+180°$ and $-180°$.

We recommend that, when finding the polar form of a complex number, you should sketch it on an Argand diagram. This will help you to avoid a common error of giving an incorrect value of the angle.

EXAMPLE 28.5

Find the modulus and argument of the complex number $3 + j4$ and express the complex number in polar form.

Let $z = 3 + j4$ which is shown in the Argand diagram in Fig. 28.10.

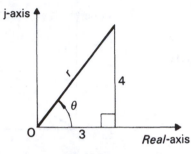

Fig. 28.10

Then $\qquad |z| = r = \sqrt{3^2 + 4^2} = 5$

and $\qquad \tan \theta = \frac{4}{3} = 1.3333$

$\therefore \qquad\qquad\quad = 53.13°$

Hence in polar form

$$z = 5(\cos 53.13° + j \sin 53.13°)$$

or $\qquad\quad z = 5\underline{/53.13°}$

EXAMPLE 28.6

Show the complex number $z = 5\underline{/-150°}$ on an Argand diagram, and find z in algebraic form.

Now z is represented by phasor $\overrightarrow{OP}$ in Fig. 28.11. It should be noted that since the angle is negative it is measured in a clockwise direction from the real-axis datum.

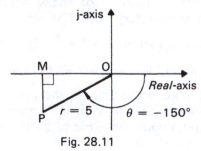

Fig. 28.11

In order to express z in algebraic form we need to find the lengths MO and MP. We use the right-angled triangle PMO in which $\widehat{POM} = 180° - 150° = 30°$.

Now $\qquad$ $MO = PO \cos \widehat{POM} = 5 \cos 30° = 4.33$

and $\qquad$ $MP = PO \sin \widehat{POM} = 5 \sin 30° = 2.50$

Hence, in algebraic form, the complex number $z = -4.33 - j2.50$.

Exercise 28.2

1) Show, indicating each one clearly, the following complex numbers on a single Argand diagram: $4 + j3$, $-2 + j$, $3 - j4$, $-3.5 - j2$, $j3$ and $-j4$.

2) Find the moduli and arguments of the complex numbers $3 + j4$ and $4 - j3$.

3) If the complex number $z_1 = -3 + j2$ find $|z_1|$ and $\arg z_1$.

4) If the complex number $z_2 = -4 - j2$ find $|z_2|$ and $\arg z_2$.

5) Express each of the following complex numbers in polar form:
(a) $4 + j3$ $\qquad$ (b) $3 - j4$ $\qquad$ (c) $-3 + j3$
(d) $-2 - j$ $\qquad$ (e) $j4$ $\qquad$ (f) $-j3.5$

6) Convert the following complex numbers, which are given in polar form, into Cartesian form:
(a) $3\underline{/45°}$ $\qquad$ (b) $5\underline{/154°}$ $\qquad$ (c) $4.6\underline{/-20°}$
(d) $3.2\underline{/-120°}$

7) The potential difference across a circuit is given by the complex number $V = 40 + j35$ volts and the current is given by the complex number $I = 6 + j3$ amperes. Sketch the appropriate phasors on an Argand diagram and find:
(a) the phase difference (i.e. the angle ϕ) between the phasors for V and I,
(b) the power, given that power $= |V| \times |I| \times \cos\phi$.

ANSWERS

ANSWERS TO CHAPTER 1

Exercise 1.1

1) 24.865 8, 24.87, 25
2) 0.008 357, 0.008 36, 0.008 4
3) 4.978 5, 4.98, 5
4) 22   5) 35.60
6) 28 388 000, 28 000 000
7) 4.149 8, 4.150, 4.15
8) 9.20
9) (a) 2.1389   (b) 2.139
   (c) 2.14
10) (a) 25.17   (b) 25.2
11) (a) 0.003 99   (b) 0.004 0
    (c) 0.004
12) (a) 7.204   (b) 7.20
    (c) 7.2
13) (a) 0.726   (b) 0.73

Exercise 1.2

1) 64.5, 63.5   2) 2474, 2464
3) 3.075, 3.065   4) 0.65, 0.55
5) 29.95, 28.85   6) 1.315, 1.205
7) 2.80, 2.60   8) 0.7515, 0.7405
9) 0.55, 0.055, 0.1, 0.005 5
10) 39.07 ± 0.005
11) 0.372 ± 0.000 5, 1.238 ± 0.000 5,
    3.222, 3.218, 0.002
12) 0.553
13) 7.00 ± 0.005 A, 0.005
14) 12.015, 11.985

ANSWERS TO CHAPTER 2

Exercise 2.1

1) 13.1   2) −11.35
3) 27.4   4) 0.001 49
5) 1.94   6) −4.26
7) 1.28   8) 18.8

9) −2.52   10) 527
11) −22.8   12) −22.8
13) 0.007 6   14) −0.348
15) 0.66   16) 0.55
17) −4.1   18) 4.0
19) 6.6   20) 0.001 30

Exercise 2.2

1) 29.0   2) 12.3
3) 0.0391   4) 0.0160
5) 0.0103   6) 94.2
7) 42.1   8) 86.3
9) 0.67   10) 2.90
11) 2.54   12) 0.51
13) 17.11   14) 2.84
15) 1.62   16) 0.52
17) 470   18) 3.22
19) 63   20) 1.92×10^8
21) 1.2   22) 9.2
23) 0.0171   24) 6.56×10^6
25) 5500   26) 610
27) 70.3   28) 55.8°
29) 45.2°   30) 18.6
31) (a) 72.4   (b) 3.22   (c) 244
    (d) 10.7

ANSWERS TO CHAPTER 3

Exercise 3.1

1) 8 km   2) 15 Mg
3) 3.8 Mm   4) 1.8 Gg
5) 7 mm   6) 1.3 μm
7) 28 g   8) 360 mm
9) 64 mg   10) 3.6 mA

Exercise 3.2

1) 0.755   2) 805
3) 54.9   4) 0.791

5) 0.391	6) 128	16) (a) 24.1	(b) 683
7) 0.227	8) 114	(c) 0.683	
9) 49.2	10) 122 000	17) 0.011	
11) 44.0	12) 330	18) (a) 88.0	(b) 26.8
13) 1270	14) 23.01 ± 0.03 mm	(c) 96.6	
15) 0.110, 0.020 9, 0.000 492		19) 42.5	

ANSWERS TO CHAPTER 4

Exercise 4.1

1) (a) 24.4°C   (b) 72.0°C   (b) 2.2 mm   (c) 410 kg
   (c) 155.7°F   (d) 11.7°F   4) (a) 2.5 ohm   (b) 7.9 ohm
2) (a) 17.0 lb   (b) 730 lb   (c) 3.1 ohm
   (c) 8.3 kg   (d) 0.19 kg   5) (a) 110   (b) 60
3) (a) 580 MN/m^2 6) (b) 34 kN, 46 kN

7)

January		February		March		April	
800	800	1000	1800	1200	3000	1400	4400
	700		1000		1200		1000
				200			
	700		1700			3100	4100

8)

		A	B	C
1		January	February	March
2				
3	Mortgage	320.00	320.00	320.00
4	Groceries	330.00	330.00	330.00
5	Car payment	220.00	220.00	220.00
6	Personal	270.00	270.00	270.00
7	Insurance	80.00	80.00	80.00
8	Telephone	141.00		
9	Electricity			186.00
10				
11	Totals	1361.00	1220.00	1406.00

Instructions: Sum(A3 ... A9), Sum(B3 ... B9), Sum(C3 ... C9)

ANSWERS TO CHAPTER 5

Exercise 5.1

1) 0	2) 0	3) 0	7) 1	8) 1	9) 1
4) 1	5) 1	6) 1	10) 1	11) 0	12) 1

13) (i)   **(ii)**   **(iii)**

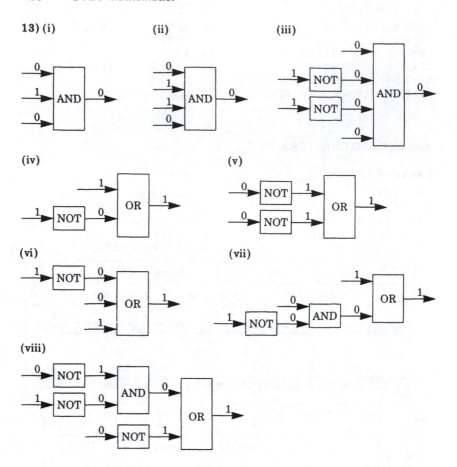

Exercise 5.3

1) (a)

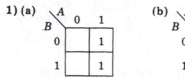

B \ A	0	1
0		1
1		1

(b)

B \ A	0	1
0		1
1	1	1

(c)

Y \ X	0	1
0		1
1	1	

2) (a)

BC \ A	0	1
00	1	1
01	1	1
11	1	1
10		1

(b)

BC \ A	0	1
00	1	1
01		1
11		
10	1	1

(c)

BC \ A	0	1
00	1	
01		
11		
10	1	1

3) (a)

YZ＼WX	00	01	11	10
00				
01			1	1
11			1	1
10				

(b)

CD＼AB	00	01	11	10
00	1	1	1	
01	1	1	1	
11	1	1		1
10	1	1		

(c)

YZ＼WX	00	01	11	10
00	1			
01	1	1	1	1
11	1	1	1	1
10				

4) (a)

A + B

(b)

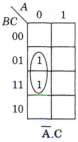

$\overline{A}.C$

(c)

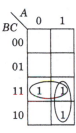

A.B + B.C

(d)

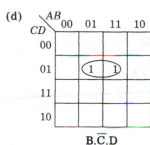

CD＼AB	00	01	11	10
00				
01		1	1	
11				
10				

B.$\overline{C}$.D

Exercise 5.4

1) (a) 51   (b) 143
   (c) 307
2) (a) 35_8 (b) 363_8
   (c) 3404_8
3) (a) 715   (b) 5614
5) (a) 10111   (b) 1111101
   (c) 1100001
6) (a) 22   (b) 57
   (c) 90
7) (a) 0.8125   (b) 0.4375
   (c) 0.1875
8) (a) 0.011   (b) 0.0101
   (c) 0.111
9) (a) 0.0010101
   (b) 10010.0111011
   (c) 1101100.1011010

10) (a) 1110   (b) 100110
    (c) 111000
11) (a) 100011   (b) 1101110
    (c) 101011111

ANSWERS TO CHAPTER 6

Exercise 6.1

1) 2^{11} 2) a^8 3) n^3
4) 3^{11} 5) b^{-3} 6) 10^4
7) z^3 8) 3^{-4} 9) m^4
10) x^{-3} 11) 9^{12} 12) y^{-6}
13) t^8 14) c^{14} 15) a^{-9}
16) 7^{-12} 17) b^{10} 18) s^{-9}
19) 8   20) 1   21) 0.5
22) 8   23) 0.25   24) 100

25) 0.25   26) 0.143   27) 0.04
28) 3   29) 7   30) 25
31) 7   32) 3.375   33) 0.003 91

Exercise 6.2

1) $x^{1/2}$ 2) $x^{4/5}$ 3) $x^{-1/2}$
4) $x^{-1/3}$ 5) $x^{-4/3}$ 6) $x^{-3/2}$
7) $x^{2/3}$ 8) $x^{0.075}$ 9) $x^{2/3}$
10) $x^{1/3}$ 11) $x^{1/3}$ 12) x
13) 5   14) 2   15) 2
16) 2   17) 4   18) 125
19) 8   20) 27   21) 2
22) 4   23) 0.167   24) 0.125
25) 0.031 25   26) 64
27) 3   28) $a^{-13/6}$ 29) $a^{-11/3}$
30) $x^{4.5}$ 31) $b^{1/2}$ 32) $m^{7/4}$
33) $z^{2.3}$ 34) 1   35) $u^{-5/2}$
36) $y^{1/4}$ 37) $n^{1/4}$ 38) $x^{11/14}$
39) $t^{-2/3}$

ANSWERS TO CHAPTER 7

Exercise 7.1

1) $3x + 12$ 2) $2a + 2b$

3) $9x + 6y$ 4) $\dfrac{x}{2} - \dfrac{1}{2}$

5) $10p - 15q$ 6) $7a - 21m$
7) $-a - b$ 8) $-a + 2b$
9) $-3p + 3q$ 10) $-7m + 6$
11) $-4x - 12$ 12) $-4x + 10$
13) $-20 + 15x$ 14) $2k^2 - 10k$
15) $-9xy - 12y$ 16) $ap - aq - ar$
17) $4abxy - 4acxy + 4dxy$
18) $3x^4 - 6x^3y + 3x^2y^2$
19) $-14P^3 + 7P^2 - 7P$
20) $2m - 6m^2 + 4mn$
21) $5x + 11$ 22) $14 - 2a$
23) $x + 7$ 24) $16 - 17x$
25) $7x - 11y$
                          26) $\dfrac{7y}{6} - \dfrac{3}{2}$
27) $-8a - 11b + 11c$
28) $7x - 2x^2$ 29) $3a - 9b$
30) $-x^3 + 18x^2 - 9x - 15$

Exercise 7.2

1) $x^2 + 3x + 2$ 2) $x^2 + 4x + 3$
3) $x^2 + 9x + 20$ 4) $2x^2 + 11x + 15$
5) $3x^2 + 25x + 42$
6) $5x^2 + 21x + 4$ 7) $6x^2 + 16x + 8$
8) $10x^2 + 17x + 3$ 9) $21x^2 + 41x + 10$

10) $x^2 - 4x + 3$ 11) $x^2 - 6x + 8$
12) $x^2 - 9x + 18$ 13) $2x^2 - 9x + 4$
14) $3x^2 - 11x + 10$
15) $4x^2 - 33x + 8$ 16) $6x^2 - 16x + 8$
17) $6x^2 - 17x + 5$ 18) $21x^2 - 29x + 10$
19) $x^2 + 2x - 3$ 20) $x^2 + 5x - 14$
21) $x^2 - 2x - 15$ 22) $2x^2 + x - 10$
23) $3x^2 + 13x - 30$
24) $3x^2 + 23x + 30$
25) $6x^2 + x - 15$ 26) $12x^2 + 4x - 21$
27) $6x^2 - x - 15$ 28) $3x^2 + 5xy + 2y^2$
29) $2p^2 - 7pq + 3q^2$
30) $6v^2 - 5uv - 6u^2$
31) $6a^2 + ab - b^2$ 32) $5a^2 - 37a + 42$
33) $6x^2 - xy - 12y^2$
34) $x^2 + 2x + 1$ 35) $4x^2 + 12x + 9$
36) $9x^2 + 42x + 49$
37) $x^2 - 2x + 1$ 38) $9x^2 - 30x + 25$
39) $4x^2 - 12x + 9$ 40) $4a^2 + 12ab + 9b^2$
41) $x^2 + 2xy + y^2$ 42) $P^2 + 6PQ + 9Q^2$
43) $a^2 - 2ab + b^2$
44) $9x^2 - 24xy + 16y^2$
45) $4x^2 - y^2$
46) $a^2 - 9b^2$
47) $4m^2 - 9n^2$
48) $x^4 - y^2$

Exercise 7.3

1) p^2q 2) ab^2 3) $3mn$
4) b 5) $3xyz$ 6) $2(x+3)$
7) $4(x-y)$ 8) $5(x-1)$
9) $4x(1-2y)$ 10) $m(x-y)$
11) $x(a+b+c)$ 12) $\dfrac{1}{2}\left(x - \dfrac{y}{4}\right)$
13) $5(a-2b+3c)$ 14) $ax(x+1)$
15) $\pi r(2r+h)$ 16) $3y(1-3y)$
17) $ab(b^2-a)$ 18) $xy(xy-a+by)$
19) $5x(x^2-2xy+3y^2)$
20) $3xy(3x^2-2xy+y^4)$
21) $I_0(1+\alpha t)$ 22) $\dfrac{1}{3}\left(x - \dfrac{y}{2} + \dfrac{z}{3}\right)$
23) $a(2a-3b)+b^2$
24) $x(x^2-x+7)$
25) $\dfrac{m^2}{pn}\left(1 - \dfrac{m}{n} + \dfrac{m^2}{pn}\right)$

Exercise 7.4

1) $(x+y)(a+b)$ 2) $(p-q)(m+n)$
3) $(ac+d)^2$ 4) $(2p+q)(r-2s)$
5) $2(a-b)(2x+3y)$
6) $(x^2+y^2)(ab-cd)$

7) $(mn-pq)(3x-1)$
8) $(k^2l-mn)(l-1)$

Exercise 7.5

1) $6a^2$ 2) $2x^2y$ 3) m^2n^2
4) $2abc^2$ 5) $2(x+1)$ 6) $x^2(a+b)^2$
7) $(a+b)(a-b)$ 8) $x(1-x)(x+1)$
9) $(x+2)(x-2)^2$
10) $(x-2)(x+2)$
11) $6(a+b)(a-b)$
12) $6(a+b)^2(a^2+b^2)$

Exercise 7.6

1) $\dfrac{1}{ab}$ 2) $\dfrac{b}{a}$ 3) $\dfrac{x^2}{y^2}$

4) $\dfrac{xy}{2}$ 5) $\dfrac{1}{ab}$ 6) $c(a+b)$

7) $1-x^2$ 8) $\dfrac{c}{(a-b)^2}$ 9) $\dfrac{x+y}{xy}$

10) $\dfrac{a+1}{a}$ 11) $\dfrac{m-n}{n}$ 12) $\dfrac{b-c^2}{c}$

13) $\dfrac{ad-bc}{bd}$ 14) $\dfrac{ac-1}{bc}$

15) $\dfrac{a^2-1}{a}$ 16) $\dfrac{1+y+xy}{xy}$

17) $\dfrac{12+x^2}{4x}$ 18) $\dfrac{3de+2ce-5cd}{cde}$

19) $\dfrac{ad+cb+bd}{bd}$ 20) $\dfrac{2h-5f-3g}{6fgh}$

21) $\dfrac{8y-16}{(y+3)(y-5)}$ 22) $\dfrac{4-2x}{x(x+2)}$

23) $\dfrac{x^2+x-1}{x^2(x-1)}$ 24) $\dfrac{x^2+1}{(1-x)(1+x)}$

25) $\dfrac{2}{2-x}$ 26) $\dfrac{2x}{x-2}$

Exercise 7.7

1) $1+\dfrac{b}{a}$ 2) $\dfrac{1}{b}-\dfrac{1}{a}$ 3) $\dfrac{1}{c}+1$

4) $\dfrac{x}{2}+\dfrac{y}{2x}$ 5) $\dfrac{a}{bc}-\dfrac{1}{c}+\dfrac{1}{b}$

6) $\dfrac{(x-1)}{(x+1)}+1$ 7) $y+\dfrac{y^2}{x(1-a)}$

8) $\dfrac{1}{(x-y)}+\dfrac{1}{x}$ 9) $\dfrac{1}{(a-b)}-\dfrac{1}{(a+b)}$

10) $\dfrac{7d-3c}{4c}$ 11) $\dfrac{x}{1+x}$

12) $\dfrac{a^2}{a^2-1}$ 13) $\dfrac{1}{2(4a+3)}$

14) $\dfrac{m+1}{1-m}$ 15) $\dfrac{3(3t+5)}{(1-6t)}$

16) $\dfrac{R_1R_2}{R_1+R_2}$ 17) $\dfrac{a(ac-b)}{c(a+c)}$

18) $\dfrac{by+cx}{c(xy+b)}$ 19) $\dfrac{b(b-a)}{ab-1}$

20) $\dfrac{a+b}{b}$ 21) $-\dfrac{2}{x}$

ANSWERS TO CHAPTER 8

Exercise 8.1

1) $x=5$ 2) $t=7$ 3) $q=2$
4) $x=20$ 5) $q=-3$ 6) $x=3$
7) $y=6$ 8) $m=12$ 9) $x=2$
10) $x=3$ 11) $p=4$ 12) $x=-2$
13) $x=-1$ 14) $x=4$ 15) $x=2$
16) $x=6$ 17) $m=2$ 18) $x=-8$
19) $d=6$ 20) $x=5$ 21) $x=3$

22) $m=5$ 23) $x=-\dfrac{29}{5}$

24) $x=2$ 25) $x=\dfrac{45}{8}$

26) $x=-2$ 27) $x=-15$

28) $x=\dfrac{50}{47}$ 29) $m=-1.5$

30) $x=\dfrac{15}{28}$ 31) $m=1$

32) $x=2.5$ 33) $t=6$
34) $x=4.2$ 35) $y=-70$

36) $x=\dfrac{5}{3}$ 37) $x=13$

38) $x=-10$ 39) $m=\dfrac{25}{26}$

40) $y=\dfrac{9}{7}$ 41) $x=\dfrac{25}{3}$

42) $x=3.5$ 43) $x=20$
44) $x=13$ 45) -53

46) $x=4$ 47) $p=\dfrac{13}{4}$

48) $m=3$ 49) $x=\dfrac{15}{4}$

50) $x=\dfrac{7}{2}$

ANSWERS TO CHAPTER 9

Exercise 9.1

1) $T = \dfrac{pV}{mR}$ 2) $h = \dfrac{Hr}{R}$

3) $u = v - at$ 4) $t = \dfrac{v-u}{a}$

5) $C = \frac{5}{9}(F - 32)$ 6) $x = \dfrac{y-c}{m}$

7) $L = \dfrac{15-A}{1.5}$ 8) $r = 1 - \dfrac{a}{S}$

9) $R = \dfrac{V}{I} - r$ 10) $h = \dfrac{S}{\pi r} - r$

11) $T = \dfrac{H}{ws} + t$ 12) $u^2 = v^2 - 2as$

13) (a) $S = \dfrac{n}{2}[2a + (n-1)d]$

(b) $a = \dfrac{S}{n} - \dfrac{d}{2}(n-1)$

14) $R_1 = \dfrac{R_2 R}{R_2 - R}$

15) $R = \dfrac{V}{\pi h^2} + \dfrac{h}{3}$

16) $R_2 = \dfrac{R_1 R}{R_1 - R}$

17) $d_1 = \dfrac{(M-L)d_2}{(i-M+L)}$

18) $m_2 = \dfrac{m_i - \tan\theta}{1 + m_1 \tan\theta}$

19) $a^2 = \dfrac{b^2}{1-e^2}$

20) $H = \dfrac{2A - (a+b)h}{a+c}$

Exercise 9.2

1) $h = \dfrac{v^2}{2g}$ 2) $r = \sqrt{\dfrac{A}{\pi}}$

3) $v = \sqrt{\dfrac{2E}{m}}$ 4) $A = \dfrac{\pi d^2}{4}$

5) $f = \sqrt{\dfrac{2EU}{V}}$ 6) $c = \sqrt{a^2 - b^2}$

7) $l = \dfrac{g}{4f^2\pi^2}$ 8) $M = \sqrt{T_e^2 - T^2}$

9) $C = \dfrac{\pi}{L}\sqrt{\dfrac{EI}{P}}$

10) $v = \sqrt{\dfrac{2}{m}(E_t - mgh)}$

11) $r = \sqrt{\dfrac{2V}{h} - \dfrac{h^2}{3}}$

12) $b = \sqrt{12k^2 - a^2}$

13) $L = \sqrt{12k^2 - 3R^2}$

14) $f = \dfrac{(D^2 + d^2)p}{D^2 - d^2}$

15) $P = \sqrt{4Q_e^2 - Q^2}$

16) $Q = \frac{1}{2}\sqrt{(2P_e - P)^2 - P^2}$

ANSWERS TO CHAPTER 10

Exercise 10.1

1) $x^2 + 3x + 2$ 2) $2x^2 + 11x + 15$
3) $6x^2 + 16x + 8$ 4) $x^2 - 6x + 8$
5) $3x^2 - 11x + 10$
6) $6x^2 - 17x + 5$ 7) $x^2 + 2x - 3$
8) $x^2 - 2x - 15$ 9) $3x^2 + 13x - 30$
10) $12x^2 + 4x - 21$
11) $2p^2 - 7pq + 3q^2$
12) $6v^2 - 5uv - 6u^2$
13) $x^2 - 9$ 14) $4x^2 - 9$
15) $x^2 + 2x + 1$ 16) $4x^2 + 12x + 9$
17) $x^2 - 2x + 1$ 18) $4x^2 - 12x + 9$
19) $4a^2 + 12ab + 9b^2$
20) $x^2 + 2xy + y^2$ 21) $a^2 - 2ab + b^2$
22) $9x^2 - 24xy + 16y^2$

Exercise 10.2

1) $(x+1)(x+3)$ 2) $(x+2)(x+4)$
3) $(x-1)(x-2)$ 4) $(x+5)(x-3)$
5) $(x+7)(x-1)$ 6) $(x+2)(x-7)$
7) $(x+y)(x-3y)$
8) $(2x+3)(x+5)$
9) $(p+1)(3p-2)$
10) $(2x+1)(2x-6)$
11) $(m+2)(3m-14)$
12) $(3x+1)(7x+10)$
13) $(2a+5)(5a-3)$
14) $(2x+5)(3x-7)$
15) $(2p+3q)(3p-q)$
16) $(4x+y)(3x-2y)$
17) $(x+y)^2$ 18) $(2x+3)^2$

19) $(p+2q)^2$ 20) $(3x+1)^2$
21) $(m-n)^2$ 22) $(5x-2)^2$
23) $(x-2)^2$ 24) $(m+n)(m-n)$
25) $(2x+y)(2x-y)$
26) $(3p+2q)(3p-2q)$
27) $(x+1/3)(x-1/3)$
28) $(1+b)(1-b)$
29) $(1/x+1/y)(1/x-1/y)$
30) $(11p+8q)(11p-8q)$

Exercise 10.3

1) $x^2-4x+3 = 0$
2) $x^2+2x-8 = 0$
3) $x^2+3x+2 = 0$
4) $x^2-2.3x+1.12 = 0$
5) $x^2-1.07x-4.53 = 0$
6) $x^2+7.32x+12.19 = 0$
7) $x^2-1.4x = 0$
8) $x^2+4.36x = 0$
9) $x^2-12.25 = 0$
10) $x^2-8x+16 = 0$

Exercise 10.4

1) ± 6 2) ± 1.25
3) ± 1.333 4) -4 or -5
5) 8 or -9 6) 2 or $\frac{1}{3}$
7) 3 8) 4 or -8
9) $\frac{4}{7}$ or $\frac{3}{2}$ 10) $\frac{7}{3}$ or $-\frac{4}{3}$
11) 1.175 or -0.425
12) $\frac{2}{6}$ or $\frac{1}{6}$
13) 0.573 or -2.907
14) 0.211 or -1.354
15) 1 or -0.2
16) 3.886 or -0.386
17) 0.956 or -1.256
18) 2.388 or 0.262
19) 0.44 or -3.775
20) 8.385 or -2.385
21) -0.225 or -1.775
22) 11.14 or -3.14
23) 1.303 or -2.303
24) ± 53.67
25) 5.24 or 0.76
26) -3.064 or -0.935

Exercise 10.5

1) 149.6 2) 92.4 mm
3) 6.5 m, 9.5 m   4) 40 mm
5) 0.685 or 23.3 m
6) 30 or 72 mm   7) 54.6 mm
8) 50 9) 2.88 m

10) 94.6×94.6 mm
11) 2.41 and -0.41 s
12) 0.412

Exercise 10.6

1) $x = 0.333, y = -1.667$;
   $x = 2, y = 5$
2) $x = -1, y = 6$;
   $x = 0.375, y = -0.875$
3) $x = 1.175, y = 2.175$;
   $x = -0.425, y = 0.575$
4) $x = -0.969, y = 0.956$;
   $x = 0.579, y = -1.256$
5) $x = 0.211, y = -1.422$;
   $x = -1.354, y = 1.708$
6) $x = 0.167, y = 0.805$;
   $x = 0.833, y = 2.278$
7) 12 or 14.7 m
8) 2 m, 8 m

ANSWERS TO CHAPTER 11

Exercise 11.1

1) $\log_a n = x$ 2) $\log_2 8 = 3$
3) $\log_5 0.04 = -2$
4) $\log_{10} 0.001 = -3$
5) $\log_x 1 = 0$ 6) $\log_{10} 10 = 1$
7) $\log_a a = 1$ 8) $\log_e 7.39 = 2$
9) $\log_{10} 1 = 0$ 10) 3
11) 3 12) 4 13) 3
14) 9 15) 64 16) 100
17) 1 18) 2 19) 3
20) $\frac{1}{2}$ 21) 1

Exercise 11.2

1) 2.08 2) 368
3) 0.795 4) $0.000\,179$
5) 0.267 6) 1.77
7) -7.38 8) -0.322
9) 4.63 10) 2.35
11) -2.28 12) -36.5
13) 22.2

Exercise 11.3

1) (a) 15.8 (b) 1.97
   (c) 1.09 (d) 0.0314
   (e) 0.581 (f) 0.925
2) (a) 853 (b) 39.3
   (c) 2.18
3) $0.003\,57$ 4) 0.0164

5) 0.590   6) 25.7
7) 132   8) 0.741
9) 44.2   10) 0.005 49
11) 0.769

Exercise 11.4

1) $y = 1.82$, $x = 0.84$
2) $x = -1.39$, -0.5
3) (a) $T = 631$ (b) $s = 1.12$
4) 4.34 mA per second
5) (a) 0.18 seconds
   (b) 1200 volts per second
6) (a) 89 800 cells
   (b) 1.35 hours
   (c) 27 900 cells/hour
7) (a) 1.10×10^{-8} grams
   (b) 10^{-10} grams
   (c) 0.578 hours
8) 0.0996

ANSWERS TO CHAPTER 12

Exercise 12.1

1) (a) 0.6109   (b) 1.457
   (c) 0.3367   (d) 0.7621
2) (a) 9°55'25" (b) 89°33'53"
   (c) 4°29'11"
3) (a) 1.05 m   (b) 22.9 mm
4) (a) 120°   (b) 10.2°
5) 89.2 mm
6) (a) 4.71 m²   (b) 508 mm²
   (c) 7620 mm²
7) 866 mm²
8) 29.3 and 80.7 mm
9) 1240 mm²   10) 185 mm²
11) 163 mm²
12) 369 mm, 20 600 mn
13) 7.96 m, 31.7 m²
14) 11.2 m²   15) 17.9 m

ANSWERS TO CHAPTER 13

Exercise 13.1

1) 67.9 mm   2) 90.02 mm
3) 115 mm   4) 167 mm
5) 39.7 mm   6) 255 mm
7) 30°33'   8) 21°33'
9) 64°45'   10) 177 mm

Exercise 13.2

1) (a) 0.8572, −0.5150, −1.6643
   (b) 0.0282, −0.9996, −0.0282
   (c) 0.9764, −0.2162, −4.5169
   (d) −0.5150, −0.8572, 0.6009
   (e) −0.8597, −0.5108, 1.6831
   (f) −0.9798, −0.2000, 4.9006
   (g) −0.6633, 0.7484, −0.8863
   (h) −0.8887, 0.4584, −1.9389
2) 1.897   3) 3.0248
4) −0.28   5) $-\frac{4}{5}$, $-\frac{3}{4}$
6) 8°14', 171°46'
7) 153°13', 206°47'
8) (a) 45°32', 134°28'
   (b) 118°46'   (c) 43°28'
   (d) 119°33'
9) (a) 232° and 308°
   (b) 304°   (c) 289°
10) 14°34', 165°26'
11) 105°42', 254°18'

Exercise 13.3

1) (a) 0.6157, 0.7880
   (b) 0.9551, 0.3090
   (c) 0.6157, −0.7880
   (d) 0.9551, −0.3090
   (e) −0.3420, −0.9397
   (f) −0.9397, −0.3420
   (g) −0.8192, 0.5736
   (h) −0.5299, 0.8480
2) 30°, 150°, −2.82
3) 78°, 282°   4) 23°, 157°

Exercise 13.4

1) (a) $C = 71°$, $b = 59.1$ mm,
       $c = 99.9$ mm
   (b) $A = 48°$, $a = 71.5$ mm,
       $c = 84.2$ mm
   (c) $B = 56°$, $a = 3.74$ m,
       $b = 9.53$ m
   (d) $B = 46°$, $b = 136$ mm,
       $c = 58.4$ mm
   (e) $C = 67°$, $a = 1.51$ m,
       $c = 2.36$ m
   (f) $C = 63°32'$
       $a = 9.486$ mm
       $b = 11.56$ mm

(g) B = 135°38',
   a = 93.93 mm,
   c = 144.4 mm
(h) B = 81°54',
   b = 9.947 m
   c = 3.609 m
(i) A = 53°39',
   a = 2124 mm,
   b = 2390 mm
(j) A = 45°30' or 134°30',
   B = 95°30' or 6°30',
   c = 23.7 m or 2.70 m
(k) A = 13°51',
   B = 144°2',
   b = 17.2 m
(l) A = 86°1' or 15°43',
   B = 54°51' or 125°9',
   a = 112 mm or 30.5 mm
(m) A = 44°46',
   C = 49°57',
   a = 10.69 m
(n) B = 93°49',
   C = 36°52',
   b = 30.3 m
(o) B = 48°31',
   C = 26°25',
   c = 4.247 m
2) (a) *c* = 10.2 m,
   A = 50°11',
   B = 69°49',
   (b) *a* = 11.8 m,
   B = 44°42',
   C = 79°18',
   (c) *b* = 4.99 m,
   A = 82°24',
   C = 60°18'
   (d) A = 38°12',
   B = 81°38',
   C = 60°10'
   (e) A = 24°42',
   B = 44°54',
   C = 110°24'
   (f) A = 34°42',
   B = 18°6',
   C = 127°12'
3) 64.00 mm   4) 37°35'
5) 40.5°   6) 41°27'
7) (a) 14.2 m   (b) 142°39'
8) 60.2 and 32.9 mm
9) 21.2 A   10) 13.4, 14°52'
11) 18.6 A   12) 52°27'
13) A = 55°44', B = 76°12', C = 48°4'
14) 14.5 m

ANSWERS TO CHAPTER 14

Exercise 14.1

1) 32.75 m   2) 3.72 m
3) 5.09 m   4) 27°53'
5) 8.91 m
6) (a) 219 m   (b) 140 m
   (c) 260 m   (d) S32°42'E
7) 14.1 m   8) 50°58'
9) 1233 m   10) 190.8 m
11) 74.9 m, 20°41'
12) 11.8 m   13) 36.1 m
14) BD = DF = AC = CE = 2.66 m,
   GB = FH = 4.41 m,
   BC = CF = 3.59 m,
   GA = EH = 3.70 m
15) 147.6 m   16) 33.3 m
17) 44.4 m, 116°
18) 15.2 m, N66°48'E

Exercise 14.2

1) *x*   0   47.55   29.39 − 29.39
   y   50.00   15.45 −40.45 −40.45
   x −47.55
   y   15.45
2) *x*   34.64   20.00 −34.64 −20.00
   y   20.00 −34.64 −20.00   34.64
3) *x*   35.24   5.22 −18.75
   y   12.83 −37.14   32.48
4) *x* −19.53   36.94
   y   56.73   47.28
5) *x*   14.28   35.72
   y   8.57   21.43

Exercise 14.3

1) 45.79 mm   2) 19.95 mm
3) 20.90 mm   4) 24.98 mm
5) 10°44'

Exercise 14.4

1) 1.64 mm
2) 1°31'; 13.04 mm; 9.59 mm
3) 65°46'; 29.71 mm
4) 53.01 mm   5) 31.99 mm
6) 4°24'; 25.51 mm
7) 104.98 mm   8) 5.18 mm

Exercise 14.5

1) 2408 mm   2) 5369 mm
3) 12.63 m   4) 1287 mm
5) 2971 mm   6) 5740 mm

7) 16.01 m   8) 2215 mm
9) 2.887 mm; 53.44 mm
10) 30.53 mm

Exercise 14.6

1) 2210 mm²   2) 3170 mm²
3) 2765 mm each side
4) 540 m²   5) 738 m²
6) (a) 7.55 m²   (b) 8.06 m²
7) 13.4 m²   8) 962 mm²
9) (a) 143 m²   (b) 53.7 m²
   (c) 43.6 m²
10) 11.9 m²   11) 28 m²
12) 8900 mm²   13) 2.12 m
14) 3060 mm²
15) (a) 13.8 m²   (b) 65.0 m²
16) (a) 3.31 m²   (b) 19.3 m²
17) 15.7 m

ANSWERS TO CHAPTER 15

Exercise 15.1

1) (b) 130 mm   (d) 153 mm
2) (a) 40 mm   (c) 71.6°
   (d) 126 mm   (e) 5060 mm²
   (f) 26 640 mm²
3) 127 mm   4) 45.3 mm
5) 8.07 m, 10.1 m
6) (a) 1524 m   (b) 43°58′
7) 19.7 m²
8) (a) 13.9 m   (b) 231 m²
9) 261 m²
10) (a) 45.2 m   (b) 927 m²
11) 34.3 m   12) 185 m
13) 108 m²   14) 43.0 m²
15) 19°29′

ANSWERS TO CHAPTER 16

Exercise 16.1

1) 8.8 mm   2) 0.0128 m²
3) (a) 1200 mm²
   (b) 276 mm²   (c) 259.5 mm²
   (d) 774 mm²   (e) 1050 mm²
4) 2800 mm²   5) 8910 mm²
6) 2.12 m   7) 3060 mm²
8) (a) 1380 mm²   (b) 6500 mm²
9) (a) 3.31 mm²   (b) 19.3 mm²
10) 157 mm   11) 29.9 mm
12) (a) 11 200 mm²   (b) 302 mm²
13) 34.1 mm   14) 2592 mm²

15) 909
16) (a) 9.05 m²   (b) 7.54 m
17) 1910 mm
18) (a) 5856 m²   (b) 792.5 m²
19) (a) 3.80 m²   (b) 4.19 m²
    (c) 0.39 m²
20) 34.6 m²

Exercise 16.2

1) 0.108 m³   2) 335 mm
3) 0.008 75 m³   4) 39 000 m³
5) 60 m³   6) 477 mm
7) 1.51 m²
8) 128 000 mm³, 11 700 mm²
9) 7.92 m³   10) 19.9 ℓ
11) 75.4 mm
12) 5.33 m³, 20.5 m²
13) 0.437 m³
14) (a) 0.366 m³   (b) 0.583 m
15) (a) 20.6 m³   (b) 33.8 m²
16) 0.004 19 m³, 0.126 m²
17) (a) 89.8 m³   (b) 76.9 m²
18) (a) 92.1 m³   (b) 50.3 m²
19) 348 m³, 119 m²   20) 1.47 ℓ
21) 943 mm², 110 mm   22) 1.80 m
23) 0.0410 m², 0.725 m
24) 1.35 × 10⁶ℓ
25) 14 m³, 22.7 m²
26) (a) 36.7 m³   (b) 80 700 kg
27) 3.03 m³, 8.36 m²   28) 16.3 ℓ
29) 93.7 m²   30) 150 mm
31) 10.5 m³, 18.8 m²   32) 232 m²

Exercise 16.3

1) 752   2) 172   3) 0.8
4) 99   5) 25.5   6) 1090 m²
7) 17.2 m³
8) 1060 m², 9.46 m/s
9) 3 × 10⁶ m³ approx.   10) 7.14 m³
11) 4430 m³
12) 0.0247 m³

ANSWERS TO CHAPTER 17

Exercise 17.1

1) 2.21 m³, 22.1 m²   2) 132 m³
3) 2.18 kg   4) 18.8 m²
5) 41.0 m³   6) 0.837 m³
7) 865 kg
8) 37.2 m², 3.88 m³, 8540 kg
9) 56 300 mm³   10) 1 248 000 mm

ANSWERS TO CHAPTER 18

Exercise 18.1

1) $m = 1, c = 3$
2) (a) $m = 1, c = 3$
   (b) $m = -3, c = 4$
   (c) $m = -3.1, c = -1.7$
   (d) $m = 4.3, c = -2.5$
3) $m = 2, c = 1$
4) $a = 0.25, b = 1.25$
5) $a = 0.29, b = -1.0$
6) 529 N
7) $E = 0.0984W + 0.72$
8) $a = 0.03, b = 0$
9) $a = 100, b = 0.43$
10) 524 N/m   11) 51 ohms

ANSWERS TO CHAPTER 19

Exercise 19.1

1) b   2) d   3) e
4) a   5) c   6) b
7) a   8) b   9) b
10) $x^2 + y^2 = 10$

Exercise 19.2

1) $a = 70, b = 50$
2) $k_m = 0.016 + \dfrac{0.023}{\mu}$
3) Gradient = 1500, Intercept = 0
4) $m = 4.5, c = 0.5$
5) $k = 0.2$
6) $a = 0.761, b = 10.1$
7) $m = 0.040, c = 0.20$
8) $m = 0.1, c = 1.4$
9) $a = 12.3, b = 0.53$

Exercise 19.3

1) $a = 3, n = 2$
2) $a = 2 \times 10^{-6}, n = 4$
3) $n = 2, R = 10$
4) $n = 4$, for $V = 80$ read $V = 70$
5) $t = 0.3 \, m^{1.5}$
6) $k = 100, n = -1.2$

ANSWERS TO CHAPTER 20

Exercise 20.1

1) 1, 2   2) 4, 5   3) 4, 1

4) 40, 75   5) 8.9, 7.3
6) 25 ohms, 0.005, 31.25 ohms
7) £0.10, £0.30   8) £96, £300
9) £0.90, £1.25   10) 0.4, 50, 170 N
11) £250, £6500
12) £14 000, £16 000
13) 2, 8   14) 18, 28
15) $x = 5400$,
    $y = 1700$; £5400, £3400
16) 3, 8

Exercise 20.2

1) 3, 4   2) 4 repeated
3) +3, −3 4) 3, −5   5) 0.667, 7
6) −5, −1.5   7) 2.414, −0.414
8) 2.181, 0.153 9) +0.745, −0.745

Exercise 20.3

1) $x = -1, y = 1; x = 4, y = 6$
2) $x = 0, y = 5; x = 3, y = 11$
3) $x = 1, y = 3; x = -0.2, y = -3$
4) $x = 2.39, y = 6.91$;
   $x = 0.26, y = 0.54$
5) 60 m × 80 m

Exercise 20.4

1) 1, −1, 4   2) 2, 0.33, −1.20
3) 1   4) −1, 1 (repeated)
5) 3, 1.18, −0.43
6) 2   7) 1, 0.57, −2.91
8) −3, 2 (repeated)
9) 2, 0.21, −1.35
10) −1   11) 7.9 m
12) 3.32 m   13) 5 m
14) 3

ANSWERS TO CHAPTER 21

Exercise 21.1

1) (a) $(8.60, 54.5°)$
   (b) $(3.61, 123.7°)$
   (c) $(3.61, 303.7°)$
   (d) $(5, 233.1°)$
2) (a) $(1.64, 1.15)$
   (b) $(-1.81, 2.39)$
   (c) $(-0.75, -1.30)$
   (d) $(0.401, -0.446)$
   (e) $(2.15, -0.824)$
   (f) $(3.21, -3.83)$

3)

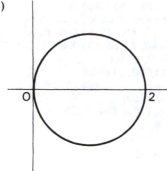

6)

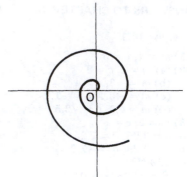

4)

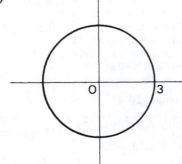

7)

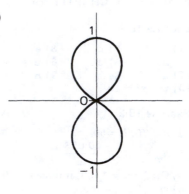

5)

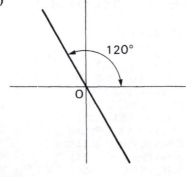

8)

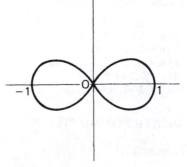

9)

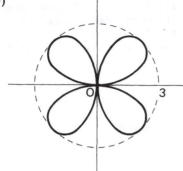

10)

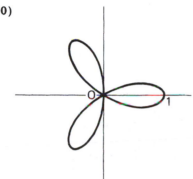

11)

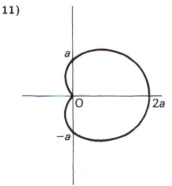

ANSWERS TO CHAPTER 22

Exercise 22.1

1) discrete (a), (e)
5) 3 ohm
7) 0.3 N/mm²

Exercise 22.2

1) 5   2) no mode
3) 3, 5 and 8   4) 121.96 ohms
5) 18.4272 mm   6) 5
7) 84   8) £183
9) Mean = 15.09 ohms,
   Median = 15.76 ohms
10) 118.1, 117.2, 119.1 ohms
11) 20.063, 20.095, 20.115 mm
12) 109.095 mm 13) 23.865 kg
14) 9.6486 m   15) 19.966 kN

Exercise 22.3

1) 10.81%, 1.2782%
2) 11.4925 N/mm²,
   0.014 52 N/mm²
3) 99.93 W, 0.17 W
4) 12.67, 2.98
5) 43.05, 4.965 gallons
6) (a) 18.54 mm and 18.66 mm
   (b) (i) 16%   (ii) 2%
7) 1920
8) $\bar{x} = 10.0000$, $\sigma = 0.0128$ mm
   (a) 1400   (b) 9800
9) $R = 1.3V + 19.8$
10) $E = 0.030t - 0.01$, 25.19 mV

ANSWERS TO CHAPTER 23

Exercise 23.1

1) $-4, 8$ 2) 3   3) 2

Exercise 23.2

1) $2x$ 2) $7x^6$ 3) $12x^2$
4) $30x^4$ 5) $1.5t^2$ 6) $2\pi R$
7) $\frac{1}{2}x^{-1/2}$ 8) $6x^{1/2}$ 9) $x^{-1/2}$
10) $2x^{-1/3}$ 11) $-2x^{-3}$ 12) $-x^{-2}$
13) $-\frac{3}{5}x^{-2}$ 14) $-6x^{-4}$ 15) $-\frac{1}{2}x^{-3/2}$
16) $-\frac{1}{3}x^{-3.2}$ 17) $-\frac{15}{2}x^{-5/2}$
18) $\frac{3}{10}t^{-1/2}$ 19) $-0.01h^{-2}$
20) $-35x^{-8}$ 21) $8x - 3$
22) $9t^2 - 4t + 5$ 23) $4u - 1$
24) $20x^3 - 21x^2 + 6x - 2$
25) $35t^4 - 6t$ 26) $\frac{1}{2}x^{-1/2} + \frac{5}{2}x^{3/2}$
27) $-3x^{-3} + 1$ 28) $\frac{1}{2}x^{-1/2} - \frac{1}{2}x^{-3/2}$
29) $3x^2 - \frac{3}{2}x^{-3/2}$
30) $1.3t^{0.3} + 0.575t^{-3.3}$
31) $\frac{9}{5}x^2 - \frac{4}{7}x + \frac{1}{2}x^{-1/2}$
32) $-0.01x^{-2}$

33) $4.65x^{0.5} - 1.44x^{-0.4}$

34) $\frac{3}{2}x^2 + 5x^{-2}$ 35) $-6 + 14t - 6t^2$

36) $-5, 19$ 37) $39.5, 5, 17$

38) $2.5, 2, 1$

Exercise 23.3

2) -2.905

3) $5\cos x - 3\sin x$

4) $4\cos\theta + 2\sin\theta$

ANSWERS TO CHAPTER 24

Exercise 24.1

1) 42 m/s 2) -6 m/s^2

3) (a) 6 m/s

   (b) 2.41 or -0.41 s

   (c) 6 m/s^2 (d) 1 s

4) $-0.074 \text{ m/s}, 0.074 \text{ m/s}^2$

5) $10 \text{ m/s}, 30 \text{ m/s}$

6) 3.46 m/s

7) (a) 4 rad/s (b) 36 rad/s^2

   (c) 0 s or 1 s

8) (a) -2.97 rad/s

   (b) 0.280 s

   (c) -8.98 rad/s^2

   (d) 1.57 s

9) 62.5 kJ

Exercise 24.2

1) 452 mm^3

2) $0.006\,00$ volt too many

3) (a) 0.785 m too few

   (b) 0.38 m too many

4) 0.201 m^2 too small

5) 2.5%

Exercise 24.3

1) (a) $11(\text{max}), -16(\text{min})$

   (b) $4(\text{max}), 0(\text{min})$

   (c) $0(\text{min}), 32(\text{max})$

2) (a) 54 (b) 2.5

   (c) $x = -2$

3) $(3, -15), (-1, 17)$

4) (a) -2 (b) 1

   (c) 9

5) (a) 12 (b) 12.48

6) $10 \text{ V}(\text{max}), 4 \text{ V}(\text{min})$

7) 8

8) $12.5 \text{ m/s}, 1.23 \text{ kW}$

9) 15 mm 10) 10 m

11) 4 12) $108\,000 \text{ mm}^3$

13) $86 \text{ mm diam.}, 86 \text{ mm long}$

14) 405 mm

15) $28.9 \text{ mm diam.}, 14.4 \text{ mm high}$

ANSWERS TO CHAPTER 25

Exercise 25.1

1) $\dfrac{x^3}{3} + c$ 2) $\dfrac{x^9}{9} + c$

3) $\dfrac{2x^{3/2}}{3} + c$ 4) $-\dfrac{1}{x} + c$

5) $-\dfrac{1}{3x^3} + c$ 6) $2x^{1/2} + c$

7) $\dfrac{3x^5}{5} + c$ 8) $\dfrac{5x^9}{9} + e^x + c$

9) $\dfrac{x^3}{3} + \dfrac{x^2}{2} + 3x + c$

10) $\dfrac{x^4}{2} - \dfrac{7x^2}{2} - 4x + c$

11) $\dfrac{x^3}{3} - \dfrac{5x^2}{2} + 2x^{1/2} - \dfrac{2}{x} + c$

12) $-\dfrac{4}{x^2} + \dfrac{2}{x} + \dfrac{2x^{3/2}}{3} + c$

13) $\dfrac{x^3}{3} - \dfrac{3x^2}{2} + 2x + c$

14) $\dfrac{x^3}{3} + 3x^2 + 9x + c$

15) $\dfrac{4x^3}{3} - 14x^2 + 49x + c$

16) $2\sin x - 3\cos x + c$

Exercise 25.2

1) $y = \dfrac{x^2}{2} + 1$ 2) 46.7

3) $y = 10 + 3x - x^2, 10$

4) $y = \dfrac{2x^3}{3} + \dfrac{3x^2}{2} + 2x - \dfrac{1}{6}$

5) $\dfrac{t^3}{3} - 3t^2 + 9t - \dfrac{17}{3}$

6) $y = x^2 - 2x + 4$

7) $y = \sin x + 1.16$

8) $y = \sin\theta$

9) $y = 4.72 - 3\cos t$

Exercise 25.3

1) 2.33	2) 8	3) 8.67
4) 4	5) 3.167	6) 1.89
7) 0.667	8) 0.667	9) 1

Exercise 25.4

1) 136	2) 87	3) 5.167
4) 0.667	5) 3.75	6) 0
7) 4	8) 2	

ANSWERS TO CHAPTER 26

Exercise 26.1

1)

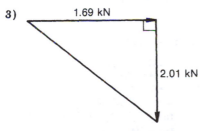

2.60 N, 1.50 N

2)

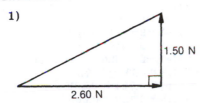

2.57 m/s, 3.06 m/s

3)

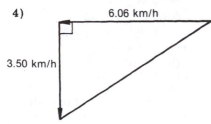

1.69 kN, 2.01 kN

4)

6.06 km/h, 3.50 km/h

5)

21.1 kN, 45.3 kN

6) 5.00 kN down the slope,
   8.66 kN at right angles to slope
7) 86.1 N horizontal,
   40.1 N vertical
8) 1.93 m/s
9) 58.3 N, 31°
10) 703.5 km/h, 5.71° east of north
11) 0.569 m/s, 38.4° west of north
12) 353 N, anticlockwise 11.0°
13) 347 N along the track,
    67.4 N between wheel flanges
    and rail
14) 3.61 V, anticlockwise 33.7°
15) 4.78 A, clockwise 13.9°
16) 10.1 V, clockwise 40.8°
17) 6.51 A, clockwise 33.0°

ANSWERS TO CHAPTER 27

Exercise 27.1

1) (a) 2×2 (b) 2×1
   (c) 3×3 (d) 2×4
2) (a) 9   (b) 8   (c) n^2
3) (a) $\begin{pmatrix} 1 & 3 \\ 2 & 4 \end{pmatrix}$ (b) $(5 \quad -6)$

(c) $\begin{pmatrix} a & 2 & x \\ b & 3 & -6 \\ 4 & 5 & 0 \end{pmatrix}$ (d) $\begin{pmatrix} 1 & 6 \\ -2 & 2 \\ -3 & 0 \\ -4 & -1 \end{pmatrix}$

4) (a) $\begin{pmatrix} 0 & 0 \\ 9 & 2 \end{pmatrix}$ (b) $\begin{pmatrix} 4 & 2 \\ -3 & 2 \end{pmatrix}$

(c) $\begin{pmatrix} \frac{5}{6} & \frac{1}{2} \\ \frac{5}{6} & 1 \end{pmatrix}$

5) $a = -8, b = 5, c = 1$

6) $\begin{pmatrix} \frac{1}{3} & \frac{1}{20} \\ \frac{1}{30} & \frac{1}{18} \end{pmatrix}$ 7) $\begin{pmatrix} 5 & 8 \\ 12 & -2 \end{pmatrix}$

8) $\begin{pmatrix} 0 \\ 5 \end{pmatrix}$

Exercise 27.2

1)(a) $\begin{pmatrix} 6 & 0 \\ -4 & 2 \end{pmatrix}$ (b) $\begin{pmatrix} -12 & 3 \\ 9 & -6 \end{pmatrix}$

(c) $\begin{pmatrix} -6 & 3 \\ 5 & -4 \end{pmatrix}$ (d) $\begin{pmatrix} 18 & -3 \\ -13 & 8 \end{pmatrix}$

2) (a) $\begin{pmatrix} 14 & 0 \\ 8 & -2 \end{pmatrix}$ (b) $\begin{pmatrix} 2 & 1 \\ 3 & 1 \end{pmatrix}$

(c) $\begin{pmatrix} 5 & 11 \\ 10 & 22 \end{pmatrix}$ (d) $\begin{pmatrix} a & b \\ c & d \end{pmatrix}$

(e) $\begin{pmatrix} ka & kb \\ kc & kd \end{pmatrix}$

3) (a) $\begin{pmatrix} 7 & 10 \\ 15 & 22 \end{pmatrix}$ (b) $\begin{pmatrix} 3 & -5 \\ 5 & 8 \end{pmatrix}$

(c) $\begin{pmatrix} 8 & 10 \\ 20 & 18 \end{pmatrix}$ (d) $\begin{pmatrix} 18 & 15 \\ 40 & 48 \end{pmatrix}$

(e) $\begin{pmatrix} 13 & 10 \\ 40 & 53 \end{pmatrix}$

Exercise 27.3

1) (a) 24   (b) −6   (c) 14
2) (a) $x = 1, y = 2$
   (b) $x = 4, y = 3$
   (c) $x = 0.5, y = 0.75$

Exercise 27.4

1) $\frac{1}{3}\begin{pmatrix} 4 & -5 \\ -1 & 2 \end{pmatrix}$ 2) $\begin{pmatrix} 3 & -5 \\ -1 & 2 \end{pmatrix}$

3) $\frac{1}{4}\begin{pmatrix} 2 & -2 \\ -1 & 3 \end{pmatrix}$ 4) No inverse

5) $\frac{1}{320}\begin{pmatrix} 4 & -24 \\ -24 & 224 \end{pmatrix}$

6) No inverse   7) $\frac{1}{7}\begin{pmatrix} 2 & -3 \\ 1 & 2 \end{pmatrix}$

8) $\frac{1}{13}\begin{pmatrix} 5 & 3 \\ -1 & 2 \end{pmatrix}$ 9) $\begin{pmatrix} 1 & -1 \\ 0 & 1 \end{pmatrix}$

10) (a) $\frac{1}{2}\begin{pmatrix} 2 & 0 \\ -3 & 1 \end{pmatrix}$ (b) $\begin{pmatrix} 2 & -5 \\ -1 & 3 \end{pmatrix}$

(c) $\frac{1}{2}\begin{pmatrix} 19 & -5 \\ -11 & 3 \end{pmatrix}$

(d) $\begin{pmatrix} 3 & 5 \\ 11 & 19 \end{pmatrix}$ (e) $\frac{1}{2}\begin{pmatrix} 19 & -5 \\ -11 & 3 \end{pmatrix}$

(f) equal

Exercise 27.5

1) $x = 6, y = -5$
2) $x = 1, y = 5$
3) $x = 3, y = -1$
4) $x = 2, y = -3$
5) $x = \frac{16}{11}, y = \frac{9}{11}$
6) $x = 2, y = -5$

ANSWERS TO CHAPTER 28

Exercise 28.1

1) (a) $18 + j10$ (b) $-j$
   (c) $8 - j10$
2) (a) $-1 + j3$ (b) $-4 - j3$
   (c) $10 - j3$
3) (a) $-9 + j21$ (b) $-36 - j32$
   (c) $-9 + j40$ (d) 34
   (e) $-21 - j20$ (f) $18 - j30$
   (g) $0.069 - j0.172$
   (h) $-0.724 + j0.690$
   (i) $-0.138 - j0.655$
   (j) $0.644 + j0.616$
   (k) $3.5 - j0.5$ (l) $0.2 + j0.6$
4) (a) $1, j0.5$ (b) $0, j5$
   (c) $-3, j3$
5) (a) $-1 \pm j$ (b) $\pm j3$
6) (a) $0.634 - j0.293$
7) (a) $3 + j4$ (b) $0.4 + j0.533$
   (c) $0.692 + j2.538$

Exercise 28.2

2) Mod 5, Arg 53.13°;
   Mod 5, Arg −36.87°
3) 3.61, 146.32°
4) 4.47, −153.43°
5) (a) $5\underline{/36.87°}$ (b) $5\underline{/-53.13°}$
   (c) $4.24\underline{/135°}$
   (d) $2.24\underline{/-153.43}$
   (e) $4\underline{/90°}$ (f) $3.5\underline{/-90°}$
6) (a) $2.12 + j2.12$
   (b) $-4.49 + j2.19$
   (c) $4.32 - j1.57$
   (d) $-1.60 - j2.77$
7) (a) 14.62°   (b) 345 watts

INDEX